LONDON MATHEMATICAL SOCIETY LECTURE NOTE SERIES

Managing Editor: Professor M. Reid, Mathematics Institute, University of Warwick, Coventry CV4 7AL, United Kingdom

The titles below are available from booksellers, or from Cambridge University Press at http://www.cambridge.org/mathematics

287 Topics on Riemann surfaces and Fuchsian groups, E. BUJALANCE, A. F. COSTA & E. MARTÍNEZ (eds)
288 Surveys in combinatorics, 2001, J. W. P. HIRSCHFELD (ed)
289 Aspects of Sobolev-type inequalities, L. SALOFF-COSTE
290 Quantum groups and Lie theory, A. PRESSLEY (ed)
291 Tits buildings and the model theory of groups, K. TENT (ed)
292 A quantum groups primer, S. MAJID
293 Second order partial differential equations in Hilbert spaces, G. DA PRATO & J. ZABCZYK
294 Introduction to operator space theory, G. PISIER
295 Geometry and integrability, L. MASON & Y. NUTKU (eds)
296 Lectures on invariant theory, I. DOLGACHEV
297 The homotopy category of simply connected 4-manifolds, H.-J. BAUES
298 Higher operads, higher categories, T. LEINSTER (ed)
299 Kleinian groups and hyperbolic 3-manifolds, Y. KOMORI, V. MARKOVIC & C. SERIES (eds)
300 Introduction to Möbius differential geometry, U. HERTRICH-JEROMIN
301 Stable modules and the D(2)-problem, F. E. A. JOHNSON
302 Discrete and continuous nonlinear Schrödinger systems, M. J. ABLOWITZ, B. PRINARI & A.D. TRUBATCH
303 Number theory and algebraic geometry, M. REID & A. SKOROBOGATOV (eds)
304 Groups St Andrews 2001 in Oxford I, C. M. CAMPBELL, E. F. ROBERTSON & G. C. SMITH (eds)
305 Groups St Andrews 2001 in Oxford II, C. M. CAMPBELL, E. F. ROBERTSON & G. C. SMITH (eds)
306 Geometric mechanics and symmetry, J. MONTALDI & T. RATIU (eds)
307 Surveys in combinatorics 2003, C. D. WENSLEY (ed.)
308 Topology, geometry and quantum field theory, U. L. TILLMANN (ed)
309 Corings and comodules, T. BRZEZINSKI & R. WISBAUER
310 Topics in dynamics and ergodic theory, S. BEZUGLYI & S. KOLYADA (eds)
311 Groups: topological, combinatorial and arithmetic aspects, T. W. MÜLLER (ed)
312 Foundations of computational mathematics, Minneapolis 2002, F. CUCKER *et al* (eds)
313 Transcendental aspects of algebraic cycles, S. MÜLLER-STACH & C. PETERS (eds)
314 Spectral generalizations of line graphs, D. CVETKOVIĆ, P. ROWLINSON & S. SIMIĆ
315 Structured ring spectra, A. BAKER & B. RICHTER (eds)
316 Linear logic in computer science, T. EHRHARD, P. RUET, J.-Y. GIRARD & P. SCOTT (eds)
317 Advances in elliptic curve cryptography, I. F. BLAKE, G. SEROUSSI & N. P. SMART (eds)
318 Perturbation of the boundary in boundary-value problems of partial differential equations, D. HENRY
319 Double affine Hecke algebras, I. CHEREDNIK
320 L-functions and Galois representations, D. BURNS, K. BUZZARD & J. NEKOVÁŘ (eds)
321 Surveys in modern mathematics, V. PRASOLOV & Y. ILYASHENKO (eds)
322 Recent perspectives in random matrix theory and number theory, F. MEZZADRI & N. C. SNAITH (eds)
323 Poisson geometry, deformation quantisation and group representations, S. GUTT *et al* (eds)
324 Singularities and computer algebra, C. LOSSEN & G. PFISTER (eds)
325 Lectures on the Ricci flow, P. TOPPING
326 Modular representations of finite groups of Lie type, J. E. HUMPHREYS
327 Surveys in combinatorics 2005, B.S. WEBB (ed)
328 Fundamentals of hyperbolic manifolds, R. CANARY, D. EPSTEIN & A. MARDEN (eds)
329 Spaces of Kleinian groups, Y. MINSKY, M. SAKUMA & C. SERIES (eds)
330 Noncommutative localization in algebra and topology, A. RANICKI (ed)
331 Foundations of computational mathematics, Santander 2005, L. M PARDO, A. PINKUS, E. SÜLI & M. J. TODD (eds)
332 Handbook of tilting theory, L. ANGELERI HÜGEL, D. HAPPEL & H. KRAUSE (eds)
333 Synthetic differential geometry (2nd Edition), A. KOCK
334 The Navier–Stokes equations, N. RILEY & P. DRAZIN
335 Lectures on the combinatorics of free probability, A. NICA & R. SPEICHER
336 Integral closure of ideals, rings, and modules, I. SWANSON & C. HUNEKE
337 Methods in Banach space theory, J. M. F. CASTILLO & W. B. JOHNSON (eds)
338 Surveys in geometry and number theory, N. YOUNG (ed)
339 Groups St Andrews 2005 I, C. M. CAMPBELL, M. R. QUICK, E. F. ROBERTSON & G. C. SMITH (eds)
340 Groups St Andrews 2005 II, C. M. CAMPBELL, M. R. QUICK, E. F. ROBERTSON & G. C. SMITH (eds)
341 Ranks of elliptic curves and random matrix theory, J. B. CONREY, D. W. FARMER, F. MEZZADRI & N. C. SNAITH (eds)
342 Elliptic cohomology, H.R. MILLER & D. C. RAVENEL (eds)
343 Algebraic cycles and motives I, J. NAGEL & C. PETERS (eds)

344 Algebraic cycles and motives II, J. NAGEL & C. PETERS (eds)
345 Algebraic and analytic geometry, A. NEEMAN
346 Surveys in combinatorics 2007, A. HILTON & J. TALBOT (eds)
347 Surveys in contemporary mathematics, N. YOUNG & Y. CHOI (eds)
348 Transcendental dynamics and complex analysis, P. J. RIPPON & G. M. STALLARD (eds)
349 Model theory with applications to algebra and analysis I, Z. CHATZIDAKIS, D. MACPHERSON, A. PILLAY & A. WILKIE (eds)
350 Model theory with applications to algebra and analysis II, Z. CHATZIDAKIS, D. MACPHERSON, A. PILLAY & A. WILKIE (eds)
351 Finite von Neumann algebras and masas, A. M. SINCLAIR & R. R. SMITH
352 Number theory and polynomials, J. MCKEE & C. SMYTH (eds)
353 Trends in stochastic analysis, J. BLATH, P. MÖRTERS & M. SCHEUTZOW (eds)
354 Groups and analysis, K. TENT (ed)
355 Non-equilibrium statistical mechanics and turbulence, J. CARDY, G. FALKOVICH & K. GAWEDZKI
356 Elliptic curves and big Galois representations, D. DELBOURGO
357 Algebraic theory of differential equations, M. A. H. MACCALLUM & A. V. MIKHAILOV (eds)
358 Geometric and cohomological methods in group theory, M. R. BRIDSON, P. H. KROPHOLLER & I. J. LEARY (eds)
359 Moduli spaces and vector bundles, L. BRAMBILA-PAZ, S. B. BRADLOW, O. GARCÍA-PRADA & S. RAMANAN (eds)
360 Zariski geometries, B. ZILBER
361 Words: Notes on verbal width in groups, D. SEGAL
362 Differential tensor algebras and their module categories, R. BAUTISTA, L. SALMERÓN & R. ZUAZUA
363 Foundations of computational mathematics, Hong Kong 2008, F. CUCKER, A. PINKUS & M. J. TODD (eds)
364 Partial differential equations and fluid mechanics, J. C. ROBINSON & J. L. RODRIGO (eds)
365 Surveys in combinatorics 2009, S. HUCZYNSKA, J. D. MITCHELL & C. M. RONEY-DOUGAL (eds)
366 Highly oscillatory problems, B. ENGQUIST, A. FOKAS, E. HAIRER & A. ISERLES (eds)
367 Random matrices: High dimensional phenomena, G. BLOWER
368 Geometry of Riemann surfaces, F. P. GARDINER, G. GONZÁLEZ-DIEZ & C. KOUROUNIOTIS (eds)
369 Epidemics and rumours in complex networks, M. DRAIEF & L. MASSOULIÉ
370 Theory of p-adic distributions, S. ALBEVERIO, A. YU. KHRENNIKOV & V. M. SHELKOVICH
371 Conformal fractals, F. PRZYTYCKI & M. URBAŃSKI
372 Moonshine: The first quarter century and beyond, J. LEPOWSKY, J. MCKAY & M.P. TUITE (eds)
373 Smoothness, regularity and complete intersection, J. MAJADAS & A. G. RODICIO
374 Geometric analysis of hyperbolic differential equations: An introduction, S. ALINHAC
375 Triangulated categories, T. HOLM, P. JØRGENSEN & R. ROUQUIER (eds)
376 Permutation patterns, S. LINTON, N. RUŠKUC & V. VATTER (eds)
377 An introduction to Galois cohomology and its applications, G. BERHUY
378 Probability and mathematical genetics, N. H. BINGHAM & C. M. GOLDIE (eds)
379 Finite and algorithmic model theory, J. ESPARZA, C. MICHAUX & C. STEINHORN (eds)
380 Real and complex singularities, M. MANOEL, M.C. ROMERO FUSTER & C.T.C WALL (eds)
381 Symmetries and integrability of difference equations, D. LEVI, P. OLVER, Z. THOMOVA & P. WINTERNITZ (eds)
382 Forcing with random variables and proof complexity, J. KRAJÍČEK
383 Motivic integration and its interactions with model theory and non-Archimedean geometry I, R. CLUCKERS, J. NICAISE & J. SEBAG (eds)
384 Motivic integration and its interactions with model theory and non-Archimedean geometry II, R. CLUCKERS, J. NICAISE & J. SEBAG (eds)
385 Entropy of hidden Markov processes and connections to dynamical systems, B. MARCUS, K. PETERSEN & T. WEISSMAN (eds)
386 Independence-friendly logic, A.L. MANN, G. SANDU & M. SEVENSTER
387 Groups St Andrews 2009 in Bath I, C.M. CAMPBELL *et al* (eds)
388 Groups St Andrews 2009 in Bath II, C.M. CAMPBELL *et al* (eds)
389 Random fields on the sphere, D. MARINUCCI & G. PECCATI
390 Localization in periodic potentials, D.E. PELINOVSKY
391 Fusion systems in algebra and topology M. ASCHBACHER, R. KESSAR & B. OLIVER
392 Surveys in combinatorics 2011, R. CHAPMAN (ed)
393 Non-abelian fundamental groups and Iwasawa theory, J. COATES *et al* (eds)
394 Variational Problems in Differential Geometry, R. BIELAWSKI, K. HOUSTON & M. SPEIGHT (eds)
395 How groups grow, A. MANN
396 Arithmetic Differential Operators over the p-adic Integers, C.C. RALPH & S.R. SIMANCA
397 Hyperbolic geometry and applications in quantum Chaos and cosmology, J. BOLTE & F. STEINER (eds)
398 Mathematical models in contact mechanics, M. SOFONEA & A. MATEI
399 Circuit double cover of graphs, C.-Q. ZHANG
400 Dense sphere packings: a blueprint for formal proofs, T. HALES
401 A double Hall algebra approach to affine quantum Schur–Weyl theory, B. DENG, J. DU & Q. FU

London Mathematical Society Lecture Note series: 401

A Double Hall Algebra Approach to Affine Quantum Schur–Weyl Theory

BANGMING DENG
Beijing Normal University

JIE DU
University of New South Wales, Sydney

QIANG FU
Tongji University, Shanghai

CAMBRIDGE
UNIVERSITY PRESS

CAMBRIDGE
UNIVERSITY PRESS

University Printing House, Cambridge CB2 8BS, United Kingdom

Cambridge University Press is part of the University of Cambridge.

It furthers the University's mission by disseminating knowledge in the pursuit of education, learning and research at the highest international levels of excellence.

www.cambridge.org
Information on this title: www.cambridge.org/9781107608603

First published 2012

A catalogue record for this publication is available from the British Library

ISBN 978-1-107-60860-3 Paperback

We dedicate the book to our teachers:

Peter Gabriel

Shaoxue Liu

Leonard Scott

Jianpan Wang

2010 Mathematics Subject Classification. Primary 17B37, 20G43, 20C08; Secondary 16G20, 20G42, 16T20

Key words and phrases. affine Hecke algebra, affine quantum Schur algebra, cyclic quiver, Drinfeld double, loop algebra, quantum group, Schur–Weyl duality, Ringel–Hall algebra, simple representation

Abstract

Over its one-hundred year history, the theory of Schur–Weyl duality and its quantum analogue have had and continue to have profound influences in several areas of mathematics such as Lie theory, representation theory, invariant theory, combinatorial theory, and so on. Recent new developments include, e.g., walled Brauer algebras and rational Schur algebras, quantum Schur superalgebras, and the integral Schur–Weyl duality for types other than A. This book takes an algebraic approach to the affine quantum Schur–Weyl theory.

The book begins with a study of extended Ringel–Hall algebras associated with the cyclic quiver of n vertices and the Green–Xiao Hopf structure on their Drinfeld double—the double Ringel–Hall algebra. This algebra is presented in terms of Chevalley type and central generators and is proved to be isomorphic to the quantum loop algebra of the general linear Lie algebra. The rest of the book investigates the affine quantum Schur–Weyl duality on three levels. This includes

- the affine quantum Schur–Weyl reciprocity;
- the bridging role played by the affine quantum Schur algebra between the quantum loop algebra and the corresponding affine Hecke algebra;
- Morita equivalence of certain representation categories;
- the presentation of affine quantum Schur algebras; and
- the realization conjecture for the double Ringel–Hall algebra which is proved to be true in the classical case.

Connections with various existing works by Lusztig, Varagnolo–Vasserot, Schiffmann, Hubery, Chari–Pressley, Frenkel–Mukhin, and others are also discussed throughout the book.

Contents

Introduction

Quantum Schur–Weyl theory refers to a three-level duality relation. At Level I, it investigates a certain double centralizer property, *the quantum Schur– Weyl reciprocity*, associated with some bimodules of quantum $\mathfrak{gl}_n$ and the Hecke algebra (of type A)—the tensor spaces of the natural representation of quantum $\mathfrak{gl}_n$ (see [**43**], [**21**], [**27**]). This is the quantum version of the well-known Schur–Weyl reciprocity which was beautifully used in H. Weyl's influential book [**77**]. The key ingredient of the reciprocity is a class of important finite dimensional endomorphism algebras, the *quantum Schur algebras* or *q-Schur algebras*, whose classical version was introduced by I. Schur over a hundred years ago (see [**69**], [**70**]). At Level II, it establishes a certain *Morita equivalence* between quantum Schur algebras and Hecke algebras. Thus, quantum Schur algebras are used to bridge representations of quantum $\mathfrak{gl}_n$ and Hecke algebras. More precisely, they link polynomial representations of quantum $\mathfrak{gl}_n$ with representations of Hecke algebras via the Morita equivalence. The third level of this duality relation is motivated by two simple questions associated with the structure of (associative) algebras. If an algebra is defined by generators and relations, the *realization problem* is to reconstruct the algebra as a vector space with hopefully explicit multiplication formulas on elements of a basis; while, if an algebra is defined in terms of a vector space such as an endomorphism algebra, it is natural to seek their generators and defining relations.

As one of the important problems in quantum group theory, the realization problem is to construct a quantum group in terms of a vector space and certain multiplication rules on basis elements. This problem is crucial to understand their structure and representations (see [**47**, p. xiii] for a similar problem for Kac–Moody Lie algebras and [**60**] for a solution in the symmetrizable case). Though the Ringel–Hall algebra realization of the ±-part of quantum enveloping algebras associated with symmetrizable Cartan matrices was an important breakthrough in the early 1990s, especially for the introduction of the geometric approach to the theory, the same problem for the entire quantum groups is far from completion. However, Beilinson–Lusztig–MacPherson (BLM) [**4**] solved the problem for quantum $\mathfrak{gl}_n$ by exploring further properties coming from the quantum Schur–Weyl reciprocity. On the other hand, as endomorphism algebras and as homomorphic images of quantum $\mathfrak{gl}_n$, it is natural to look for presentations for quantum Schur algebras via the presentation of quantum $\mathfrak{gl}_n$. This problem was first considered in [**18**] (see also [**26**]). Thus, as a particular feature in the type A theory, realizing quantum $\mathfrak{gl}_n$ and presenting quantum Schur algebras form Level III of this duality relation. For a complete account of the quantum Schur–Weyl theory and further references, see Parts 3 and 5 of [**12**] (see also [**17**] for more applications).

There are several developments in the establishment of an affine analogue of the quantum Schur–Weyl theory. Soon after BLM's work, Ginzburg and Vasserot [**32, 75**] used a geometric and K-theoretic approach to investigate *affine quantum Schur algebras*[1] as homomorphic images of *quantum loop algebra* $\mathbf{U}(\widehat{\mathfrak{gl}}_n)$ of $\mathfrak{gl}_n$ in the sense of Drinfeld's new presentation [**20**], called quantum affine $\mathfrak{gl}_n$ (at level 0) in this book. This establishes at Level I the first centralizer property for the affine analogue of the quantum Schur–Weyl reciprocity. Six years later, investigations around affine quantum Schur algebras focused on their different definitions and, hence, different applications. For example, Lusztig [**56**] generalized the fundamental multiplication formulas [**4**, 3.4] for quantum Schur algebras to the affine case and showed that the "extended" quantum affine $\mathfrak{sl}_n$, $\mathbf{U}_\Delta(n)$, does not map onto affine quantum Schur algebras; Varagnolo–Vasserot [**73**] investigated Ringel–Hall algebra actions on tensor spaces and described the geometrically defined affine quantum Schur algebras in terms of the endomorphism algebras of tensor spaces. Moreover, they proved that the tensor space definition coincides with Green's definition [**35**] via q-permutation modules. Some progress on the second centralizer property has also been made recently by Pouchin [**61**]. The approaches used in these works are mainly geometric. However, like the non-affine case, there would be more favorable algebraic and combinatorial approaches.

At Level II, representations at non-roots-of-unity of quantum affine $\mathfrak{sl}_n$ and $\mathfrak{gl}_n$ over the complex number field $\mathbb{C}$, including classifications of finite dimensional simple modules, have been thoroughly investigated by Chari–Pressley [**6, 7, 8**], and Frenkel–Mukhin [**28**] in terms of Drinfeld polynomials. Moreover, an equivalence between the module category of the Hecke algebra

[1] Perhaps they should be called quantum affine Schur algebras. Since our purpose is to establish an affine analogue of the quantum Schur–Weyl theory, this terminology seems more appropriate to reflect this.

$\mathcal{H}_\Delta(r)_\mathbb{C}$ and a certain full subcategory of quantum affine $\mathfrak{sl}_n$ (resp., $\mathfrak{gl}_n$) has also been established algebraically by Chari–Pressley [9] (resp., geometrically by Ginzburg–Reshetikhin–Vasserot [**31**]) under the condition $n > r$ (resp., $n \geqslant r$). Note that the approach in [**31**] uses intersection cohomology complexes. It would be interesting to know how affine quantum Schur algebras would play a role in these works.

Much less progress has been made at Level III. When $n > r$, Doty–Green [**18**] and McGerty [**58**] have found a presentation for affine quantum Schur algebras, while the last two authors of this book have investigated the realization problem in [**24**], where they first developed an approach without using the stabilization property, a key property used in the BLM approach, and presented an ideal candidate for the realization of quantum affine $\mathfrak{gl}_n$.

This book attempts to establish the affine quantum Schur–Weyl theory as a whole and is an outcome of *algebraically* understanding the works mentioned above.

First, building on Schiffmann [**67**] and Hubery [**40**], our starting point is to present the double Ringel–Hall algebra $\mathfrak{D}_\Delta(n)$ of the cyclic quiver with n vertices in terms of Chevalley type generators together with infinitely many central generators. Thus, we obtain a central subalgebra $\mathbf{Z}_\Delta(n)$ such that $\mathfrak{D}_\Delta(n) = \mathbf{U}_\Delta(n)\mathbf{Z}_\Delta(n) \cong \mathbf{U}_\Delta(n) \otimes \mathbf{Z}_\Delta(n)$. We then establish an isomorphism between $\mathfrak{D}_\Delta(n)$ and Drinfeld's quantum affine $\mathfrak{gl}_n$ in the sense of [**20**]. In this way, we easily obtain an action on the tensor space which upon restriction coincides with the Ringel–Hall algebra action defined geometrically by Varagnolo–Vasserot [**73**] and commutes with the affine Hecke algebra action.

Second, by a thorough investigation of a BLM type basis for affine quantum Schur algebras, we introduce certain triangular relations for the corresponding structure constants and, hence, a triangular decomposition for affine quantum Schur algebras. With this decomposition, we establish explicit algebra epimorphisms $\xi_r = \xi_{r,\mathbb{Q}(v)}$ from the double Ringel–Hall algebra $\mathfrak{D}_\Delta(n)$ to affine quantum Schur algebras $\boldsymbol{\mathcal{S}}_\Delta(n,r) := \mathcal{S}_\Delta(n,r)_{\mathbb{Q}(v)}$ for all $r \geqslant 0$. This algebraic construction has several nice applications, especially at Levels II and III. For example, the homomorphic image of commutator formulas for semisimple generators gives rise to a beautiful polynomial identity whose combinatorial proof remains mysterious.

Like the quantum Schur algebra case, we will establish for $n \geqslant r$ a Morita equivalence between affine quantum Schur algebras $\mathcal{S}_\Delta(n,r)_\mathbb{F}$ and affine Hecke algebras $\mathcal{H}_\Delta(r)_\mathbb{F}$ of type A over a field $\mathbb{F}$ with a non-root-of-unity parameter. As a by-product, we prove that every simple $\mathcal{S}_\Delta(n,r)_\mathbb{F}$-module is finite dimensional. Thus, applying the classification of simple $\mathcal{H}_\Delta(r)_\mathbb{C}$-modules by Zelevinsky [**81**] and Rogawski [**66**] yields a classification of simple

$\mathcal{S}_\Delta(n, r)_\mathbb{C}$-modules. Hence, inflation via the epimorphisms $\xi'_{r,\mathbb{C}}$ gives many finite dimensional simple $\mathrm{U}_\mathbb{C}(\widehat{\mathfrak{gl}}_n)$-modules. We will also use $\xi'_{r,\mathbb{C}}$ together with the action on tensor spaces and a result of Chari–Pressley to prove that finite dimensional simple polynomial representations of $\mathrm{U}_\mathbb{C}(\widehat{\mathfrak{gl}}_n)$ are all inflations of simple $\mathcal{S}_\Delta(n, r)_\mathbb{C}$-modules. In this way, we can see the bridging role played by affine quantum Schur algebras between representations of quantum affine $\mathfrak{gl}_n$ and those of affine Hecke algebras. Moreover, we obtain a classification of simple $\mathcal{S}_\Delta(n, r)_\mathbb{C}$-modules in terms of Drinfeld polynomials and, when $n > r$, we identify them with those arising from simple $\mathcal{H}_\Delta(r)_\mathbb{C}$-modules.

Our findings also show that, if we regard the category $\mathcal{S}_\Delta(n, r)_\mathbb{C}$-Mod of $\mathcal{S}_\Delta(n, r)_\mathbb{C}$-modules as a full subcategory of $\mathrm{U}_\mathbb{C}(\widehat{\mathfrak{gl}}_n)$-modules, this category is quite different from the category $\mathcal{C}^{hi} \cap \mathcal{C}'$ considered in [**54**, §6.2]. For example, the latter is completely reducible and simple objects are usually infinite dimensional, while $\mathcal{S}_\Delta(n, r)_\mathbb{C}$-Mod is not completely reducible and all simple objects are finite dimensional. As observed in [**23**, Rem. 9.4(2)] for quantum $\mathfrak{gl}_\infty$ and infinite quantum Schur algebras, this is another kind of phenomenon of infinite type in contrast to the finite type case.

The discussion of the realization and presentation problems is also based on the algebra epimorphisms ξ_r and relies on the use of semisimple generators and indecomposable generators for $\mathfrak{D}_\Delta(n)$ which are crucial to understand the integral structure and multiplication formulas. We first use the new presentation for $\mathfrak{D}_\Delta(n)$ to give a decomposition for $\boldsymbol{\mathcal{S}}_\Delta(n, r) = \mathbf{U}_\Delta(n, r)\mathbf{Z}_\Delta(n, r)$ into a product of two subalgebras, where $\mathbf{Z}_\Delta(n, r)$ is a central subalgebra and $\mathbf{U}_\Delta(n, r)$ is the homomorphic image of $\mathbf{U}_\Delta(n)$, the extended quantum affine $\mathfrak{sl}_n$. By taking a close look at this structure, we manage to get a presentation for $\boldsymbol{\mathcal{S}}_\Delta(r, r)$ for all $r \geqslant 1$ and acknowledge that the presentation problem is very complicated in the $n < r$ case. On the other hand, we formulate a realization conjecture suggested by the work [**24**] and prove the conjecture in the classical ($v = 1$) case.

We remark that, unlike the geometric approach in which the ground ring must be a field or mostly the complex number field $\mathbb{C}$, the algebraic, or rather, the representation-theoretic approach we use in this book works largely over a ring or mostly the integral Laurent polynomial ring $\mathbb{Z}[v, v^{-1}]$.

We have organized the book as follows.

In the first preliminary chapter, we introduce in §1.4 three different types of generators and their associated monomial bases for the Ringel–Hall algebras of cyclic quivers, and display in §1.5 the Green–Xiao Hopf structure on the extended version of these algebras.

Chapter 2 introduces a new presentation using Chevalley generators for Drinfeld's quantum loop algebra $\mathbf{U}(\widehat{\mathfrak{gl}}_n)$ of $\mathfrak{gl}_n$. This is achieved by constructing the presentation for the double Ringel–Hall algebra $\mathfrak{D}_\Delta(n)$ associated with cyclic quivers (Theorem 2.3.1), based on the work of Schiffmann and Hubery, and by lifting Beck's algebra monomorphism from the quantum $\widehat{\mathfrak{sl}}_n$ with a Drinfeld–Jimbo presentation into $\mathbf{U}(\widehat{\mathfrak{gl}}_n)$ to obtain an isomorphism between $\mathfrak{D}_\Delta(n)$ and $\mathbf{U}(\widehat{\mathfrak{gl}}_n)$ (Theorem 2.5.3).

Chapter 3 investigates the structure of affine quantum Schur algebras. We first recall the geometric definition by Ginzburg–Vasserot and Lusztig, the Hecke algebra definition by R. Green, and the tensor space definition by Varagnolo–Vasserot. Using the Chevalley generators of $\mathfrak{D}_\Delta(n)$, we easily obtain an action on the $\mathbb{Q}(v)$-space $\boldsymbol{\Omega}$ with a basis indexed by $\mathbb{Z}$ and, hence, an action of $\mathfrak{D}_\Delta(n)$ on $\boldsymbol{\Omega}^{\otimes r}$ (§3.5). We prove that this action commutes with the affine Hecke algebra action defined in [**73**]. Moreover, we show that the restriction of the action to the negative part of $\mathfrak{D}_\Delta(n)$ (i.e., to the corresponding Ringel–Hall algebra) coincides with the Ringel–Hall algebra action geometrically defined by Varagnolo–Vasserot (Theorem 3.6.3). As an application of this coincidence, the commutator formula associated with semisimple generators, arising from the skew-Hopf pairing, gives rise to a certain polynomial identity associated with a pair of elements $\lambda, \mu \in \mathbb{N}_\Delta^n$ (Corollary 3.9.6). The main result of the chapter is an elementary proof of the surjective homomorphism ξ_r from the double Ringel–Hall algebra $\mathfrak{D}_\Delta(n)$, i.e., the quantum loop algebra $\mathbf{U}(\widehat{\mathfrak{gl}}_n)$, onto the affine quantum Schur algebra $\boldsymbol{\mathcal{S}}_\Delta(n, r)$ (Theorem 3.8.1). The approach we used is the establishment of a triangular decomposition of $\boldsymbol{\mathcal{S}}_\Delta(n, r)$ (Theorem 3.7.7) through an analysis of the BLM type bases.

In Chapter 4, we discuss the representation theory of affine quantum Schur algebras over $\mathbb{C}$ and its connection to polynomial representations of quantum affine $\mathfrak{gl}_n$ and representations of affine Hecke algebras. We first establish a category equivalence between the module categories $\mathcal{S}_\Delta(n, r)_\mathbb{C}$-Mod and $\mathcal{H}_\Delta(r)_\mathbb{C}$-Mod for $n \geqslant r$ (Theorem 4.1.3). As an application, we will reinterpret Chari–Pressley's category equivalence ([**9**, Th. 4.2]) between (level r) representations of $\mathrm{U}_\mathbb{C}(\widehat{\mathfrak{sl}}_n)$ and those of affine Hecke algebras $\mathcal{H}_\Delta(r)_\mathbb{C}$, where $n > r$, in terms of representations of $\mathcal{S}_\Delta(n, r)_\mathbb{C}$ (Proposition 4.2.1). We then develop two approaches to the classification of simple $\mathcal{S}_\Delta(n, r)_\mathbb{C}$-modules. In the so-called upward approach, we use the classification of simple $\mathcal{H}_\Delta(r)_\mathbb{C}$-modules of Zelevinsky and Rogawski to classify simple $\mathcal{S}_\Delta(n, r)_\mathbb{C}$-modules (Theorems 4.3.4 and 4.5.3), while in the downward approach, we determine the classification of simple $\mathcal{S}_\Delta(n, r)_\mathbb{C}$-modules (Theorem 4.6.8) in terms of simple polynomial representations of $\mathrm{U}_\mathbb{C}(\widehat{\mathfrak{gl}}_n)$. When $n > r$, we prove an identification theorem (Theorem 4.4.2) for the two classifications. Finally, in §4.7, a classification of finite dimensional simple $U_\Delta(n, r)_\mathbb{C}$-modules is also completed and its connections to finite dimensional simple $\mathrm{U}_{\mathbb{C}}(\widehat{\mathfrak{sl}}_n)$-modules and finite dimensional simple (polynomial) $\mathrm{U}_{\mathbb{C}}(\widehat{\mathfrak{gl}}_n)$-modules are also discussed.

We move on to look at the presentation and realization problems in Chapter 5. We first observe $\boldsymbol{\mathcal{S}}_\Delta(n,r) = \mathbf{U}_\Delta(n,r)\mathbf{Z}_\Delta(n,r)$, where $\mathbf{U}_\Delta(n,r)$ and $\mathbf{Z}_\Delta(n,r)$ are homomorphic images of $\mathbf{U}_\Delta(n)$ and $\mathbf{Z}_\Delta(n)$, respectively, and that $\mathbf{Z}_\Delta(n,r) \subseteq \mathbf{U}_\Delta(n,r)$ if and only if $n > r$. A presentation for $\mathbf{U}_\Delta(n,r)$ is given in [**58**] (see also [**19**] for the $n > r$ case). Building on McGerty's presentation, we first give a Drinfeld–Jimbo type presentation for the subalgebra $\mathbf{U}_\Delta(n,r)$ (Theorem 5.1.3). We then describe a presentation for the central subalgebra $\mathbf{Z}_\Delta(n,r)$ as a Laurent polynomial ring in one indeterminate over a polynomial ring in $r-1$ indeterminates over $\mathbb{Q}(v)$. We manage to describe a presentation for $\boldsymbol{\mathcal{S}}_\Delta(r,r)$ for all $r \geqslant 1$ (Theorem 5.3.5) by adding an extra generator (and its inverse) together with an additional set of relations on top of the relations given in Theorem 5.1.3. What we will see from this case is that the presentation for $\boldsymbol{\mathcal{S}}_\Delta(n,r)$ with $r > n$ can be very complicated.

We discuss the realization problem from §5.4 onwards. We first describe the modified BLM approach developed in [**24**]. With some supporting evidence, we then formulate the realization conjecture (Conjecture 5.4.2) as suggested in [**24**, 5.5(2)], and state its classical ($v = 1$) version. We end the chapter with a closer look at Lusztig's transfer maps [**57**] by displaying some explicit formulas for their action on the semisimple generators for $\boldsymbol{\mathcal{S}}_\Delta(n,r)$ (Corollary 5.5.2). These formulas also show that the homomorphism from $\mathbf{U}(\widehat{\mathfrak{sl}}_n)$ to $\varprojlim \boldsymbol{\mathcal{S}}_\Delta(n,n+m)$ induced by the transfer maps cannot either be extended to the double Ringel–Hall algebra $'\mathfrak{D}_\Delta(n)$. (Lusztig already pointed out that it cannot be extended to $\mathbf{U}_\Delta(n)$.) This somewhat justifies why a direct product is used in the realization conjecture.

In the final Chapter 6, we prove the realization conjecture for the classical ($v = 1$) case. The key step in the proof is the establishment of more multiplication formulas (Proposition 6.2.3) between homogeneous indecomposable generators and an arbitrary BLM type basis element. As a by-product, we display a basis for the universal enveloping algebra of the loop algebra of $\mathfrak{gl}_n$ (Theorem 6.3.4) together with explicit multiplication formulas between generators and arbitrary basis elements (Corollary 6.3.5).

There are two appendices in §§3.10 and 6.4 which collect a number of lengthy calculations used in some proofs.

Conjectures and problems. There are quite a few conjectures and problems throughout the book. The conjectures are mostly natural generalizations to the affine case, for example, the realization conjecture 5.4.2 and the conjectures in §3.8 on an integral form for double Ringel–Hall algebras and the second centralizer property in the affine quantum Schur–Weyl reciprocity. Some problems are designed to seek further solutions to certain questions such as "quantum Serre relations" for semisimple generators (Problem 2.6.4), the Affine Branching Rule (Problem 4.3.6), and further identification of simple modules from different classifications (Problem 4.6.11). There are also problems for seeking different proofs. Problems 3.4.3 and 6.4.2 form a key step towards the proof of the realization conjecture.

Notational scheme. For most of the notation used throughout the book, if it involves a subscript $_\triangle$ or a superscript $^\triangle$, it indicates that the same notation without $\triangle$ has been used in the non-affine case, say, in [**4**], [**12**], [**33**], etc. Here the triangle $\triangle$ depicts the cyclic Dynkin diagram of affine type A.

For a ground ring $\mathcal{Z}$ and a $\mathcal{Z}$-module (or a $\mathcal{Z}$-algebra) $\mathcal{A}$, we often use the notation $\mathcal{A}_{\mathbb{F}} := \mathcal{A} \otimes \mathbb{F}$ to represent the object obtained by *base change* to a field $\mathbb{F}$, which itself is a $\mathcal{Z}$-module. In particular, if $\mathcal{Z} = \mathbb{Z}[v, v^{-1}]$, then we write $\boldsymbol{\mathcal{A}}$ for $\mathcal{A}_{\mathbb{Q}(v)}$.

Acknowledgements. The main results of the book have been presented by the authors at the following conferences and workshops. We would like to thank the organizers for the opportunities of presenting our work.

- Conference on Perspectives in Representation Theory, Cologne, September 2009;
- International Workshop on Combinatorial and Geometric Approach to Representation Theory, Seoul National University, September 2009;
- 2010 ICM Satellite Conference, Bangalore, August 2010;
- 12th National Algebra Conference, Lanzhou, June 2010;
- Southeastern Lie Theory Workshop: Finite and Algebraic Groups and Leonard Scott Day, Charlottesville, June 2011;
- 55th Annual Meeting of the Australian Mathematical Society, Wollongong, September 2011.

The research was partially supported by the Australian Research Council, the Natural Science Foundation of China, the 111 Program of China, the Program NCET, the Fok Ying Tung Education Foundation, the Fundamental Research Funds for the Central Universities of China, and the UNSW Goldstar Award. The first three and last two chapters were written while Deng and Fu were visiting the University of New South Wales at various times. The hospitality and support of UNSW are gratefully acknowledged.

The second author would like to thank Alexander Kleshchev and Arun Ram for helpful comments on the Affine Branching Rule (4.3.5.1), and Vyjayanthi

Chari for several discussions and explanations on the paper [9] and some related topics. He would also like to thank East China Normal University, and the Universities of Mainz, Virginia, and Auckland for their hospitality during his sabbatical leave in the second half of 2009.

Finally, for all their help, encouragement, and infinite patience, we thank our wives and children: Wenlian Guo and Zhuoran Deng; Chunli Yu, Andy Du, and Jason Du; Shanshan Xia.

Bangming Deng
Jie Du
Qiang Fu

Sydney
5 December 2011

1
Preliminaries

We start with the loop algebra of $\mathfrak{gl}_n(\mathbb{C})$ and its interpretation in terms of matrix Lie algebras. We use the subalgebra of integer matrices of the latter to introduce several important index sets which will be used throughout the book. Ringel–Hall algebras $\mathfrak{H}_\Delta(n)$ associated with cyclic quivers $\Delta(n)$ and their geometric construction are introduced in §1.2. In §1.3, we discuss the composition subalgebra $\mathfrak{C}_\Delta(n)$ of $\mathfrak{H}_\Delta(n)$ and relate it to the quantum loop algebra $\mathbf{U}(\widehat{\mathfrak{sl}}_n)$. We then describe in §1.4 three types of generators for $\mathfrak{H}_\Delta(n)$, which consist of all simple modules together with, respectively, the Schiffmann–Hubery central elements, homogeneous semisimple modules, and homogeneous indecomposable modules, and their associated monomial bases (Corollaries 1.4.2 and 1.4.6). These generating sets will play different roles in what follows. Finally, extended Ringel–Hall algebras and their Hopf structure are discussed in §1.5.

1.1. The loop algebra $\widehat{\mathfrak{gl}}_n$ and some notation

For a positive integer n, let $\mathfrak{gl}_n(\mathbb{C})$ be the complex general linear Lie algebra, and let

$$\widehat{\mathfrak{gl}}_n(\mathbb{C}) := \mathfrak{gl}_n(\mathbb{C}) \otimes \mathbb{C}[t, t^{-1}]$$

be the loop algebra of $\mathfrak{gl}_n(\mathbb{C})$; see [**47**]. Thus, $\widehat{\mathfrak{gl}}_n(\mathbb{C})$ is spanned by $E_{i,j} \otimes t^m$ for all $1 \leqslant i, j \leqslant n$, and $m \in \mathbb{Z}$, where $E_{i,j}$ is the matrix $(\delta_{k,i}\delta_{j,l})_{1\leqslant k,l\leqslant n}$. The (Lie) multiplication is the bracket product associated with the multiplication

$$(E_{i,j} \otimes t^m)(E_{k,l} \otimes t^{m'}) = \delta_{j,k} E_{i,l} \otimes t^{m+m'}.$$

We may interpret the Lie algebra $\widehat{\mathfrak{gl}}_n(\mathbb{C})$ as a matrix Lie algebra. Let $M_{\Delta,n}(\mathbb{C})$ be the set of all $\mathbb{Z}\times\mathbb{Z}$ complex matrices $A = (a_{i,j})_{i,j\in\mathbb{Z}}$ with $a_{i,j} \in \mathbb{C}$

such that

(a) $a_{i,j} = a_{i+n,j+n}$ for $i, j \in \mathbb{Z}$, and
(b) for every $i \in \mathbb{Z}$, the set $\{j \in \mathbb{Z} \mid a_{i,j} \neq 0\}$ is finite.

Clearly, conditions (a) and (b) imply that there are only finitely many non-zero entries in each column of A. For $A, B \in M_{\triangle,n}(\mathbb{C})$, let $[A, B] = AB - BA$. Then $(M_{\triangle,n}(\mathbb{C}), [\ ,\ ])$ becomes a Lie algebra over $\mathbb{C}$.

Denote by $M_{n,\bullet}(\mathbb{C})$ the set of $n \times \mathbb{Z}$ matrices $A = (a_{i,j})$ over $\mathbb{C}$ satisfying (b) with $i \in [1, n] := \{1, 2, \ldots, n\}$. Then there is a bijection

$$\flat_1 : M_{\triangle,n}(\mathbb{C}) \longrightarrow M_{n,\bullet}(\mathbb{C}), \quad (a_{i,j})_{i,j\in\mathbb{Z}} \longmapsto (a_{i,j})_{1\leqslant i\leqslant n, j\in\mathbb{Z}}. \tag{1.1.0.1}$$

For $i, j \in \mathbb{Z}$, let $E^{\triangle}_{i,j} \in M_{\triangle,n}(\mathbb{C})$ be the matrix $(e^{i,j}_{k,l})_{k,l\in\mathbb{Z}}$ defined by

$$e^{i,j}_{k,l} = \begin{cases} 1, & \text{if } k = i + sn, l = j + sn \text{ for some } s \in \mathbb{Z}; \\ 0, & \text{otherwise.} \end{cases}$$

The set $\{E^{\triangle}_{i,j} | 1 \leqslant i \leqslant n, j \in \mathbb{Z}\}$ is a $\mathbb{C}$-basis of $M_{\triangle,n}(\mathbb{C})$. Since

$$E^{\triangle}_{i,j+ln} E^{\triangle}_{p,q+kn} = \delta_{j,p} E^{\triangle}_{i,q+(l+k)n},$$

for all $i, j, p, q, l, k \in \mathbb{Z}$ with $1 \leqslant j, p \leqslant n$, it follows that the map

$$M_{\triangle,n}(\mathbb{C}) \longrightarrow \widehat{\mathfrak{gl}}_n(\mathbb{C}), \quad E^{\triangle}_{i,j+ln} \longmapsto E_{i,j} \otimes t^l, \ 1 \leqslant i, j \leqslant n, l \in \mathbb{Z}$$

is a Lie algebra isomorphism. We will identify the loop algebra $\widehat{\mathfrak{gl}}_n(\mathbb{C})$ with $M_{\triangle,n}(\mathbb{C})$ in the sequel.

In Chapter 6, we will consider the loop algebra $\widehat{\mathfrak{gl}}_n := \widehat{\mathfrak{gl}}_n(\mathbb{Q}) = M_{\triangle,n}(\mathbb{Q})$ defined over $\mathbb{Q}$ and its universal enveloping algebra $\mathcal{U}(\widehat{\mathfrak{gl}}_n)$ and triangular parts $\mathcal{U}(\widehat{\mathfrak{gl}}_n)^+, \mathcal{U}(\widehat{\mathfrak{gl}}_n)^-$, and $\mathcal{U}(\widehat{\mathfrak{gl}}_n)^0$. Here $\mathcal{U}(\widehat{\mathfrak{gl}}_n)^+$ (resp., $\mathcal{U}(\widehat{\mathfrak{gl}}_n)^-, \mathcal{U}(\widehat{\mathfrak{gl}}_n)^0$) is the subalgebra of $\mathcal{U}(\widehat{\mathfrak{gl}}_n)$ generated by $E^{\triangle}_{i,j}$ for all $i < j$ (resp., $E^{\triangle}_{i,j}$ for all $i > j$, $E^{\triangle}_{i,i}$ for all i). We will also relate these algebras in §6.1 with the specializations at $v = 1$ of the Ringel–Hall algebra $\mathfrak{H}_{\triangle}(n)$ and the double Ringel–Hall algebra $\mathfrak{D}_{\triangle}(n)$.

We now introduce some **notation** which will be used throughout the book.

Consider the subset $M_{\triangle,n}(\mathbb{Z})$ of $M_{\triangle,n}(\mathbb{C})$ consisting of matrices with integer entries. For each $A \in M_{\triangle,n}(\mathbb{Z})$, let

$$\mathrm{ro}(A) = \Big(\sum_{j\in\mathbb{Z}} a_{i,j}\Big)_{i\in\mathbb{Z}} \quad \text{and} \quad \mathrm{co}(A) = \Big(\sum_{i\in\mathbb{Z}} a_{i,j}\Big)_{j\in\mathbb{Z}}.$$

We obtain functions

$$\mathrm{ro}, \mathrm{co} : M_{\triangle,n}(\mathbb{Z}) \longrightarrow \mathbb{Z}^n_{\triangle},$$

where

$$\mathbb{Z}^n_\Delta := \{(\lambda_i)_{i\in\mathbb{Z}} \mid \lambda_i \in \mathbb{Z},\ \lambda_i = \lambda_{i-n} \text{ for } i \in \mathbb{Z}\}.$$

For $\lambda = (\lambda_i)_{i\in\mathbb{Z}} \in \mathbb{Z}^n_\Delta$, $A \in M_{\Delta,n}(\mathbb{Z})$, and $i_0 \in \mathbb{Z}$, let

$$\sigma(\lambda) = \sum_{i_0+1\leqslant i\leqslant i_0+n} \lambda_i \quad\text{and}\quad \sigma(A) = \sum_{\substack{i_0+1\leqslant i\leqslant i_0+n\\ j\in\mathbb{Z}}} a_{i,j} = \sum_{\substack{i_0+1\leqslant j\leqslant i_0+n\\ i\in\mathbb{Z}}} a_{i,j}.$$

Clearly, both $\sigma(\lambda)$ and $\sigma(A)$ are defined and independent of i_0. We sometimes identify $\mathbb{Z}^n_\Delta$ with $\mathbb{Z}^n$ via the following bijection

$$\flat_2 : \mathbb{Z}^n_\Delta \longrightarrow \mathbb{Z}^n, \quad \lambda \longmapsto \flat_2(\lambda) = (\lambda_1, \ldots, \lambda_n). \tag{1.1.0.2}$$

For example, we define a "dot product" on $\mathbb{Z}^n_\Delta$ by $\lambda \centerdot \mu := \flat_2(\lambda) \centerdot \flat_2(\mu) = \sum_{i=1}^n \lambda_i\mu_i$, and define the order relation $\leqslant$ on $\mathbb{Z}^n_\Delta$ by setting

$$\lambda \leqslant \mu \iff \flat_2(\lambda) \leqslant \flat_2(\mu) \iff \lambda_i \leqslant \mu_i \text{ for all } 1 \leqslant i \leqslant n. \tag{1.1.0.3}$$

Also, let $\boldsymbol{e}^\Delta_i \in \mathbb{Z}^n_\Delta$ be defined by $\flat_2(\boldsymbol{e}^\Delta_i) = \boldsymbol{e}_i = (0, \ldots, 0, \underset{(i)}{1}, 0, \ldots, 0)$.

Let

$$\begin{aligned}\Theta_\Delta(n) &:= \{A = (a_{i,j}) \in M_{\Delta,n}(\mathbb{Z}) \mid a_{i,j} \in \mathbb{N}\} = M_{\Delta,n}(\mathbb{N}),\\ \mathbb{N}^n_\Delta &:= \{(\lambda_i)_{i\in\mathbb{Z}} \in \mathbb{Z}^n_\Delta \mid \lambda_i \geqslant 0\},\end{aligned}$$

and, for $r \geqslant 0$, let

$$\begin{aligned}\Theta_\Delta(n,r) &:= \{A \in \Theta_\Delta(n) \mid \sigma(A) = r\} \quad\text{and}\\ \Lambda_\Delta(n,r) &:= \{\lambda \in \mathbb{N}^n_\Delta \mid \sigma(\lambda) = r\}.\end{aligned}$$

The set $M_n(\mathbb{Z})$ can be naturally regarded as a subset of $M_{n,\bullet}(\mathbb{Z})$ by sending $(a_{i,j})_{1\leqslant i,j\leqslant n}$ to $(a_{i,j})_{1\leqslant i\leqslant n, j\in\mathbb{Z}}$, where $a_{i,j} = 0$ if $j \in \mathbb{Z}\backslash[1,n]$. Thus, (the inverse of) $\flat_1$ induces an embedding

$$\flat_1' : M_n(\mathbb{Z}) \longrightarrow M_{\Delta,n}(\mathbb{Z}). \tag{1.1.0.4}$$

By removing the subscripts Δ, we define similarly the subsets $\Theta(n)$, $\Theta(n,r)$ of $M_n(\mathbb{Z})$ and subset $\Lambda(n,r)$ of $\mathbb{N}^n$, etc. Note that $\flat_2(\Lambda_\Delta(n,r)) = \Lambda(n,r)$.

Let $\mathcal{Z} = \mathbb{Z}[v, v^{-1}]$, where v is an indeterminate, and let $\mathbb{Q}(v)$ be the fraction field of $\mathcal{Z}$. For integers N, t with $t \geqslant 0$, let

$$\begin{bmatrix} N\\ t\end{bmatrix} = \prod_{1\leqslant i\leqslant t} \frac{v^{N-i+1} - v^{-(N-i+1)}}{v^i - v^{-i}} \in \mathcal{Z} \quad\text{and}\quad \begin{bmatrix} N\\ 0\end{bmatrix} = 1. \tag{1.1.0.5}$$

If we put $[m] = \frac{v^m - v^{-m}}{v - v^{-1}} = \left[\begin{smallmatrix} m\\ 1\end{smallmatrix}\right]$ and $[N]^! := [1][2]\cdots[N]$, then $\left[\begin{smallmatrix} N\\ t\end{smallmatrix}\right] = \frac{[N]^!}{[t]^![N-t]^!}$ for all $1 \leqslant t \leqslant N$. Given a polynomial $f \in \mathcal{Z}$ and $z \in \mathbb{C}^* := \mathbb{C}\backslash\{0\}$, we sometimes write f_z for $f(z)$, e.g., $[m]_z$, $[m]^!_z$, etc.

When counting occurs, we often use

$$\left[\!\left[{N \atop t} \right]\!\right] := v^{t(N-t)} \left[{N \atop t} \right]$$

to denote the Gaussian polynomials in v^2.

Also, for any $\mathbb{Q}(v)$-algebra $\mathscr{A}$ and an invertible element $X \in \mathscr{A}$, let

$$\left[{X; a \atop t} \right] = \prod_{s=1}^{t} \frac{Xv^{a-s+1} - X^{-1}v^{-a+s-1}}{v^s - v^{-s}} \quad \text{and} \quad \left[{X; a \atop 0} \right] = 1, \qquad (1.1.0.6)$$

for all $a, t \in \mathbb{Z}$ with $t \geqslant 1$.

1.2. Representations of cyclic quivers and Ringel–Hall algebras

Let $\triangle(n)$ $(n \geqslant 2)$ be the cyclic quiver

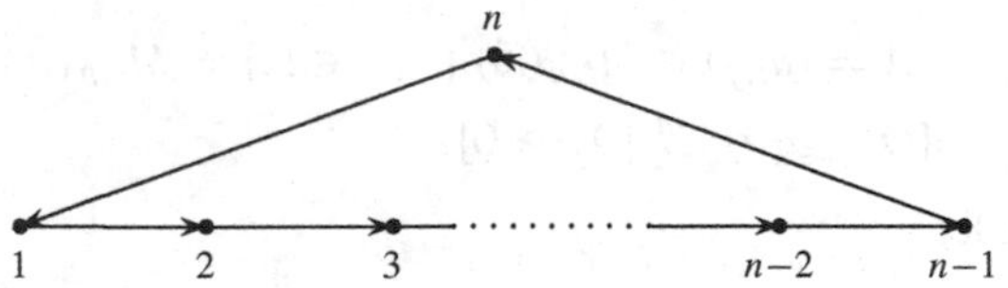

with vertex set $I = \mathbb{Z}/n\mathbb{Z} = \{1, 2, \ldots, n\}$ and arrow set $\{i \to i+1 \mid i \in I\}$. Let $\mathbb{F}$ be a field. By $\mathbf{Rep}^0\triangle(n) = \mathbf{Rep}^0_{\mathbb{F}}\triangle(n)$ we denote the category of finite dimensional *nilpotent representations* of $\triangle(n)$ over $\mathbb{F}$, i.e., representations $V = (V_i, f_i)_{i\in I}$ of $\triangle(n)$ such that all V_i are finite dimensional and the composition $f_n \cdots f_2 f_1 : V_1 \to V_1$ is nilpotent. The vector $\mathbf{dim}\, V = (\dim_{\mathbb{F}} V_i) \in \mathbb{N}I = \mathbb{N}^n$ is called the dimension vector of V. (We shall sometimes identify $\mathbb{N}I$ with $\mathbb{N}^n_\triangle$ under (1.1.0.2).) For each vertex $i \in I$, there is a one-dimensional representation S_i in $\mathbf{Rep}^0\triangle(n)$ satisfying $(S_i)_i = \mathbb{F}$ and $(S_i)_j = 0$ for $j \neq i$. It is known that the S_i form a complete set of simple objects in $\mathbf{Rep}^0\triangle(n)$. Hence, each *semisimple representation* $S_{\mathbf{a}}$ in $\mathbf{Rep}^0\triangle(n)$ is given by $S_{\mathbf{a}} = \oplus_{i\in I} a_i S_i$, where $\mathbf{a} = (a_1, \ldots, a_n) \in \mathbb{N}I$. A semisimple representation $S_{\mathbf{a}}$ is called *sincere* if $\mathbf{a}$ is sincere, namely, all a_i are positive. In particular, the vector

$$\delta := (1, \ldots, 1) \in \mathbb{N}^n$$

will often be used.

Moreover, up to isomorphism, all indecomposable representations in $\mathbf{Rep}^0\triangle(n)$ are given by $S_i[l]$ $(i \in I$ and $l \geqslant 1)$ of length l with top S_i. Thus, the isoclasses of representations in $\mathbf{Rep}^0\triangle(n)$ are indexed by multisegments

$\pi = \sum_{i\in I, l\geqslant 1} \pi_{i,l}[i; l)$, where the representation $M(\pi)$ corresponding to π is defined by

$$M(\pi) = M_{\mathbb{F}}(\pi) = \bigoplus_{1\leqslant i\leqslant n, l\geqslant 1} \pi_{i,l} S_i[l].$$

Since the set Π of all multisegments can be identified with the set

$$\Theta_\Delta^+(n) = \{A = (a_{i,j}) \in \Theta_\Delta(n) \mid a_{i,j} = 0 \text{ for } i \geqslant j\}$$

of all strictly upper triangular matrices via

$$\flat_3 : \Theta_\Delta^+(n) \longrightarrow \Pi, \quad A = (a_{i,j})_{i,j\in\mathbb{Z}} \longmapsto \sum_{i<j, 1\leqslant i\leqslant n} a_{i,j}[i; j-i),$$

we will use $\Theta_\Delta^+(n)$ to index the finite dimensional nilpotent representations. In particular, for any $i, j \in \mathbb{Z}$ with $i < j$, we have

$$M^{i,j} := M(E^\Delta_{i,j}) = S_i[j-i], \text{ and } M^{i+n,j+n} = M^{i,j}.$$

Thus, for any $A = (a_{i,j}) \in \Theta_\Delta^+(n)$ and $i_0 \in \mathbb{Z}$,

$$M(A) = M_{\mathbb{F}}(A) = \bigoplus_{1\leqslant i\leqslant n, i<j} a_{i,j} M^{i,j} = \bigoplus_{i_0+1\leqslant i\leqslant i_0+n, i<j} a_{i,j} M^{i,j}.$$

For $A = (a_{i,j}) \in \Theta_\Delta^+(n)$, set

$$\mathfrak{d}(A) = \sum_{i<j, 1\leqslant i\leqslant n} a_{i,j}(j-i).$$

Then $\dim_{\mathbb{F}} M(A) = \mathfrak{d}(A)$. Moreover, for each $\lambda = (\lambda_i) \in \mathbb{N}I$, set $A_\lambda = (a_{i,j})$ with $a_{i,j} = \delta_{j,i+1}\lambda_i$, i.e., $A_\lambda = \sum_{i=1}^n \lambda_i E^\Delta_{i,i+1}$. Then

$$M(A_\lambda) = \bigoplus_{1\leqslant i\leqslant n} \lambda_i S_i =: S_\lambda \tag{1.2.0.1}$$

is semisimple. Also, for $A \in \Theta_\Delta^+(n)$, we write $\mathbf{d}(A) = \mathbf{dim}\, M(A) \in \mathbb{Z}I$, the dimension vector of $M(A)$. Hence, $\mathbb{Z}I$ is identified with the Grothendieck group of $\mathbf{Rep}^0\!\Delta(n)$.

A matrix $A = (a_{i,j}) \in \Theta_\Delta^+(n)$ is called *aperiodic* if, for each $l \geqslant 1$, there exists $i \in \mathbb{Z}$ such that $a_{i,i+l} = 0$. Otherwise, A is called *periodic*. A nilpotent representation $M(A)$ is called aperiodic (resp., periodic) if A is aperiodic (resp., periodic).

It is well known that there exist Auslander–Reiten sequences in $\mathbf{Rep}^0\!\Delta(n)$; see [**1**]. More precisely, for each $i \in I$ and each $l \geqslant 1$, there is an Auslander–Reiten sequence

$$0 \longrightarrow S_{i+1}[l] \longrightarrow S_i[l+1] \oplus S_{i+1}[l-1] \longrightarrow S_i[l] \longrightarrow 0,$$

where we set $S_{i+1}[0] = 0$ by convention. $S_{i+1}[l]$ is called the Auslander–Reiten translate of $S_i[l]$, denoted by $\tau S_i[l]$. In this case, τ indeed defines an equivalence from $\mathbf{Rep}^0\!\triangle(n)$ to itself, called the Auslander–Reiten translation. For each $A = (a_{i,j}) \in \Theta_\triangle^+(n)$, we define $\tau(A) \in \Theta_\triangle^+(n)$ by $M(\tau(A)) = \tau M(A)$. Thus, if we write $\tau(A) = (b_{i,j}) \in \Theta_\triangle^+(n)$, then $b_{i,j} = a_{i-1,j-1}$ for all i, j.

We now introduce the degeneration order on $\Theta_\triangle^+(n)$ and generic extensions of nilpotent representations. These notions play an important role in the study of bases for both the Ringel–Hall algebra $\mathfrak{H}_\triangle(n)$ of $\triangle(n)$ and its composition subalgebra $\mathfrak{C}_\triangle(n)$; see, for example, [**11, 13**]. For two nilpotent representations M, N in $\mathbf{Rep}^0\!\triangle(n)$ with $\mathbf{dim}\, M = \mathbf{dim}\, N$, define

$$N \leqslant_{\mathrm{dg}} M \iff \dim_{\mathbb{F}} \mathrm{Hom}(X, N) \geqslant \dim_{\mathbb{F}} \mathrm{Hom}(X, M), \text{ for all } X \in \mathbf{Rep}^0\!\triangle(n); \tag{1.2.0.2}$$

see [**82**]. This gives rise to a partial order on the set of isoclasses of representations in $\mathbf{Rep}^0\!\triangle(n)$, called the *degeneration order*. Thus, it also induces a partial order on $\Theta_\triangle^+(n)$ by letting

$$A \leqslant_{\mathrm{dg}} B \iff M(A) \leqslant_{\mathrm{dg}} M(B).$$

By [**62**] and [**11**, §3], for any two nilpotent representations M and N, there exists a unique extension G (up to isomorphism) of M by N with minimal $\dim \mathrm{End}(G)$. This representation G is called the *generic extension* of M by N and will be denoted by $M * N$ in the sequel. Moreover, for nilpotent representations M_1, M_2, M_3,

$$(M_1 * M_2) * M_3 \cong M_1 * (M_2 * M_3).$$

Also, taking generic extensions preserves the degeneration order. More precisely, if $N_1 \leqslant_{\mathrm{dg}} M_1$ and $N_2 \leqslant_{\mathrm{dg}} M_2$, then $N_1 * N_2 \leqslant_{\mathrm{dg}} M_1 * M_2$. For $A, B \in \Theta_\triangle^+(n)$, let $A * B \in \Theta_\triangle^+(n)$ be defined by $M(A * B) \cong M(A) * M(B)$.

As above, let $\mathcal{Z} = \mathbb{Z}[v, v^{-1}]$ be the Laurent polynomial ring in indeterminate v. By [**65**] and [**36**], for $A, B_1, \ldots, B_m \in \Theta_\triangle^+(n)$, there is a polynomial $\varphi^A_{B_1,\ldots,B_m} \in \mathbb{Z}[v^2]$ in v^2, called the *Hall polynomial*, such that for any finite field $\mathbb{F}$ of q elements, $\varphi^A_{B_1,\ldots,B_m}|_{v^2=q}$ (the evaluation of $\varphi^A_{B_1,\ldots,B_m}$ at $v^2 = q$) equals the number $F^{M_{\mathbb{F}}(A)}_{M_{\mathbb{F}}(B_1),\ldots,M_{\mathbb{F}}(B_m)}$ of the filtrations

$$0 = M_m \subseteq M_{m-1} \subseteq \cdots \subseteq M_1 \subseteq M_0 = M_{\mathbb{F}}(A)$$

such that $M_{t-1}/M_t \cong M_{\mathbb{F}}(B_t)$ for all $1 \leqslant t \leqslant m$.

Moreover, for each $A = (a_{i,j}) \in \Theta_\triangle^+(n)$, there is a polynomial $\mathfrak{a}_A = \mathfrak{a}_A(v^2) \in \mathcal{Z}$ in v^2 such that, for each finite field $\mathbb{F}$ with q elements, $\mathfrak{a}_A|_{v^2=q} = |\mathrm{Aut}(M_{\mathbb{F}}(A))|$; see, for example, [**59**, Cor. 2.1.1]. For later use, we give an explicit formula for $\mathfrak{a}_A$. Let m_A denote the dimension of rad End$(M_{\mathbb{F}}(A))$, which is known to be independent of the field $\mathbb{F}$. We also have

$$\mathrm{End}(M_{\mathbb{F}}(A))/\mathrm{rad}\,\mathrm{End}(M_{\mathbb{F}}(A)) \cong \prod_{1\leqslant i\leqslant n,\, a_{i,j}>0} M_{a_{i,j}}(\mathbb{F}),$$

where $M_{a_{i,j}}(\mathbb{F})$ denotes the full matrix algebra of $a_{i,j} \times a_{i,j}$ matrices over $\mathbb{F}$. Hence,

$$|\,\mathrm{Aut}(M_{\mathbb{F}}(A))| = |\mathbb{F}|^{m_A} \prod_{1\leqslant i\leqslant n,\, a_{i,j}>0} |GL_{a_{i,j}}(\mathbb{F})|.$$

Consequently,

$$\mathfrak{a}_A = v^{2m_A} \prod_{1\leqslant i\leqslant n,\, a_{i,j}>0} (v^{2a_{i,j}}-1)(v^{2a_{i,j}}-v^2)\cdots(v^{2a_{i,j}}-v^{2a_{i,j}-2}). \quad (1.2.0.3)$$

In particular, if $z \in \mathbb{C}^*$ is not a root of unity, then $\mathfrak{a}_A|_{v^2=z} \neq 0$.

Let $\mathfrak{H}_\Delta(n)$ be the *(generic) Ringel–Hall algebra* of the cyclic quiver $\Delta(n)$, which is by definition the free $\mathcal{Z}$-module with basis $\{u_A = u_{[M(A)]} \mid A \in \Theta_\Delta^+(n)\}$. The multiplication is given by

$$u_A u_B = v^{\langle \mathbf{d}(A),\mathbf{d}(B)\rangle} \sum_{C\in\Theta_\Delta^+(n)} \varphi_{A,B}^C u_C$$

for $A, B \in \Theta_\Delta^+(n)$, where

$$\langle \mathbf{d}(A), \mathbf{d}(B)\rangle = \dim \mathrm{Hom}(M(A), M(B)) - \dim \mathrm{Ext}^1(M(A), M(B)) \quad (1.2.0.4)$$

is the *Euler form* associated with the cyclic quiver $\Delta(n)$. If we write $\mathbf{d}(A) = (a_i)$ and $\mathbf{d}(B) = (b_i)$, then

$$\langle \mathbf{d}(A), \mathbf{d}(B)\rangle = \sum_{i\in I} a_i b_i - \sum_{i\in I} a_i b_{i+1}. \quad (1.2.0.5)$$

Since both $\dim_{\mathbb{F}} \mathrm{End}(M_{\mathbb{F}}(A))$ and $\dim_{\mathbb{F}} M_{\mathbb{F}}(A) = \mathfrak{d}(A)$ are independent of the ground field, we put for each $A \in \Theta_\Delta^+(n)$,

$$d'_A = \dim_{\mathbb{F}} \mathrm{End}(M_{\mathbb{F}}(A)) - \dim_{\mathbb{F}} M_{\mathbb{F}}(A) \text{ and } \widetilde{u}_A = v^{d'_A} u_A; \quad (1.2.0.6)$$

cf. [**13**, (8.1)].[1] As seen in [**13**], it is sometimes convenient to work with the PBW type basis $\{\widetilde{u}_A \mid A \in \Theta_\Delta^+(n)\}$ of $\mathfrak{H}_\Delta(n)$.

The degeneration order gives rise to the following "triangular relation" in $\mathfrak{H}_\Delta(n)$: for $A_1, \ldots, A_t \in \Theta_\Delta^+(n)$,

$$u_{A_1}\cdots u_{A_t} = v^{\sum_{1\leqslant r<s\leqslant t}\langle \mathbf{d}(A_r),\mathbf{d}(A_s)\rangle} \sum_{B\leqslant_{\mathrm{dg}} A_1*\cdots*A_t} \varphi_{A_1,\ldots,A_t}^B u_B. \quad (1.2.0.7)$$

[1] There is another notation, called d_A, which will be defined in (3.1.3.2) for all $A \in \Theta_\Delta(n)$.

There is a natural $\mathbb{N}I$-grading on $\mathfrak{H}_\Delta(n)$:

$$\mathfrak{H}_\Delta(n) = \bigoplus_{\mathbf{d}\in\mathbb{N}I} \mathfrak{H}_\Delta(n)_{\mathbf{d}}, \tag{1.2.0.8}$$

where $\mathfrak{H}_\Delta(n)_{\mathbf{d}}$ is spanned by all u_A with $\mathbf{d}(A) = \mathbf{d}$. Moreover, we will frequently consider in the sequel the algebra $\boldsymbol{\mathfrak{H}}_\Delta(n) = \mathfrak{H}_\Delta(n) \otimes_{\mathcal{Z}} \mathbb{Q}(v)$ obtained by base change to the fraction field $\mathbb{Q}(v)$.

In order to relate Ringel–Hall algebras $\mathfrak{H}_\Delta(n)$ of cyclic quivers with affine quantum Schur algebras later on, we recall Lusztig's geometric construction of $\mathfrak{H}_\Delta(n)$ (specializing v to a square root of a prime power) developed in [**53, 54**] (cf. also [**55**]).

Let $\mathbb{F} = \mathbb{F}_q$ be the finite field of q elements and $\mathbf{d} \in \mathbb{N}I$. Fix an I-graded $\mathbb{F}$-vector space $V = \oplus_{i\in I} V_i$ of dimension vector $\mathbf{d}$, i.e., $\dim_{\mathbb{F}} V_i = d_i$ for all $i \in I$. Then each element in

$$E_V = \Big\{(f_i) \in \bigoplus_{i\in I} \mathrm{Hom}_{\mathbb{F}}(V_i, V_{i+1}) \mid f_n \cdots f_1 \text{ is nilpotent}\Big\}$$

can be viewed as a nilpotent representation of $\Delta(n)$ over $\mathbb{F}$ of dimension vector $\mathbf{d}$. The group $G_V = \prod_{i\in I} GL(V_i)$ acts on E_V by conjugation. Then there is a bijection between the G_V-orbits in E_V and the isoclasses of nilpotent representations of $\Delta(n)$ of dimension vector $\mathbf{d}$. For each $A \in \Theta_\Delta^+(n)$ with $\mathbf{d}(A) = \mathbf{d}$, we will denote by $\mathfrak{O}_A$ the orbit in E_V corresponding to the isoclass of $M(A)$.

Define $\mathtt{H}_{\mathbf{d}} = \mathbb{C}_{G_V}(E_V)$ to be the vector space of G_V-invariant functions from E_V to $\mathbb{C}$. Now let $\mathbf{a}, \mathbf{b} \in \mathbb{N}_\Delta^n$ with $\mathbf{d} = \mathbf{a} + \mathbf{b}$ and fix I-graded $\mathbb{F}$-vector spaces $U = \oplus_{i\in I} U_i$ and $W = \oplus_{i\in I} W_i$ with dimension vectors $\mathbf{a}$ and $\mathbf{b}$, respectively. Let E be the set of triples (x, ϕ, ψ) such that (1) $x \in E_V$, (2) the sequence

$$0 \longrightarrow W \xrightarrow{\phi} V \xrightarrow{\psi} U \longrightarrow 0$$

of I-graded spaces is exact, and (3) $\phi(W)$ is stable by x. Let F be the set of pairs (x, W'), where $x \in E_V$ and $W' \subseteq V$ is an x-stable I-graded subspace of dimension vector $\mathbf{b}$. Consider the diagram

$$E_U \times E_W \xleftarrow{p_1} E \xrightarrow{p_2} F \xrightarrow{p_3} E_V,$$

where p_1, p_2, p_3 are projections defined in an obvious way. Given $f \in \mathbb{C}_{G_U}(E_U) = \mathtt{H}_{\mathbf{a}}$ and $g \in \mathbb{C}_{G_W}(E_W) = \mathtt{H}_{\mathbf{b}}$, define the convolution product of f and g by

$$fg = q^{-\frac{1}{2}m(\mathbf{a},\mathbf{b})}(p_3)_! h \in \mathbb{C}_{G_V}(E_V) = \mathtt{H}_{\mathbf{d}},$$

where $h \in \mathbb{C}(F)$ is the function such that $p_2^* h = p_1^*(fg)$ and

$$m(\mathbf{a}, \mathbf{b}) = \sum_{1 \leqslant i \leqslant n} a_i b_i + \sum_{1 \leqslant i \leqslant n} a_i b_{i+1}.$$

Consequently, $\mathrm{H} = \bigoplus_{\mathbf{d} \in \mathbb{N}_\Delta^n} \mathrm{H}_\mathbf{d}$ becomes an associative algebra over $\mathbb{C}$.[2] For each $A \in \Theta_\Delta^+(n)$, let $\chi_{\mathfrak{O}_A}$ be the characteristic function of $\mathfrak{O}_A$ and put

$$\langle \mathfrak{O}_A \rangle = q^{-\frac{1}{2} \dim \mathfrak{O}_A} \chi_{\mathfrak{O}_A}.$$

By [**13**, 8.1]), there is an algebra isomorphism

$$\mathfrak{H}_\Delta(n) \otimes_{\mathcal{Z}} \mathbb{C} \longrightarrow \mathrm{H}, \quad u_A \longmapsto v^{\mathfrak{d}(A) - \mathbf{d}(A).\mathbf{d}(A)} \chi_{\mathfrak{O}_A}, \quad \text{for all } A \in \Theta_\Delta^+(n), \tag{1.2.0.9}$$

where $\mathbb{C}$ is viewed as a $\mathcal{Z}$-module by specializing v to $q^{\frac{1}{2}}$. In particular, this isomorphism takes $\widetilde{u}_A$ to $\langle \mathfrak{O}_A \rangle$.

1.3. The quantum loop algebra $\mathbf{U}(\widehat{\mathfrak{sl}}_n)$

As mentioned in the introduction, an important breakthrough for the structure of quantum groups associated with semisimple complex Lie algebras is Ringel's Hall algebra realization of the $\pm$-part of the quantum enveloping algebra associated with the same quiver; see [**63, 64**]. For the Ringel–Hall algebra $\mathfrak{H}_\Delta(n)$ associated with a cyclic quiver, it is known from [**65**] that a subalgebra, the composition algebra, is isomorphic to the $\pm$-part of a quantum affine $\mathfrak{sl}_n$. We now describe this algebra and use it to display a certain monomial basis.

The $\mathcal{Z}$-subalgebra $\mathfrak{C}_\Delta(n)$ of $\mathfrak{H}_\Delta(n)$ generated by $u_{[mS_i]}$ ($i \in I$ and $m \geqslant 1$) is called the *composition algebra* of $\Delta(n)$. By (1.2.0.8), $\mathfrak{C}_\Delta(n)$ inherits an $\mathbb{N}I$-grading by dimension vectors:

$$\mathfrak{C}_\Delta(n) = \bigoplus_{\mathbf{d} \in \mathbb{N}I} \mathfrak{C}_\Delta(n)_\mathbf{d},$$

where $\mathfrak{C}_\Delta(n)_\mathbf{d} = \mathfrak{C}_\Delta(n) \cap \mathfrak{H}_\Delta(n)_\mathbf{d}$. Let $u_i := u_{E_{i,i+1}^\Delta} = u_{[S_i]}$ and define the divided power

$$u_i^{(m)} = \frac{1}{[m]^!} u_i^m \in \mathfrak{C}_\Delta(n),$$

for $i \in I$ and $m \geqslant 1$. In fact,

$$u_i^{(m)} = v^{m(m-1)} u_{[mS_i]} \in \mathfrak{C}_\Delta(n) \subset \mathfrak{H}_\Delta(n).$$

[2] The algebra H defined also geometrically in [**73**, 3.2] has a multiplication opposite to the one for H here.

We now use the strong monomial basis property developed in **[11, 13]** to construct an explicit monomial basis of $\mathfrak{C}_\Delta(n)$. For each $A = (a_{i,j}) \in \Theta^+_\Delta(n)$, define

$$\ell = \ell_A = \max\{j - i \mid a_{i,j} \neq 0\}.$$

In other words, ℓ is the Loewy length of the representation $M(A)$.

Suppose now A is aperiodic. Then there is $i_1 \in [1, n]$ such that $a_{i_1,i_1+\ell} \neq 0$, but $a_{i_1+1,i_1+1+\ell} = 0$. If there are some $a_{i_1+1,j} \neq 0$, we let $p \geqslant 1$ satisfy $a_{i_1+1,i_1+1+p} \neq 0$ and $a_{i_1+1,j} = 0$ for all $j > i_1 + 1 + p$; if $a_{i_1+1,j} = 0$ for all $j > i_1 + 1$, let $p = 0$. Thus, $\ell > p$. Now set

$$t_1 = a_{i_1,i_1+1+p} + \cdots + a_{i_1,i_1+\ell}$$

and define $A_1 = (b_{i,j}) \in \Theta^+_\Delta(n)$ by letting

$$b_{i,j} = \begin{cases} 0, & \text{if } i = i_1,\ j \geqslant i_1 + 1 + p; \\ a_{i_1+1,j} + a_{i_1,j}, & \text{if } i = i_1 + 1 < j,\ j \geqslant i_1 + 1 + p; \\ a_{i,j}, & \text{otherwise.} \end{cases}$$

Then, A_1 is again aperiodic. Applying the above process to A_1, we get i_2 and t_2. Repeating the above process (ending with the zero matrix), we finally get two sequences $i_1, \dots, i_m$ and $t_1, \dots, t_m$. This gives a word

$$w_A = i_1^{t_1} i_2^{t_2} \cdots i_m^{t_m},$$

where $i_1, \dots, i_m$ are viewed as elements in $I = \mathbb{Z}/n\mathbb{Z}$, and define the monomial

$$u^{(A)} = u_{i_1}^{(t_1)} u_{i_2}^{(t_2)} \cdots u_{i_m}^{(t_m)} \in \mathfrak{C}_\Delta(n).$$

The algorithm above can be easily modified to get a similar algorithm for quantum $\mathfrak{gl}_n$. We illustrate the algorithm with an example in this case.

Example 1.3.1. If $A = \begin{smallmatrix}1&2&3&4\\ &5&0&0\\ &&6&0\\ &&&7\end{smallmatrix}$, then $\ell = 4$, $i_1 = 1$, $p = 1$ and $t_1 = 2+3+4 = 9$. Here we ignore all zero entries on and below the diagonal for simplicity. Thus,

$$A_1 = \begin{smallmatrix}1&0&0&0\\ &7&3&4\\ &&6&0\\ &&&7\end{smallmatrix}, \quad \ell = 3, i_2 = 2, p = 1, \text{ and } t_2 = 3 + 4 = 7,$$

$$A_2 = \begin{smallmatrix}1&0&0&0\\ &7&0&0\\ &&9&4\\ &&&7\end{smallmatrix}, \quad \ell = 2, i_3 = 3, p = 1, \text{ and } t_3 = 4, \quad \text{and}$$

$$A_3 = \begin{smallmatrix}1&0&0&0\\ &7&0&0\\ &&9&0\\ &&&11\end{smallmatrix}, \quad \ell = 1, i_4 = 4, p = 0, \text{ and } t_4 = 11.$$

Now, for a matrix defining a semisimple representation, we have all $\ell = 1$ and $p = 0$. So the remaining cases are $i_5 = 3, t_5 = 9$; $i_6 = 2, t_6 = 7$; and $i_7 = 1, t_7 = 1$. Hence,

$$u^{(A)} = u_1^{(9)} u_2^{(7)} u_3^{(4)} u_4^{(11)} u_3^{(9)} u_2^{(7)} u_1.$$

Proposition 1.3.2. *The set*

$$\{u^{(A)} \mid A \in \Theta_\Delta^+(n) \text{ aperiodic}\}$$

is a $\mathcal{Z}$-basis of $\mathfrak{C}_\Delta(n)$.

Proof. Let $A \in \Theta_\Delta^+(n)$ be aperiodic and let $w_A = i_1^{t_1} i_2^{t_2} \cdots i_m^{t_m}$ be the corresponding word constructed as above. By [**11**, Th. 5.5], w_A is distinguished, that is, $\varphi_{A_1,\ldots,A_m}^A = 1$, where $A_s = t_s E_{i_s,i_s+1}^\Delta$ for $1 \leqslant s \leqslant m$. By [**13**, Th. 7.5(i)], the $u^{(A)}$ with $A \in \Theta_\Delta^+(n)$ aperiodic form a $\mathcal{Z}$-basis of $\mathfrak{C}_\Delta(n)$. □

We now define the quantum enveloping algebra of the loop algebra (the *quantum loop algebra* for short) of $\mathfrak{sl}_n$. Let $C = C_{\Delta(n)} = (c_{i,j})_{i,j \in I}$ be the generalized Cartan matrix of type $\widetilde{A}_{n-1}$, where $I = \mathbb{Z}/n\mathbb{Z}$. We always assume that if $n \geqslant 3$, then $c_{i,i} = 2$, $c_{i,i+1} = c_{i+1,i} = -1$ and $c_{i,j} = 0$ otherwise. If $n = 2$, then $c_{1,1} = c_{2,2} = 2$ and $c_{1,2} = c_{2,1} = -2$. In other words,

$$C = \begin{pmatrix} 2 & -2 \\ -2 & 2 \end{pmatrix} \text{ or } C = \begin{pmatrix} 2 & -1 & 0 & \cdots & 0 & -1 \\ -1 & 2 & -1 & \cdots & 0 & 0 \\ 0 & -1 & 2 & \cdots & 0 & 0 \\ \vdots & \vdots & \vdots & \ddots & \vdots & \vdots \\ 0 & 0 & 0 & \cdots & 2 & -1 \\ -1 & 0 & 0 & \cdots & -1 & 2 \end{pmatrix} \ (n \geqslant 3). \tag{1.3.2.1}$$

The quantum group associated to C is denoted by $\mathbf{U}(\widehat{\mathfrak{sl}}_n)$.

Definition 1.3.3. Let $n \geqslant 2$ and $I = \mathbb{Z}/n\mathbb{Z}$. The quantum loop algebra $\mathbf{U}(\widehat{\mathfrak{sl}}_n)$ is the algebra over $\mathbb{Q}(\upsilon)$ presented by generators

$$E_i,\ F_i,\ \widetilde{K}_i,\ \widetilde{K}_i^{-1},\ i \in I,$$

and relations, for $i, j \in I$,

(QSL0) $\widetilde{K}_1 \widetilde{K}_2 \cdots \widetilde{K}_n = 1$;
(QSL1) $\widetilde{K}_i \widetilde{K}_j = \widetilde{K}_j \widetilde{K}_i$, $\widetilde{K}_i \widetilde{K}_i^{-1} = 1$;
(QSL2) $\widetilde{K}_i E_j = \upsilon^{c_{i,j}} E_j \widetilde{K}_i$;
(QSL3) $\widetilde{K}_i F_j = \upsilon^{-c_{i,j}} F_j \widetilde{K}_i$;
(QSL4) $E_i E_j = E_j E_i$, $F_i F_j = F_j F_i$ if $i \neq j \pm 1$;
(QSL5) $E_i F_j - F_j E_i = \delta_{i,j} \frac{\widetilde{K}_i - \widetilde{K}_i^{-1}}{\upsilon - \upsilon^{-1}}$;
(QSL6) $E_i^2 E_j - (\upsilon + \upsilon^{-1}) E_i E_j E_i + E_j E_i^2 = 0$ if $i = j \pm 1$ and $n \geqslant 3$;

(QSL7) $F_i^2F_j - (v+v^{-1})F_iF_jF_i + F_jF_i^2 = 0$ if $i = j \pm 1$ and $n \geqslant 3$;
(QSL6′) $E_i^3E_j - (v^2+1+v^{-2})E_i^2E_jE_i + (v^2+1+v^{-2})E_iE_jE_i^2 - E_jE_i^3 = 0$ if $i \neq j$ and $n = 2$;
(QSL7′) $F_i^3F_j - (v^2+1+v^{-2})F_i^2F_jF_i + (v^2+1+v^{-2})F_iF_jF_i^2 - F_jF_i^3 = 0$ if $i \neq j$ and $n = 2$.

For later use in representation theory, let $\mathrm{U}_{\mathbb{C}}(\widehat{\mathfrak{sl}}_n)$ be the quantum loop algebra defined by the same generators and relations (QSL0)–(QSL7) with v replaced by a non-root-of-unity $z \in \mathbb{C}^*$ and $\mathbb{Q}(v)$ by $\mathbb{C}$.

A new presentation for $\mathbf{U}(\widehat{\mathfrak{sl}}_n)$ and $\mathrm{U}_{\mathbb{C}}(\widehat{\mathfrak{sl}}_n)$, known as Drinfeld's new presentation, will be discussed in §2.5.

In this book, *quantum affine* $\mathfrak{sl}_n$ always refers to the *quantum loop (Hopf) algebra* $\mathbf{U}(\widehat{\mathfrak{sl}}_n)$.[3] We will mainly work with $\mathbf{U}(\widehat{\mathfrak{sl}}_n)$ or quantum groups defined over $\mathbb{Q}(v)$ and mention from time to time a parallel theory over $\mathbb{C}$.

Let $\mathbf{U}(\widehat{\mathfrak{sl}}_n)^+$ (resp., $\mathbf{U}(\widehat{\mathfrak{sl}}_n)^-$, $\mathbf{U}(\widehat{\mathfrak{sl}}_n)^0$) be the positive (resp., negative, zero) part of the quantum enveloping algebra $\mathbf{U}(\widehat{\mathfrak{sl}}_n)$. In other words, $\mathbf{U}(\widehat{\mathfrak{sl}}_n)^+$ (resp., $\mathbf{U}(\widehat{\mathfrak{sl}}_n)^-$, $\mathbf{U}(\widehat{\mathfrak{sl}}_n)^0$) is a $\mathbb{Q}(v)$-subalgebra generated by E_i (resp., F_i, $\widetilde{K}_i^{\pm1}$), $i \in I$.

Let

$$\mathfrak{C}_\Delta(n) = \mathfrak{C}_\Delta(n) \otimes_{\mathcal{Z}} \mathbb{Q}(v).$$

Thus, $\mathfrak{C}_\Delta(n)$ identifies with the $\mathbb{Q}(v)$-subalgebra $\mathfrak{H}_\Delta(n)$ generated by $u_i = u_{[S_i]}$ for $i \in I$.

Theorem 1.3.4. ([**65**]) *There are $\mathbb{Q}(v)$-algebra isomorphisms*

$$\mathfrak{C}_\Delta(n) \longrightarrow \mathbf{U}(\widehat{\mathfrak{sl}}_n)^+,\ u_i \longmapsto E_i \ \text{ and } \ \mathfrak{C}_\Delta(n)^{\mathrm{op}} \longrightarrow \mathbf{U}(\widehat{\mathfrak{sl}}_n)^-,\ u_i \longmapsto F_i.$$

By this theorem and the triangular decomposition

$$\mathbf{U}(\widehat{\mathfrak{sl}}_n) = \mathbf{U}(\widehat{\mathfrak{sl}}_n)^+ \otimes \mathbf{U}(\widehat{\mathfrak{sl}}_n)^0 \otimes \mathbf{U}(\widehat{\mathfrak{sl}}_n)^-,$$

the basis displayed in Proposition 1.3.2 gives rise to a monomial basis for $\mathbf{U}(\widehat{\mathfrak{sl}}_n)$.

1.4. Three types of generators and associated monomial bases

In this section, we display three distinct minimal sets of generators for $\mathfrak{H}_\Delta(n)$, each of which contains the generators $\{u_i\}_{i\in I}$ for $\mathfrak{C}_\Delta(n)$. We also describe their associated monomial bases for $\mathfrak{H}_\Delta(n)$ in the respective generators.

[3] If (QSL0) is dropped, it also defines a quantum affine $\mathfrak{sl}_n$ with the *central extension*; see, e.g., [**9**].

The first minimal set of generators contains simple modules and certain central elements. These generators are convenient for a presentation for the double Ringel–Hall algebras over $\mathbb{Q}(v)$ (or a specialization at a non-root-of-unity) associated to cyclic quivers (see Chapter 2).

In [**67**] Schiffmann first described the structure of $\mathfrak{H}_\Delta(n)$ as a tensor product of $\mathfrak{C}_\Delta(n)$ and a polynomial algebra in infinitely many indeterminates. Later Hubery explicitly constructed these central elements in [**39**]. More precisely, for each $m \geqslant 1$, let

$$c_m = (-1)^m v^{-2nm} \sum_A (-1)^{\dim \mathrm{End}(M(A))} \mathfrak{a}_A u_A \in \mathfrak{H}_\Delta(n), \qquad (1.4.0.1)$$

where the sum is taken over all $A \in \Theta_\Delta^+(n)$ such that $\mathbf{d}(A) = \mathbf{dim}\, M(A) = m\delta$ with $\delta = (1, \dots, 1) \in \mathbb{N}^n$, and $\mathrm{soc}\, M(A)$ is square-free, i.e., $\mathbf{dim}\ \mathrm{soc}\, M(A) \leqslant \delta$ in the order defined in (1.1.0.3). Note that in this case, $\mathrm{soc}\, M(A)$ is square-free if and only if $\mathrm{top}\, M(A) := M(A)/\mathrm{rad}\, M(A)$ is square-free. The following result is proved in [**67, 39**].

Theorem 1.4.1. *The elements c_m are central in $\mathfrak{H}_\Delta(n)$. Moreover, there is a decomposition*

$$\mathfrak{H}_\Delta(n) = \mathfrak{C}_\Delta(n) \otimes_{\mathbb{Q}(v)} \mathbb{Q}(v)[c_1, c_2, \dots],$$

where $\mathbb{Q}(v)[c_1, c_2, \dots]$ is the polynomial algebra in c_m for $m \geqslant 1$. In particular, $\mathfrak{H}_\Delta(n)$ is generated by u_i and c_m for $i \in I$ and $m \geqslant 1$.

We will call the central elements c_m the *Schiffmann–Hubery generators.*

Let $A = (a_{i,j}) \in \Theta_\Delta^+(n)$. For each $s \geqslant 1$, define

$$m_s = m_s(A) = \min\{a_{i,j} \mid j - i = s\} \ \text{ and } \ A' = A - \sum_{1 \leqslant i \leqslant n,\, i<j} m_{j-i} E_{i,j}^\Delta. \qquad (1.4.1.1)$$

Then A' is aperiodic. Moreover, for $A, B \in \Theta_\Delta^+(n)$,

$$A = B \iff A' = B' \ \text{ and } \ m_s(A) = m_s(B),\ \forall\, s \geqslant 1.$$

The next corollary is a direct consequence of Theorem 1.4.1 and Proposition 1.3.2.

Corollary 1.4.2. *The set*

$$\{u^{(A')} \prod_{s \geqslant 1} c_s^{m_s(A)} \mid A \in \Theta_\Delta^+(n)\}$$

is a $\mathbb{Q}(v)$-basis of $\mathfrak{H}_\Delta(n)$.

Next, we look at the minimal set of generators consisting of simple modules and *homogeneous* semisimple modules. It is known from [**73**, Prop. 3.5] (or [**13**, Th. 5.2(i)]) that $\mathfrak{H}_\Delta(n)$ is also generated by $u_{\mathbf{a}} = u_{[S_{\mathbf{a}}]}$ for $\mathbf{a} \in \mathbb{N}I$; see also (1.4.4.1) below. If $\mathbf{a}$ is not sincere, say $a_i = 0$, then

$$u_{\mathbf{a}} = \prod_{j\in I,\, j\neq i} \frac{v^{a_j(1-a_j)}}{[a_j]^!} u_{i-1}^{a_{i-1}} \cdots u_1^{a_1} u_n^{a_n} \cdots u_{i+1}^{a_{i+1}} \in \mathfrak{C}_\Delta(n). \tag{1.4.2.1}$$

Thus, $\mathfrak{H}_\Delta(n)$ is generated by u_i and $u_{\mathbf{a}}$, for $i \in I$ and sincere $\mathbf{a} \in \mathbb{N}I$. Indeed, this result can be strengthened as follows; see also [**67**, p. 421].

Proposition 1.4.3. *The Ringel–Hall algebra $\mathfrak{H}_\Delta(n)$ is generated by u_i and $u_{m\delta}$, for $i \in I$ and $m \geqslant 1$.*

Proof. Let $\mathfrak{H}'$ be the $\mathbb{Q}(v)$-subalgebra generated by u_i and $u_{m\delta}$ for $i \in I$ and $m \geqslant 1$. To show $\mathfrak{H}' = \mathfrak{H}_\Delta(n)$, it suffices to prove $u_{\mathbf{a}} = u_{[S_{\mathbf{a}}]} \in \mathfrak{H}'$ for all $\mathbf{a} \in \mathbb{N}I$.

Take an arbitrary $\mathbf{a} \in \mathbb{N}I$. We proceed by induction on $\sigma(\mathbf{a}) = \sum_{i\in I} a_i$ to show $u_{\mathbf{a}} \in \mathfrak{H}'$. If $\sigma(\mathbf{a}) = 0$ or 1, then clearly $u_{\mathbf{a}} \in \mathfrak{H}'$. Now let $\sigma(\mathbf{a}) > 1$. If $\mathbf{a}$ is not sincere, then by (1.4.2.1), $u_{\mathbf{a}} \in \mathfrak{H}'$. So we may assume $\mathbf{a}$ is sincere. The case where $a_1 = \cdots = a_n$ is trivial. Suppose now there exists $i \in I$ such that $a_i \neq a_{i+1}$. Define $\mathbf{a}' = (a'_j)$, $\mathbf{a}'' = (a''_j) \in \mathbb{N}I$ by

$$a'_j = \begin{cases} a_i - 1, & \text{if } j = i; \\ a_i, & \text{otherwise,} \end{cases} \quad \text{and} \quad a''_j = \begin{cases} a_{i+1} - 1, & \text{if } j = i+1; \\ a_i, & \text{otherwise.} \end{cases}$$

Then, in $\mathfrak{H}_\Delta(n)$,

$$u_i u_{\mathbf{a}'} = v^{a_i - a_{i+1} - 1}(u_X + v^{a_i - 1}[a_i]u_{\mathbf{a}}) \quad \text{and}$$

$$u_{\mathbf{a}''} u_{i+1} = v^{a_{i+1} - a_i - 1}(u_X + v^{a_{i+1} - 1}[a_{i+1}]u_{\mathbf{a}}),$$

where $X \in \Theta_\Delta^+(n)$ is given by

$$M(X) \cong \bigoplus_{j\neq i,i+1} a_j S_j \oplus (a_i - 1)S_i \oplus (a_{i+1} - 1)S_{i+1} \oplus S_i[2].$$

Therefore,

$$u_i u_{\mathbf{a}'} - v^{2a_i - 2a_{i+1}} u_{\mathbf{a}''} u_{i+1} = v^{a_i - a_{i+1} - 1}(v^{a_i - 1}[a_i] - v^{a_{i+1} - 1}[a_{i+1}])u_{\mathbf{a}}.$$

The inequality $a_i \neq a_{i+1}$ implies $v^{a_i-1}[a_i] - v^{a_{i+1}-1}[a_{i+1}] \neq 0$. Thus, we obtain

$$u_{\mathbf{a}} = \frac{v^{a_{i+1} - a_i + 1}}{v^{a_i - 1}[a_i] - v^{a_{i+1} - 1}[a_{i+1}]} u_i u_{\mathbf{a}'} - \frac{v^{a_i - a_{i+1} + 1}}{v^{a_i - 1}[a_i] - v^{a_{i+1} - 1}[a_{i+1}]} u_{\mathbf{a}''} u_{i+1}.$$

Since $\sigma(\mathbf{a}') = \sigma(\mathbf{a}'') = \sigma(\mathbf{a}) - 1$, we have by the inductive hypothesis that both $u_{\mathbf{a}'}$ and $u_{\mathbf{a}''}$ belong to $\mathfrak{H}'$. Hence, $u_{\mathbf{a}} \in \mathfrak{H}'$. This finishes the proof. □

Remark 1.4.4. We will see that semisimple modules as generators are convenient for the description of Lusztig type integral forms. First, by [**13**, Th. 5.2(ii)], they generate the *integral* Ringel–Hall algebra $\mathfrak{H}_\Delta(n)$ over $\mathcal{Z}$. Second, there are in §2.6 explicit commutator formulas between semisimple generators in the double Ringel–Hall algebra. Thus, a natural candidate for the Lusztig type form of quantum affine $\mathfrak{gl}_n$ is proposed in §3.8.

Finally, we introduce a set of generators for $\mathfrak{H}_\Delta(n)$ consisting of simple and *homogeneous* indecomposable modules in $\mathbf{Rep}^0\Delta(n)$. Since indecomposable modules correspond to the simplest non-diagonal matrices, these generators are convenient for deriving explicit multiplication formulas; see §§3.4, 5.4, and 6.2.

For each $A \in \Theta_\Delta^+(n)$, consider the radical filtration of $M(A)$

$$M(A) \supseteq \operatorname{rad} M(A) \supseteq \cdots \supseteq \operatorname{rad}^{t-1} M(A) \supseteq \operatorname{rad}^t M(A) = 0,$$

where t is the Loewy length of $M(A)$. For $1 \leqslant s \leqslant t$, we write

$$\operatorname{rad}^{s-1} M(A)/\operatorname{rad}^s M(A) = S_{\mathbf{a}_s}, \quad \text{for some } \mathbf{a}_s \in \mathbb{N}I.$$

Write $\mathfrak{m}_A = u_{\mathbf{a}_1} \cdots u_{\mathbf{a}_t}$. Applying (1.2.0.7) gives that

$$\mathfrak{m}_A = \sum_{B \leqslant_{\mathrm{dg}} A} f(B) u_B,$$

where $f(B) \in \mathbb{Q}(v)$ with $f(A) = v^{\sum_{l<s}\langle \mathbf{a}_l, \mathbf{a}_s\rangle}$. In other words,

$$u_A = f(A)^{-1}\mathfrak{m}_A - \sum_{B <_{\mathrm{dg}} A} f(A)^{-1} f(B) u_B.$$

Repeating the above construction for maximal B with $B <_{\mathrm{dg}} A$ and continuing this process, we finally get that

$$u_A = \sum_{B \leqslant_{\mathrm{dg}} A} \phi_A^B \mathfrak{m}_B, \tag{1.4.4.1}$$

where $\phi_A^B \in \mathbb{Q}(v)$ with $\phi_A^A = f(A)^{-1} \neq 0$.

Proposition 1.4.5. *For each $s \geqslant 1$, fix an arbitrary $i_s \in \mathbb{Z}^+$. Then $\mathfrak{H}_\Delta(n)$ is generated by u_i and $u_{E^\Delta_{i_s, i_s+sn}}$ for $i \in I$ and $s \geqslant 1$.*

Proof. For each $m \geqslant 1$, let $\mathfrak{H}_\Delta(n)^{(m)}$ be the $\mathbb{Q}(v)$-subalgebra of $\mathfrak{H}_\Delta(n)$ generated by u_i and $u_{s\delta}$ for $i \in I$ and $1 \leqslant s \leqslant m$. By convention, we set $\mathfrak{H}_\Delta(n)^{(0)} = \mathfrak{C}_\Delta(n)$. Clearly, each $\mathfrak{H}_\Delta(n)^{(m)}$ is also $\mathbb{N}I$-graded. By Theorem 1.4.1, $\mathfrak{H}_\Delta(n)^{(m)}$ is also generated by u_i and c_s, for $i \in I$ and $1 \leqslant s \leqslant m$. Moreover, $\mathfrak{H}_\Delta(n)^{(m-1)} \subsetneqq \mathfrak{H}_\Delta(n)^{(m)}$ for all $m \geqslant 1$.

We use induction on m to show the following

Claim: $\mathfrak{H}_\Delta(n)^{(m)}$ is generated by u_i and $u_{E^\Delta_{i_s,i_s+sn}}$ for $i \in I$ and $1 \leqslant s \leqslant m$.

Let $m \geqslant 1$ and suppose the claim is true for $\mathfrak{H}_\Delta(n)^{(m-1)}$. Applying (1.4.4.1) to $E = E^\Delta_{i_m,i_m+mn}$ gives

$$u_E = \sum_{B \leqslant_{\rm dg} E} \phi^B_E \mathfrak{m}_B = \phi^E_E \mathfrak{m}_E + \sum_{\substack{B <_{\rm dg} E \\ B \neq C}} \phi^B_E \mathfrak{m}_B + \phi^C_E \mathfrak{m}_C, \tag{1.4.5.1}$$

where $C = A_{m\delta} = \sum_{i \in I} m E^\Delta_{i,i+1}$, i.e., $M(C) \cong S_{m\delta}$ and $\mathfrak{m}_C = u_{m\delta}$. For each B with $B \leqslant_{\rm dg} E$ and $B \neq C$, the Loewy length of $M(B)$ is strictly greater than 1. Using an argument similar to the proof of Proposition 1.4.3, we have $\mathfrak{m}_B \in \mathfrak{H}_\Delta(n)^{(m-1)}$. We now prove that $\phi^C_E \neq 0$. Suppose $\phi^C_E = 0$. Then $u_E \in \mathfrak{H}_\Delta(n)^{(m-1)}$. Furthermore, for each $j \in I$, there is $1 \leqslant p \leqslant n$ such that $E^\Delta_{j,j+mn} = \tau^p(E)$, where τ is the Auslander–Reiten translation defined in §1.2. Thus, applying τ^p to (1.4.5.1) gives

$$u_{E^\Delta_{j,j+mn}} = u_{\tau^p(E)} = \phi^E_E \mathfrak{m}_{\tau^p E} + \sum_{\substack{B <_{\rm dg} E \\ B \neq C}} \phi^B_E \mathfrak{m}_{\tau^p B}.$$

Using a similar argument as above, we get $u_{E^\Delta_{j,j+mn}} \in \mathfrak{H}_\Delta(n)^{(m-1)}$. By definition, we have

$$\mathfrak{H}_\Delta(n)^{(m-1)}_{\mathbf{d}} = \mathfrak{H}_\Delta(n)_{\mathbf{d}} \text{ for all } \mathbf{d} \in \mathbb{N}I \text{ with } \sigma(\mathbf{d}) < mn.$$

In particular, $u_{E^\Delta_{i,j}} \in \mathfrak{H}_\Delta(n)^{(m-1)}$ for all $i < j$ with $j - i < mn$. Consequently, we get

$$u_{E^\Delta_{i,j}} \in \mathfrak{H}_\Delta(n)^{(m-1)} \text{ for all } i < j \text{ with } j - i \leqslant mn.$$

By [**37**, Th. 3.1], each element in $\mathfrak{H}_\Delta(n)$ can be written as a linear combination of products of u_A's with $M(A)$ indecomposable. This together with the above discussion implies that $\mathfrak{H}_\Delta(n)^{(m-1)}_{m\delta} = \mathfrak{H}_\Delta(n)_{m\delta}$. Thus, $u_{m\delta} \in \mathfrak{H}_\Delta(n)^{(m-1)}$. This contradicts the fact that $\mathfrak{H}_\Delta(n)^{(m-1)} \subsetneqq \mathfrak{H}_\Delta(n)^{(m)}$. Therefore, $\phi^C_A \neq 0$. We conclude from (1.4.5.1) that

$$u_E - \phi^C_E u_{m\delta} = \phi^E_E \mathfrak{m}_E + \sum_{\substack{B <_{\rm dg} E \\ B \neq C}} \phi^B_E \mathfrak{m}_B \in \mathfrak{H}_\Delta(n)^{(m-1)},$$

which shows the claim for $\mathfrak{H}_\Delta(n)^{(m)}$. This finishes the proof. □

By the proof of the above proposition, we see that for $m \geqslant 1$, each of the following three sets

$$\{u_i, c_s \mid i \in I, 1 \leqslant s \leqslant m\}, \quad \{u_i, u_{s\delta} \mid i \in I, 1 \leqslant s \leqslant m\}, \quad \text{and}$$
$$\{u_i, u_{E^{\Delta}_{i_s, i_s+sn}} \mid i \in I, 1 \leqslant s \leqslant m\}$$

generates $\mathfrak{H}_\Delta(n)^{(m)}$. Hence, for each $m \geqslant 1$, there are non-zero elements $x_m, y_m \in \mathbb{Q}(v)$ such that

$$c_m \equiv x_m u_{m\delta} \bmod \mathfrak{H}_\Delta(n)^{(m-1)} \quad \text{and} \quad c_m \equiv y_m u_{E^{\Delta}_{i_s, i_s+sn}} \bmod \mathfrak{H}_\Delta(n)^{(m-1)}.$$

This together with Corollary 1.4.2 gives the following result.

Corollary 1.4.6. *The set*

$$\Big\{u^{(A')} \prod_{s \geqslant 1} (u_{s\delta})^{m_s(A)} \mid A \in \Theta^+_\Delta(n)\Big\}$$

is a $\mathbb{Q}(v)$*-basis of* $\mathfrak{H}_\Delta(n)$*, where* A' *is defined in* (1.4.1.1). *For each* $s \geqslant 1$*, choose* $i_s \in \mathbb{Z}^+$*. Then the set*

$$\Big\{u^{(A')} \prod_{s \geqslant 1} (u_{E^{\Delta}_{i_s, i_s+sn}})^{m_s(A)} \mid A \in \Theta^+_\Delta(n)\Big\}$$

is also a $\mathbb{Q}(v)$*-basis of* $\mathfrak{H}_\Delta(n)$.

1.5. Hopf structure on extended Ringel–Hall algebras

It is known that *generic* Ringel–Hall algebras exist only for Dynkin or cyclic quivers. In the finite type case, these algebras give a realization for the $\pm$-parts of quantum groups. It is natural to expect that this is also true for cyclic quivers. In other words, we look for a quantum group such that $\mathfrak{H}_\Delta(n)$ is isomorphic to its $\pm$-part. We will see in Chapter 2 that this quantum group is the quantum loop algebra of $\mathfrak{gl}_n$ in the sense of Drinfeld [**20**], which, in fact, is isomorphic to the so-called double Ringel–Hall algebra defined as the Drinfeld double of two Hopf algebras $\mathfrak{H}_\Delta(n)^{\geqslant 0}$ and $\mathfrak{H}_\Delta(n)^{\leqslant 0}$ together with a skew-Hopf pairing. In this section, we first introduce the pair $\mathfrak{H}_\Delta(n)^{\geqslant 0}$ and $\mathfrak{H}_\Delta(n)^{\leqslant 0}$.

We need some preparation. If we define the *symmetrization* of the Euler form (1.2.0.4) by

$$(\alpha, \beta) = \langle \alpha, \beta \rangle + \langle \beta, \alpha \rangle,$$

then I together with (,) becomes a Cartan datum in the sense of [**54**, 1.1.1]. To a Cartan datum, there are associated *root data* in the sense of [**54**, 2.2.1] which play an important role in the theory of quantum groups. We shall fix the following root datum throughout the book.

Definition 1.5.1. Let $X = \mathbb{Z}^n$, $Y = \mathrm{Hom}(X, \mathbb{Z})$, and let $\langle\ ,\ \rangle^{\mathrm{rd}} : Y \times X \to \mathbb{Z}$ be the natural perfect pairing. If we denote the standard basis of X by $\boldsymbol{e}_1, \ldots, \boldsymbol{e}_n$ and the dual basis by $\boldsymbol{f}_1, \ldots, \boldsymbol{f}_n$, then $\langle \boldsymbol{f}_i, \boldsymbol{e}_j \rangle^{\mathrm{rd}} = \delta_{i,j}$. Thus, the embeddings

$$I \longrightarrow Y, i \longmapsto \tilde{i} := \boldsymbol{f}_i - \boldsymbol{f}_{i+1} \text{ and } I \longrightarrow X, i \longmapsto i' = \boldsymbol{e}_i - \boldsymbol{e}_{i+1} \quad (1.5.1.1)$$

with $\boldsymbol{e}_{n+1} = \boldsymbol{e}_1$ and $\boldsymbol{f}_{n+1} = \boldsymbol{f}_1$ define a root datum $(Y, X, \langle , \rangle^{\mathrm{rd}}, \ldots)$.

For notational simplicity, we shall identify both X and Y with $\mathbb{Z}I$ by setting $\boldsymbol{e}_i = i = \boldsymbol{f}_i$ for all $i \in I$. Under this identification, the form $\langle\ ,\ \rangle^{\mathrm{rd}} : \mathbb{Z}I \times \mathbb{Z}I \to \mathbb{Z}$ becomes a symmetric bilinear form, which is different from the Euler forms $\langle\ ,\ \rangle$ and its symmetrization $(\ ,\ )$. However, they are related as follows.

Lemma 1.5.2. *For* $\mathbf{a} = \sum a_i i \in \mathbb{Z}I$, *if we put* $\tilde{\mathbf{a}} = \sum a_i \tilde{i}$ *and* $\mathbf{a}' = \sum a_i i'$, *then*

(1) $(\mathbf{a}, \mathbf{b}) = \langle \tilde{\mathbf{a}}, \mathbf{b}' \rangle^{\mathrm{rd}}$; (2) $\langle \mathbf{a}, \mathbf{b} \rangle = \langle \mathbf{a}', \mathbf{b} \rangle^{\mathrm{rd}}$, *for all* $\mathbf{a}, \mathbf{b} \in \mathbb{Z}I$.

Proof. Since all forms are bilinear, (1) follows from $(i, j) = \langle \tilde{i}, j' \rangle^{\mathrm{rd}}$ for all $i, j \in I$, and (2) from $\langle i, j \rangle = \delta_{i,j} - \delta_{i+1,j} = \langle i', j \rangle^{\mathrm{rd}}$. □

We may use the following commutative diagrams to describe the two relations:

$$\begin{array}{ccc} \mathbb{Z}I \times \mathbb{Z}I & \xrightarrow{(\ ,\ )} & \\ \tilde{(\,)} \times (\,)' \downarrow & & \mathbb{Z} \\ \mathbb{Z}I \times \mathbb{Z}I & \xrightarrow{\langle\ ,\ \rangle^{\mathrm{rd}}} & \end{array} \quad \text{and} \quad \begin{array}{ccc} \mathbb{Z}I \times \mathbb{Z}I & \xrightarrow{\langle\ ,\ \rangle} & \\ (\,)' \times 1 \downarrow & & \mathbb{Z}. \\ \mathbb{Z}I \times \mathbb{Z}I & \xrightarrow{\langle\ ,\ \rangle^{\mathrm{rd}}} & \end{array}$$

We also record the following fact which will be used below and in §2.1.

Let $\mathbb{F}$ be a field. A Hopf algebra $\mathscr{A}$ over $\mathbb{F}$ is an $\mathbb{F}$-vector space together with multiplication $\mu_{\mathscr{A}}$, unit $\eta_{\mathscr{A}}$, comultiplication $\Delta_{\mathscr{A}}$, counit $\varepsilon_{\mathscr{A}}$, and antipode $\sigma_{\mathscr{A}}$ which satisfy certain axioms; see, e.g., **[72]** and **[12**, §5.1**]**.

Lemma 1.5.3. *If* $\mathscr{A} = (\mathscr{A}, \mu, \eta, \Delta, \varepsilon, \sigma)$ *is a Hopf algebra with multiplication* μ, *unit* η, *comultiplication* Δ, *counit* ε, *and antipode* σ, *then* $\mathscr{A}^{\mathrm{op}} = (\mathscr{A}, \mu^{\mathrm{op}}, \eta, \Delta^{\mathrm{op}}, \varepsilon, \sigma)$ *is also a Hopf algebra. This is called the* opposite *Hopf algebra of* $\mathscr{A}$. *Moreover, if* σ *is invertible, then both* $(\mathscr{A}, \mu^{\mathrm{op}}, \eta, \Delta, \varepsilon, \sigma^{-1})$ *and* $(\mathscr{A}, \mu, \eta, \Delta^{\mathrm{op}}, \varepsilon, \sigma^{-1})$ *are also Hopf algebras, which are called* semi-opposite *Hopf algebras.*

Let

$$\mathfrak{H}_\Delta(n)^{\geqslant 0} = \mathfrak{H}_\Delta(n) \otimes_{\mathbb{Q}(v)} \mathbb{Q}(v)[K_1^{\pm 1}, \ldots, K_n^{\pm 1}]. \tag{1.5.3.1}$$

Putting $x = x\otimes 1$ and $y = 1\otimes y$ for $x \in \mathfrak{H}_\Delta(n)$ and $y \in \mathbb{Q}(v)[K_1^{\pm 1}, \ldots, K_n^{\pm 1}]$, $\mathfrak{H}_\Delta(n)^{\geqslant 0}$ is a $\mathbb{Q}(v)$-space with basis $\{u_A^+ K_\alpha \mid \alpha \in \mathbb{Z}I, A \in \Theta_\Delta^+(n)\}$. We are now ready to introduce the Ringel–Green–Xiao Hopf structure on $\mathfrak{H}_\Delta(n)^{\geqslant 0}$. Let

$$\Theta_\Delta^+(n)^* := \Theta_\Delta^+(n)\backslash\{0\}.$$

Proposition 1.5.4. *The $\mathbb{Q}(v)$-space $\mathfrak{H}_\Delta(n)^{\geqslant 0}$ with basis $\{u_A^+ K_\alpha \mid \alpha \in \mathbb{Z}I, A \in \Theta_\Delta^+(n)\}$ becomes a Hopf algebra with the following algebra, coalgebra, and antipode structures.*

(a) Multiplication and unit (Ringel [**64**])*: for all $A, B \in \Theta_\Delta^+(n)$ and α, $\beta \in \mathbb{Z}I$,*

$$\begin{aligned} u_A^+ u_B^+ &= \sum_{C \in \Theta_\Delta^+(n)} v^{\langle \mathbf{d}(A), \mathbf{d}(B)\rangle} \varphi_{A,B}^C u_C^+, \\ K_\alpha u_A^+ &= v^{\langle \mathbf{d}(A), \alpha\rangle} u_A^+ K_\alpha, \\ K_\alpha K_\beta &= K_{\alpha+\beta}, \quad \text{and} \\ 1 &= u_0^+ = K_0. \end{aligned}$$

(b) Comultiplication and counit (Green [**34**])*: for all $C \in \Theta_\Delta^+(n)$ and $\alpha \in \mathbb{Z}I$,*

$$\begin{aligned} \Delta(u_C^+) &= \sum_{A,B \in \Theta_\Delta^+(n)} v^{\langle \mathbf{d}(A), \mathbf{d}(B)\rangle} \frac{\mathfrak{a}_A \mathfrak{a}_B}{\mathfrak{a}_C} \varphi_{A,B}^C u_B^+ \otimes u_A^+ \widetilde{K}_{\mathbf{d}(B)}, \\ \Delta(K_\alpha) &= K_\alpha \otimes K_\alpha, \\ \varepsilon(u_C^+) &= 0 \ (C \neq 0), \quad \text{and} \quad \varepsilon(K_\alpha) = 1. \end{aligned}$$

Here, if $\alpha = (a_i)$, then $\widetilde{K}_\alpha$ denotes $(\widetilde{K}_1)^{a_1} \cdots (\widetilde{K}_n)^{a_n}$ with $\widetilde{K}_i = K_i K_{i+1}^{-1}$.

(c) Antipode (Xiao [**78**])*: for all $C \in \Theta_\Delta^+(n)$ and $\alpha \in \mathbb{Z}I$,*

$$\begin{aligned} \sigma(u_C^+) &= \delta_{C,0} + \sum_{m \geqslant 1} (-1)^m \sum_{\substack{D \in \Theta_\Delta^+(n) \\ C_1, \ldots, C_m \in \Theta_\Delta^+(n)^*}} \frac{\mathfrak{a}_{C_1} \cdots \mathfrak{a}_{C_m}}{\mathfrak{a}_C} \\ &\qquad \times \varphi_{C_1,\ldots,C_m}^C \varphi_{C_m,\ldots,C_1}^D u_D^+ \widetilde{K}_{-\mathbf{d}(C)} \quad \text{and} \\ \sigma(K_\alpha) &= K_{-\alpha}. \end{aligned}$$

Moreover, the inverse of σ *is given by*

$$\sigma^{-1}(u_C^+) = \delta_{C,0} + \sum_{m\geqslant 1}(-1)^m \sum_{\substack{D\in\Theta_\triangle^+(n)\\ C_1,\cdots,C_m\in\Theta_\triangle^+(n)^*}} v^{2\sum_{i<j}\langle \mathbf{d}(C_i),\mathbf{d}(C_j)\rangle} \frac{\mathfrak{a}_{C_1}\cdots\mathfrak{a}_{C_m}}{\mathfrak{a}_C} \times \varphi^C_{C_1,\ldots,C_m}\varphi^D_{C_1,\ldots,C_m}\widetilde{K}_{-\mathbf{d}(C)}u_D^+ \quad \text{and}$$

$$\sigma^{-1}(K_\alpha) = K_{-\alpha}.$$

Proof. The Hopf structure on $\mathfrak{H}_\triangle(n)^{\geqslant 0}$ is almost identical to the Hopf algebra H' defined in the proof of [**78**, Prop. 4.8] except that we used $\widetilde{K}_\alpha$ instead of K_α in the comultiplication and antipode. Thus, the comultiplication of $\mathfrak{H}_\triangle(n)^{\geqslant 0}$ defined here is opposite to that defined in [**78**, Th. 4.5], while the antipode is the inverse. Hence, by Lemma 1.5.3, $\mathfrak{H}_\triangle(n)^{\geqslant 0}$ is the Hopf algebra semi-opposite to a variant of the Hopf algebras considered in [**78**, loc cit]. (Of course, one can directly check by mimicking the proof of [**78**, Th. 4.5] that $\mathfrak{H}_\triangle(n)^{\geqslant 0}$ with the operations defined above satisfies the axioms of a Hopf algebra.) □

Remarks 1.5.5. (1) Because of Lemma 1.5.2, we are able to make the root datum used in the second relation in (a) invisible in the definition of $\mathfrak{H}_\triangle(n)^{\geqslant 0}$.

(2) Besides the modification of changing K_α to $\widetilde{K}_\alpha$ in comultiplication and antipode, we also used the Euler form $\langle\,,\,\rangle$ (or rather the form $\langle\,,\,\rangle^{\mathrm{rd}}$ for the root datum given in Definition 1.5.1) instead of the symmetric Euler form $(\,,\,)$ used in Xiao's definition of H' ([**78**, p. 129]) for the commutator formulas between K_α and u_A^+. This means that H' is not isomorphic to $\mathfrak{H}_\triangle(n)^{\geqslant 0}$. However, there is a Hopf algebra homomorphism from H' to $\mathfrak{H}_\triangle(n)^{\geqslant 0}$ by sending $u_A^+K_\alpha$ to $u_A^+\widetilde{K}_\alpha$ whose image is the (Hopf) subalgebra $'\mathfrak{H}_\triangle(n)^{\geqslant 0}$ generated by $\widetilde{K}_i$ and u_A^+, for all $i \in I$ and $A \in \Theta_\triangle^+(n)$. Note that it sends the central element $K_1\cdots K_n \neq 1$ in H' to $\widetilde{K}_1\cdots\widetilde{K}_n = 1$ in $\mathfrak{H}_\triangle(n)^{\geqslant 0}$.

(3) The above modifications are necessary for compatibility with Lusztig's construction for quantum groups in [**54**] and with the corresponding relations in affine quantum Schur algebras.

It is clear that the subalgebra of $\mathfrak{H}_\triangle(n)^{\geqslant 0}$ generated by u_A^+ ($A \in \Theta_\triangle^+(n)$) is isomorphic to $\mathfrak{H}_\triangle(n)$. The subalgebra generated by K_α, $\alpha \in \mathbb{Z}I$, is isomorphic to the Laurent polynomial ring $\mathbb{Q}(v)[K_1^{\pm 1},\ldots,K_n^{\pm 1}]$.

Corollary 1.5.6. *The* $\mathbb{Q}(v)$*-space* $\mathfrak{H}_\triangle(n)^{\leqslant 0}$ *with basis* $\{K_\alpha u_A^- \mid \alpha \in \mathbb{Z}I, A \in \Theta_\triangle^+(n)\}$ *becomes a Hopf algebra with the following algebra, coalgebra, and antipode structures.*

(a′) Multiplication and unit: for all $A, B \in \Theta_\Delta^+(n)$ and $\alpha, \beta \in \mathbb{Z}I$,

$$u_A^- u_B^- = \sum_{C\in\Theta_\Delta^+(n)} v^{\langle \mathbf{d}(B),\mathbf{d}(A)\rangle} \varphi_{B,A}^C u_C^-,$$
$$u_A^- K_\alpha = v^{\langle \mathbf{d}(A),\alpha\rangle} K_\alpha u_A^-,$$
$$K_\alpha K_\beta = K_{\alpha+\beta}, \quad \textit{and}$$
$$1 = u_0^- = K_0.$$

(b′) Comultiplication and counit: for all $C \in \Theta_\Delta^+(n)$ and $\alpha \in \mathbb{Z}I$,

$$\Delta(u_C^-) = \sum_{A,B\in\Theta_\Delta^+(n)} v^{-\langle \mathbf{d}(B),\mathbf{d}(A)\rangle} \frac{\mathfrak{a}_A\mathfrak{a}_B}{\mathfrak{a}_C} \varphi_{A,B}^C \widetilde{K}_{-\mathbf{d}(A)} u_B^- \otimes u_A^-,$$
$$\Delta(K_\alpha) = K_\alpha \otimes K_\alpha,$$
$$\varepsilon(u_C^-) = 0\ (C \neq 0), \quad \textit{and} \quad \varepsilon(K_\alpha) = 1.$$

(c′) Antipode: for all $C \in \Theta_\Delta^+(n)$ and $\alpha \in \mathbb{Z}I$,

$$\sigma(u_C^-) = \delta_{C,0} + \sum_{m\geqslant 1}(-1)^m \sum_{\substack{D\in\Theta_\Delta^+(n)\\ C_1,\ldots,C_m\in\Theta_\Delta^+(n)^*}} v^{2\sum_{i<j}\langle \mathbf{d}(C_i),\mathbf{d}(C_j)\rangle} \frac{\mathfrak{a}_{C_1}\cdots\mathfrak{a}_{C_m}}{\mathfrak{a}_C}$$
$$\times \varphi_{C_1,\ldots,C_m}^C \varphi_{C_1,\ldots,C_m}^D \widetilde{K}_{\mathbf{d}(C)} u_D^- \quad \textit{and}$$
$$\sigma(K_\alpha) = K_{-\alpha}.$$

Proof. Let $\mathfrak{H}'$ be the $\mathbb{Q}(v)$-space with basis $\{K'_\alpha (u_A^-)' \mid \alpha \in \mathbb{Z}I, A \in \Theta_\Delta^+(n)\}$ and define the following operations on $\mathfrak{H}'$:

(a″) for all $A, B \in \Theta_\Delta^+(n)$ and $\alpha, \beta \in \mathbb{Z}I$,

$$(u_A^-)'(u_B^-)' = \sum_{C\in\Theta_\Delta^+(n)} v^{\langle \mathbf{d}(B),\mathbf{d}(A)\rangle} \varphi_{B,A}^C (u_C^-)',$$
$$(u_A^-)' K'_\alpha = v^{-\langle \mathbf{d}(A),\alpha\rangle} K'_\alpha (u_A^-)',$$
$$K'_\alpha K'_\beta = K'_{\alpha+\beta},$$

and let $1 = (u_0^-)' = K'_0$;

(b″) for all $C \in \Theta_\Delta^+(n)$ and $\alpha \in \mathbb{Z}I$,

$$\Delta((u_C^-)') = \sum_{A,B\in\Theta_\Delta^+(n)} v^{\langle \mathbf{d}(A),\mathbf{d}(B)\rangle} \frac{\mathfrak{a}_A\mathfrak{a}_B}{\mathfrak{a}_C} \varphi_{A,B}^C (u_B^-)' \otimes \widetilde{K}'_{-\mathbf{d}(B)} (u_A^-)',$$
$$\Delta(K'_\alpha) = K'_\alpha \otimes K'_\alpha,$$
$$\varepsilon((u_A^-)') = 0\ (A \neq 0), \quad \text{and} \quad \varepsilon(K'_\alpha) = 1,$$

where $\widetilde{K}'_i = K'_i (K'_{i+1})^{-1}$;

(c″) for all $C \in \Theta_\Delta^+(n)$ and $\alpha \in \mathbb{Z}I$,

$$\sigma((u_C^-)') = \delta_{C,0} + \sum_{m\geqslant 1}(-1)^m \sum_{\substack{D\in\Theta_\Delta^+(n)\\ C_1,\dots,C_m\in\Theta_\Delta^+(n)^*}} v^{2\sum_{i<j}\langle \mathbf{d}(C_i),\mathbf{d}(C_j)\rangle}\frac{\mathfrak{a}_{C_1}\cdots\mathfrak{a}_{C_m}}{\mathfrak{a}_C} \times \varphi^C_{C_1,\dots,C_m}\varphi^D_{C_1,\dots,C_m}(u_D^-)'\widetilde{K}'_{\mathbf{d}(C)} \quad \text{and}$$

$$\sigma(K'_\alpha) = K'_{-\alpha}.$$

By Lemma 1.5.3, if we replace the multiplication of $\mathfrak{H}_\Delta(n)^{\geqslant 0}$ by its opposite one and σ by σ^{-1} and keep other structure maps unchanged, then we obtain the semi-opposite Hopf algebra $\mathfrak{H}^{\mathrm{op}}$ of $\mathfrak{H}_\Delta(n)^{\geqslant 0}$. It is clear that the $\mathbb{Q}(v)$-linear isomorphism $\mathfrak{H}^{\mathrm{op}} \to \mathfrak{H}'$ taking $u_A^+K_\alpha \mapsto (u_A^-)'K'_{-\alpha}$ preserves all the operations. Thus, $\mathfrak{H}'$ is a Hopf algebra with the operations (a″)–(c″).

Now, for $\alpha \in \mathbb{Z}I$ and $A \in \Theta_\Delta^+(n)$, set in $\mathfrak{H}'$,

$$K_\alpha := K'_{-\alpha} \quad \text{and} \quad u_A^- := v^{-\langle \mathbf{d}(A),\mathbf{d}(A)\rangle}\widetilde{K}'_{\mathbf{d}(A)}(u_A^-)'. \tag{1.5.6.1}$$

Then $\{K_\alpha u_A^- \mid \alpha \in \mathbb{Z}I, A \in \Theta_\Delta^+(n)\}$ is a new basis of $\mathfrak{H}'$. It is easy to check that applying the operations (a″)–(c″) to the basis elements $K_\alpha u_A^-$ gives (a′)–(c′). Consequently, $\mathfrak{H}_\Delta(n)^{\leqslant 0}$ with the operations (a′)–(c′) is a Hopf algebra. □

The proof above shows that $\mathfrak{H}_\Delta(n)^{\leqslant 0}$ is the semi-opposite Hopf algebra of $\mathfrak{H}_\Delta(n)^{\geqslant 0}$ in the sense that multiplication and antipode are replaced by the opposite and inverse ones, respectively. Note that the inverse σ^{-1} of σ in $\mathfrak{H}_\Delta(n)^{\leqslant 0}$ is defined by

$$\sigma^{-1}(u_C^-) = \delta_{C,0} + \sum_{m\geqslant 1}(-1)^m \sum_{\substack{D\in\Theta_\Delta^+(n)\\ C_1,\dots,C_m\in\Theta_\Delta^+(n)^*}} \frac{\mathfrak{a}_{C_1}\cdots\mathfrak{a}_{C_m}}{\mathfrak{a}_C}\varphi^C_{C_1,\dots,C_m}\varphi^D_{C_m,\dots,C_1}u_D^-\widetilde{K}_{\mathbf{d}(C)}.$$

By (a′), the subalgebra of $\mathfrak{H}_\Delta(n)^{\leqslant 0}$ generated by u_A^- ($A \in \Theta_\Delta^+(n)$) is isomorphic to $\mathfrak{H}_\Delta(n)^{\mathrm{op}}$. Moreover, there is a $\mathbb{Q}(v)$-vector space isomorphism

$$\mathfrak{H}_\Delta(n)^{\leqslant 0} \cong \mathbb{Q}(v)[K_1^{\pm 1},\dots,K_n^{\pm 1}] \otimes_{\mathbb{Q}(v)} \mathfrak{H}_\Delta(n)^{\mathrm{op}}. \tag{1.5.6.2}$$

Remark 1.5.7. If we put $\mathcal{A} = \mathcal{Z}[(v^m-1)^{-1}]_{m\geqslant 1}$, then (1.2.0.3) guarantees that extended Ringel–Hall algebras $\mathfrak{H}_\Delta(n)_{\mathcal{A}}^{\geqslant 0}$ and $\mathfrak{H}_\Delta(n)_{\mathcal{A}}^{\leqslant 0}$ over $\mathcal{A}$ are well-defined. Thus, if v is specialized to z in a field (or a ring) $\mathbb{F}$ which is not a root of unity, then extended Ringel–Hall algebras $\mathfrak{H}_\Delta(n)_{\mathbb{F}}^{\geqslant 0} := \mathfrak{H}_\Delta(n)_{\mathcal{A}}^{\geqslant 0} \otimes \mathbb{F}$ and $\mathfrak{H}_\Delta(n)_{\mathbb{F}}^{\leqslant 0}$ over $\mathbb{F}$ are defined.

2

Double Ringel–Hall algebras of cyclic quivers

A Drinfeld double refers to a construction of gluing two Hopf algebras via a skew-Hopf pairing between them to obtain a new Hopf algebra. We apply this construction in §2.1 to the extended Ringel–Hall algebras $\mathfrak{H}_\Delta(n)^{\geqslant 0}$ and $\mathfrak{H}_\Delta(n)^{\leqslant 0}$ discussed in §1.5 to obtain double Ringel–Hall algebras $\mathfrak{D}_\Delta(n)$ of cyclic quivers.

The algebras $\mathfrak{D}_\Delta(n)$ possess a rich structure. First, by using the Schiffmann–Hubery generators and the connection with the quantum enveloping algebra associated with a Borcherds–Cartan matrix, we obtain a presentation for $\mathfrak{D}_\Delta(n)$ (Theorem 2.3.1). Second, by using Drinfeld's new presentation for the quantum loop algebra $\mathbf{U}(\widehat{\mathfrak{gl}}_n)$, we extend Beck's (and Jing's) embedding of quantum affine $\widehat{\mathfrak{sl}}_n$ into $\mathbf{U}(\widehat{\mathfrak{gl}}_n)$ to obtain an isomorphism between $\mathfrak{D}_\Delta(n)$ and $\mathbf{U}(\widehat{\mathfrak{gl}}_n)$ (Theorem 2.5.3). Finally, applying the skew-Hopf pairing to semisimple generators yields certain commutator relations (Theorem 2.6.3). Thus, we propose a possible presentation using semisimple generators; see Problem 2.6.4.

2.1. Drinfeld doubles and the Hopf algebra $\mathfrak{D}_\Delta(n)$

In this section, we first recall from [**46**] the notion of a skew-Hopf pairing and define the associated Drinfeld double. We then apply this general construction to obtain the Drinfeld double $\mathfrak{D}_\Delta(n)$ of the Ringel–Hall algebra $\mathfrak{H}_\Delta(n)$; see [**78**] for a general construction.

Let $\mathscr{A} = (\mathscr{A}, \mu_{\mathscr{A}}, \eta_{\mathscr{A}}, \Delta_{\mathscr{A}}, \varepsilon_{\mathscr{A}}, \sigma_{\mathscr{A}})$ and $\mathscr{B} = (\mathscr{B}, \mu_{\mathscr{B}}, \eta_{\mathscr{B}}, \Delta_{\mathscr{B}}, \varepsilon_{\mathscr{B}}, \sigma_{\mathscr{B}})$ be two Hopf algebras over a field $\mathbb{F}$. A *skew-Hopf pairing* of $\mathscr{A}$ and $\mathscr{B}$ is an $\mathbb{F}$-bilinear form $\psi : \mathscr{A} \times \mathscr{B} \to \mathbb{F}$ satisfying:

(HP1) $\psi(1, b) = \varepsilon_{\mathscr{B}}(b)$, $\psi(a, 1) = \varepsilon_{\mathscr{A}}(a)$, for all $a \in \mathscr{A}$, $b \in \mathscr{B}$;
(HP2) $\psi(a, bb') = \psi(\Delta_{\mathscr{A}}(a), b \otimes b')$, for all $a \in \mathscr{A}$, $b, b' \in \mathscr{B}$;

(HP3) $\psi(aa', b) = \psi(a \otimes a', \Delta^{\mathrm{op}}_{\mathscr{B}}(b))$, for all $a, a' \in \mathscr{A}$, $b \in \mathscr{B}$;
(HP4) $\psi(\sigma_{\mathscr{A}}(a), b) = \psi(a, \sigma^{-1}_{\mathscr{B}}(b))$, for all $a \in \mathscr{A}$, $b \in \mathscr{B}$,

where $\psi(a \otimes a', b \otimes b') = \psi(a, b)\psi(a', b')$, and $\Delta^{\mathrm{op}}_{\mathscr{B}}$ is defined by $\Delta^{\mathrm{op}}_{\mathscr{B}}(b) = \sum b_2 \otimes b_1$ if $\Delta_{\mathscr{B}}(b) = \sum b_1 \otimes b_2$. Note that we have assumed here that $\sigma_{\mathscr{B}}$ is invertible.

Let $\mathscr{A} * \mathscr{B}$ be the free product of $\mathbb{F}$-algebras $\mathscr{A}$ and $\mathscr{B}$ with identity. Then $\mathscr{A} * \mathscr{B}$ is the coproduct of $\mathscr{A}$ and $\mathscr{B}$ in the category of $\mathbb{F}$-algebras. More precisely, for any fixed bases $B_{\mathscr{A}}$ and $B_{\mathscr{B}}$ for $\mathscr{A}$, $\mathscr{B}$, respectively, where both $B_{\mathscr{A}}$ and $B_{\mathscr{B}}$ contain the identity element, $\mathscr{A} * \mathscr{B}$ is the $\mathbb{F}$-vector space spanned by the basis consisting of all words $b_1 b_2 \cdots b_m$ ($b_i \in (B_{\mathscr{A}} \backslash \{1\}) \cup (B_{\mathscr{B}} \backslash \{1\})$) of any length $m \geqslant 0$ such that $b_i b_{i+1}$ is not defined (in other words, b_i, b_{i+1} are not in the same $\mathscr{A}$ or $\mathscr{B}$) with multiplication given by "contracted juxtaposition"

$$(b_1 \cdots b_m) * (b'_1 \cdots b'_{m'}) = \begin{cases} b_1 \cdots b_m b'_1 \cdots b'_{m'}, & \text{if } b_m b'_1 \text{ is not defined;} \\ b_1 \cdots b_{m-1} c b'_2 \cdots b'_{m'}, & \text{if } b_m b'_1 \text{ is defined,} \\ & \quad b_m b'_1 \neq 0; \\ 0, & \text{otherwise.} \end{cases}$$

Note that, since $c = b_m b'_1 = \sum_k \lambda_k a_k$, where all $a_k \in B_{\mathscr{A}}$ or all $a_k \in B_{\mathscr{B}}$, is a linear combination of basis elements, the element $b_1 \cdots b_{m-1} c b'_2 \cdots b'_{m'}$ is a linear combination of words and is defined inductively. Thus, 1 is replaced by the empty word.

Let $\mathcal{I} = \mathcal{I}_{\mathscr{A},\mathscr{B}}$ be the ideal of $\mathscr{A} * \mathscr{B}$ generated by

$$\sum (b_2 * a_2)\psi(a_1, b_1) - \sum (a_1 * b_1)\psi(a_2, b_2) \quad (a \in \mathscr{A}, b \in \mathscr{B}), \tag{2.1.0.1}$$

where $\Delta_{\mathscr{A}}(a) = \sum a_1 \otimes a_2$ and $\Delta_{\mathscr{B}}(b) = \sum b_1 \otimes b_2$. Moreover, $\mathcal{I}$ is also generated by

$$b * a - \sum \psi(a_1, \sigma_{\mathscr{B}}(b_1))(a_2 * b_2)\psi(a_3, b_3) \quad (a \in \mathscr{A}, b \in \mathscr{B}), \tag{2.1.0.2}$$

where $\Delta^{(2)}_{\mathscr{A}}(a) = \sum a_1 \otimes a_2 \otimes a_3$ and $\Delta^{(2)}_{\mathscr{B}}(b) = \sum b_1 \otimes b_2 \otimes b_3$ (see, for example, [**46**, p. 72]).

The Drinfeld double of the pair $\mathscr{A}$ and $\mathscr{B}$ is by definition the quotient algebra

$$D(\mathscr{A}, \mathscr{B}) := \mathscr{A} * \mathscr{B} / \mathcal{I}.$$

By (2.1.0.2), each element in $D(\mathscr{A}, \mathscr{B})$ can be expressed as a linear combination of elements of the form $a * b + \mathcal{I}$ for $a \in \mathscr{A}$ and $b \in \mathscr{B}$. Note that there is an $\mathbb{F}$-vector space isomorphism

$$D(\mathscr{A},\mathscr{B}) \longrightarrow \mathscr{A}\otimes_{\mathbb{F}}\mathscr{B},\ a*b+\mathcal{I}\longmapsto a\otimes b;$$

see [**71**, Lem. 3.1]. For notational simplicity, we write $a*b+\mathcal{I}$ as $a*b$. Both $\mathscr{A}$ and $\mathscr{B}$ can be viewed as subalgebras of $D(\mathscr{A},\mathscr{B})$ via $a\mapsto a*1$ and $b\mapsto 1*b$, respectively. Thus, if x and y lie in $\mathscr{A}$ or $\mathscr{B}$, we write xy instead of $x*y$.

The algebra $D(\mathscr{A},\mathscr{B})$ admits a Hopf algebra structure induced by those of $\mathscr{A}$ and $\mathscr{B}$; see [**46**, 3.2.3]. More precisely, comultiplication, counit, and antipode in $D(\mathscr{A},\mathscr{B})$ are defined by

$$\begin{aligned}\Delta(a*b)&=\sum(a_1*b_1)\otimes(a_2*b_2),\\ \varepsilon(a*b)&=\varepsilon_{\mathscr{A}}(a)\varepsilon_{\mathscr{B}}(b),\quad\text{and}\\ \sigma(a*b)&=\sigma_{\mathscr{B}}(b)*\sigma_{\mathscr{A}}(a),\end{aligned}\tag{2.1.0.3}$$

where $a\in\mathscr{A}, b\in\mathscr{B}$, $\Delta_{\mathscr{A}}(a)=\sum a_1\otimes a_2$, and $\Delta_{\mathscr{B}}(b)=\sum b_1\otimes b_2$.

By modifying [**71**, Lem. 3.2], we obtain that the above ideal $\mathcal{I}$ can be generated by the elements in certain generating sets of $\mathscr{A}$ and $\mathscr{B}$ as described in the following result which will be used in §2.6.

Lemma 2.1.1. *Let $\mathscr{A},\mathscr{B}$ be Hopf algebras over $\mathbb{F}$ and let $\psi:\mathscr{A}\times\mathscr{B}\to\mathbb{F}$ be a skew-Hopf pairing. Assume that $X_{\mathscr{A}}\subseteq\mathscr{A}$ and $X_{\mathscr{B}}\subseteq\mathscr{B}$ are generating sets of $\mathscr{A}$ and $\mathscr{B}$, respectively. If $\Delta_{\mathscr{A}}(X_{\mathscr{A}})\subseteq \operatorname{span}_{\mathbb{F}}X_{\mathscr{A}}\otimes\operatorname{span}_{\mathbb{F}}X_{\mathscr{A}}$ and $\Delta_{\mathscr{B}}(X_{\mathscr{B}})\subseteq \operatorname{span}_{\mathbb{F}}X_{\mathscr{B}}\otimes\operatorname{span}_{\mathbb{F}}X_{\mathscr{B}}$, then the ideal $\mathcal{I}=\mathcal{I}_{\mathscr{A},\mathscr{B}}$ is generated by the following elements*

$$\sum(b_2*a_2)\psi(a_1,b_1)-\sum(a_1*b_1)\psi(a_2,b_2),\quad \textit{for all } a\in X_{\mathscr{A}}, b\in X_{\mathscr{B}}.\tag{2.1.1.1}$$

Proof. For $a\in\mathscr{A}$ and $b\in\mathscr{B}$, put

$$h_{a,b}:=\sum(b_2*a_2)\psi(a_1,b_1)-\sum(a_1*b_1)\psi(a_2,b_2).$$

Let $\mathcal{I}'$ be the ideal of $\mathscr{A}*\mathscr{B}$ generated by $h_{a,b}$ for all $a\in X_{\mathscr{A}}, b\in X_{\mathscr{B}}$, and set $\mathscr{H}=\mathscr{A}*\mathscr{B}/\mathcal{I}'$. We need to show $\mathcal{I}=\mathcal{I}'$. Since $\mathcal{I}'\subseteq\mathcal{I}$, it remains to show that for all $a\in\mathscr{A}$ and $b\in\mathscr{B}$, $h_{a,b}\in\mathcal{I}'$, or equivalently, $h_{a,b}=0$ in $\mathscr{H}$.

First, suppose $a\in X_{\mathscr{A}}$ and $b=y_1\cdots y_t$ with $y_j\in X_{\mathscr{B}}$. We proceed by induction on t to show that $h_{a,b}\in\mathcal{I}'$. Let $b'=y_1\cdots y_{t-1}$ and $b''=y_t$. Write $\Delta_{\mathscr{A}}(a)=\sum a_1\otimes a_2$, $\Delta_{\mathscr{A}}(a_1)=\sum a_{1,1}\otimes a_{1,2}$ and $\Delta_{\mathscr{A}}(a_2)=\sum a_{2,1}\otimes a_{2,2}$. The coassociativity of $\Delta_{\mathscr{A}}$ implies that

$$\sum a_{1,1}\otimes a_{1,2}\otimes a_2=\sum\Delta_{\mathscr{A}}(a_1)\otimes a_2=\sum a_1\otimes\Delta_{\mathscr{A}}(a_2)=\sum a_1\otimes a_{2,1}\otimes a_{2,2}.$$

Further, write $\Delta_{\mathscr{B}}(b') = \sum b'_1 \otimes b'_2$ and $\Delta_{\mathscr{B}}(b'') = \sum b''_1 \otimes b''_2$. Then $\Delta_{\mathscr{B}}(b) = \sum b'_1 b''_1 \otimes b'_2 b''_2$. Thus, we obtain in $\mathscr{H}$ that

$$
\begin{aligned}
&\sum (b'_2 b''_2 * a_2)\psi(a_1, b'_1 b''_1)\\
&= \sum b'_2 (b''_2 * a_2)\psi(a_{1,1}, b'_1)\psi(a_{1,2}, b''_1)\\
&= \sum b'_2 (b''_2 * a_{2,2})\psi(a_1, b'_1)\psi(a_{2,1}, b''_1)\\
&= \sum (b'_2 * a_{2,1}) * b''_1 \psi(a_1, b'_1)\psi(a_{2,2}, b''_2)\\
&\qquad\qquad\qquad\qquad (\text{since } a_2 \in \operatorname{span}_{\mathbb{F}} X_{\mathscr{A}} \text{ and } b'' \in X_{\mathscr{B}})\\
&= \sum (b'_2 * a_{1,2}) * b''_1 \psi(a_{1,1}, b'_1)\psi(a_2, b''_2)\\
&= \sum (a_{1,1} * b'_1) b''_1 \psi(a_{1,2}, b'_2)\psi(a_2, b''_2) \quad (\text{by induction for } a_1 \text{ and } b')\\
&= \sum (a_1 * b'_1) b''_1 \psi(a_{2,1}, b'_2)\psi(a_{2,2}, b''_2)\\
&= \sum (a_1 * b'_1 b''_1)\psi(a_2, b'_2 b''_2),
\end{aligned}
$$

that is, $h_{a,b} \in \mathcal{I}'$.

Now suppose $a = x_1 \cdots x_s \in \mathscr{A}$ with $x_i \in X_{\mathscr{A}}$ and $b \in \mathscr{B}$. We proceed by induction on s. The case $s = 1$ has already been treated above. So assume $s > 1$. Let $a' = x_1 \cdots x_{s-1}$ and $a'' = x_s$. Then

$$
\begin{aligned}
&\sum (b_2 * a'_2 a''_2)\psi(a'_1 a''_1, b_1)\\
&= \sum (b^{(2)}_3 * a'_2) a''_2 \psi(a'_1, b^{(2)}_2)\psi(a''_1, b^{(2)}_1)\\
&= \sum a'_1 * (b^{(2)}_2 * a''_2)\psi(a'_2, b^{(2)}_3)\psi(a''_1, b^{(2)}_1) \quad (\text{by induction})\\
&= \sum a'_1 a''_1 * b^{(2)}_1 \psi(a'_2, b^{(2)}_3)\psi(a''_2, b^{(2)}_2) \qquad (\text{since } a'' \in X_{\mathscr{A}})\\
&= \sum a'_1 a''_1 * b_1 \psi(a'_2 a''_2, b_2),
\end{aligned}
$$

where $\Delta^{(2)}_{\mathscr{B}}(b) = \sum b^{(2)}_1 \otimes b^{(2)}_2 \otimes b^{(2)}_3$. Hence, $h_{a,b} \in \mathcal{I}'$. This completes the proof. □

A skew-Hopf pairing can be passed on to opposite Hopf algebras (see Lemma 1.5.3). The following lemma can be checked directly.

Lemma 2.1.2. *Let* $\mathscr{A} = (\mathscr{A}, \mu_{\mathscr{A}}, \eta_{\mathscr{A}}, \Delta_{\mathscr{A}}, \varepsilon_{\mathscr{A}}, \sigma_{\mathscr{A}})$, $\mathscr{B} = (\mathscr{B}, \mu_{\mathscr{B}}, \eta_{\mathscr{B}}, \Delta_{\mathscr{B}}, \varepsilon_{\mathscr{B}}, \sigma_{\mathscr{B}})$ *be two Hopf algebras over a field* $\mathbb{F}$ *together with a skew-Hopf pairing* $\psi : \mathscr{A} \times \mathscr{B} \to \mathbb{F}$. *Assume that* $\sigma_{\mathscr{A}}$ *and* $\sigma_{\mathscr{B}}$ *are both invertible. Then* ψ *is also a skew-Hopf pairing of the semi-opposite Hopf algebras* $(\mathscr{A}, \mu_{\mathscr{A}}, \eta_{\mathscr{A}}, \Delta^{\mathrm{op}}_{\mathscr{A}}, \varepsilon_{\mathscr{A}}, \sigma^{-1}_{\mathscr{A}})$ *and* $(\mathscr{B}, \mu^{\mathrm{op}}_{\mathscr{B}}, \eta_{\mathscr{B}}, \Delta_{\mathscr{B}}, \varepsilon_{\mathscr{B}}, \sigma^{-1}_{\mathscr{B}})$ *(resp.,* $(\mathscr{A}, \mu^{\mathrm{op}}_{\mathscr{A}}, \eta_{\mathscr{A}}, \Delta_{\mathscr{A}}, \varepsilon_{\mathscr{A}}, \sigma^{-1}_{\mathscr{A}})$ *and* $(\mathscr{B}, \mu_{\mathscr{B}}, \eta_{\mathscr{B}}, \Delta^{\mathrm{op}}_{\mathscr{B}}, \varepsilon_{\mathscr{B}}, \sigma^{-1}_{\mathscr{B}})$*).*

We end this section with the construction of the Drinfeld double associated with the Ringel–Hall algebras $\mathfrak{H}_\Delta(n)^{\geqslant 0}$ and $\mathfrak{H}_\Delta(n)^{\leqslant 0}$ introduced in §1.5.

First, we need a skew-Hopf pairing. Applying Lemma 2.1.2 to [**78**, Prop. 5.3] yields the following result. For completeness, we sketch a proof. We introduce some notation which is used in the proof. For each $\alpha = \sum_{i\in I} a_i i \in \mathbb{Z}I$, write $\tau\alpha = \sum_{i\in I} a_{i-1} i$. In particular, for each $A \in \Theta_\Delta^+(n)$, we have $\tau\mathbf{d}(A) = \mathbf{d}(\tau(A))$. Then, for $\alpha, \beta \in \mathbb{Z}I$,

$$\langle \alpha, \beta\rangle = (\alpha - \tau\alpha)\centerdot\beta = -\langle \beta, \tau\alpha\rangle \quad\text{and}\quad \widetilde{K}_\alpha = K_{\alpha-\tau\alpha} = \widetilde{K}_1^{a_1}\cdots\widetilde{K}_n^{a_n}.$$

Proposition 2.1.3. *The $\mathbb{Q}(v)$-bilinear form $\psi : \mathfrak{H}_\Delta(n)^{\geqslant 0} \times \mathfrak{H}_\Delta(n)^{\leqslant 0} \to \mathbb{Q}(v)$ defined by*

$$\psi(u_A^+ K_\alpha, K_\beta u_B^-) = v^{\alpha\centerdot\beta - \langle \mathbf{d}(A), \mathbf{d}(A)+\alpha\rangle + 2\mathfrak{d}(A)} \mathfrak{a}_A^{-1}\delta_{A,B}, \tag{2.1.3.1}$$

where $\alpha, \beta \in \mathbb{Z}I$ and $A, B \in \Theta_\Delta^+(n)$, is a skew-Hopf pairing.

Proof. Condition (HP1) is obvious. We now check condition (HP2). Without loss of generality, we take $a = u_A^+ K_\alpha$, $b = K_\beta u_B^-$, and $b' = K_\gamma u_C^-$ for $\alpha, \beta, \gamma \in \mathbb{Z}I$ and $A, B, C \in \Theta_\Delta^+(n)$. Then

$$\begin{aligned}
&\psi(u_A^+ K_\alpha, K_\beta u_B^- K_\gamma u_C^-) = \psi(u_A^+ K_\alpha, v^{\langle \mathbf{d}(B),\gamma\rangle} K_{\beta+\gamma} u_B^- u_C^-)\\
&= \psi(u_A^+ K_\alpha, v^{\langle \mathbf{d}(B),\gamma\rangle} K_{\beta+\gamma} \sum_D v^{\langle \mathbf{d}(C),\mathbf{d}(B)\rangle}\varphi_{C,B}^D u_D^-)\\
&= v^{x_1}\varphi_{C,B}^A \mathfrak{a}_A^{-1},
\end{aligned}$$

where $x_1 = \langle \mathbf{d}(B), \gamma\rangle + \langle \mathbf{d}(C), \mathbf{d}(B)\rangle + \alpha\centerdot(\beta+\gamma) - \langle \mathbf{d}(A), \mathbf{d}(A)+\alpha\rangle + 2\mathfrak{d}(A)$.

On the other hand,

$$\begin{aligned}
&\psi(\Delta(u_A^+ K_\alpha), K_\beta u_B^- \otimes K_\gamma u_C^-)\\
&= \psi\Big(\sum_{B',C'} v^{\langle \mathbf{d}(B'),\mathbf{d}(C')\rangle}\frac{\mathfrak{a}_{B'}\mathfrak{a}_{C'}}{\mathfrak{a}_A}\varphi_{B',C'}^A u_{C'}^+ K_\alpha \otimes u_{B'}^+ \widetilde{K}_{\mathbf{d}(C')}K_\alpha, K_\beta u_B^- \otimes K_\gamma u_C^-\Big)\\
&= \sum_{B',C'} v^{\langle \mathbf{d}(B'),\mathbf{d}(C')\rangle}\frac{\mathfrak{a}_{B'}\mathfrak{a}_{C'}}{\mathfrak{a}_A}\varphi_{B',C'}^A \psi(u_{C'}^+ K_\alpha, K_\beta u_B^-)\psi(u_{B'}^+ \widetilde{K}_{\mathbf{d}(C')}K_\alpha, K_\gamma u_C^-)\\
&= v^{x_2}\varphi_{C,B}^A \mathfrak{a}_A^{-1},
\end{aligned}$$

where $x_2 = \langle \mathbf{d}(C), \mathbf{d}(B)\rangle + \alpha\centerdot\beta - \langle \mathbf{d}(B), \mathbf{d}(B)+\alpha\rangle + (\mathbf{d}(B) - \tau\mathbf{d}(B) + \alpha)\centerdot\gamma - \langle \mathbf{d}(C), \mathbf{d}(C)+\mathbf{d}(B) - \tau\mathbf{d}(B)+\alpha\rangle + 2\mathfrak{d}(A)$. Here we have assumed $\mathbf{d}(A) = \mathbf{d}(B) + \mathbf{d}(C)$ since $\varphi_{C,B}^A = 0$ otherwise. A direct calculation shows $x_1 = x_2$. Hence, (HP2) holds.

Condition (HP3) can be checked similarly as (HP2). It remains to check (HP4). Take $a = u_A^+ K_\alpha$ and $b = K_\beta u_B^-$. We may suppose $A \neq 0 \neq B$. Then

$$\begin{aligned}&\psi(\sigma(u_A^+ K_\alpha), K_\beta u_B^-)\\ &= \sum_{m\geqslant 1}(-1)^m v^{y_1} \sum_{C_1,\dots,C_m\in\Theta_\Delta^+(n)^*} \frac{\mathfrak{a}_{C_1}\cdots\mathfrak{a}_{C_m}}{\mathfrak{a}_A\mathfrak{a}_B}\varphi^A_{C_1,\dots,C_m}\varphi^B_{C_m,\dots,C_1}\end{aligned}$$

and

$$\begin{aligned}&\psi(u_A^+ K_\alpha, \sigma^{-1}(K_\beta u_B^-))\\ &= \sum_{m\geqslant 1}(-1)^m v^{y_2} \sum_{C_1,\dots,C_m\in\Theta_\Delta^+(n)^*} \frac{\mathfrak{a}_{C_1}\cdots\mathfrak{a}_{C_m}}{\mathfrak{a}_A\mathfrak{a}_B}\varphi^A_{C_1,\dots,C_m}\varphi^B_{C_m,\dots,C_1},\end{aligned}$$

where

$$y_1 = -\langle \mathbf{d}(B), \alpha\rangle + (-\alpha - \mathbf{d}(A) + \tau\mathbf{d}(A)) . \beta - \langle \mathbf{d}(B), \mathbf{d}(B) - \alpha - \mathbf{d}(A) + \tau\mathbf{d}(A)\rangle$$

and

$$y_2 = \langle \mathbf{d}(A), \mathbf{d}(B) - \tau\mathbf{d}(B) - \beta\rangle + \alpha . (\mathbf{d}(B) - \tau\mathbf{d}(B) - \beta) - \langle \mathbf{d}(A), \mathbf{d}(A) + \alpha\rangle.$$

Clearly, if $\mathbf{d}(A) \neq \mathbf{d}(B)$, then $\varphi^A_{C_1,\dots,C_m}\varphi^B_{C_m,\dots,C_1} = 0$. Hence, we may suppose $\mathbf{d}(A) = \mathbf{d}(B)$. Then

$$y_1 = -\alpha . \beta - \langle \mathbf{d}(A), \beta\rangle + \langle \mathbf{d}(A), \mathbf{d}(A)\rangle = y_2.$$

Therefore, $\psi(\sigma(u_A^+ K_\alpha), K_\beta u_B^-) = \psi(u_A^+ K_\alpha, \sigma^{-1}(K_\beta u_B^-))$, that is, (HP4) holds. □

Second, the Hopf algebras $\mathfrak{H}_\Delta(n)^{\geqslant 0}$ and $\mathfrak{H}_\Delta(n)^{\leqslant 0}$ together with the skew-Hopf pairing ψ give rise to the Drinfeld double

$$\widehat{\mathfrak{D}_\Delta(n)} := D(\mathfrak{H}_\Delta(n)^{\geqslant 0}, \mathfrak{H}_\Delta(n)^{\leqslant 0}).$$

Since as $\mathbb{Q}(v)$-vector spaces, we have

$$\widehat{\mathfrak{D}_\Delta(n)} = D(\mathfrak{H}_\Delta(n)^{\geqslant 0}, \mathfrak{H}_\Delta(n)^{\leqslant 0}) \cong \mathfrak{H}_\Delta(n)^{\geqslant 0} \otimes \mathfrak{H}_\Delta(n)^{\leqslant 0},$$

we sometimes write the elements in $\widehat{\mathfrak{D}_\Delta(n)}$ as linear combinations of $a \otimes b$ for $a \in \mathfrak{H}_\Delta(n)^{\geqslant 0}$ and $b \in \mathfrak{H}_\Delta(n)^{\leqslant 0}$. Moreover, it follows from (1.5.3.1) and (1.5.6.2) that there is a $\mathbb{Q}(v)$-vector space isomorphism

$$\begin{aligned}\widehat{\mathfrak{D}_\Delta(n)} \cong &\mathfrak{H}_\Delta(n) \otimes_{\mathbb{Q}(v)} \mathbb{Q}(v)[K_1^{\pm 1}, \dots, K_n^{\pm 1}]\\ &\otimes_{\mathbb{Q}(v)} \mathbb{Q}(v)[K_1^{\pm 1}, \dots, K_n^{\pm 1}] \otimes_{\mathbb{Q}(v)} \mathfrak{H}_\Delta(n)^{\mathrm{op}}.\end{aligned}$$

Finally, we define the *reduced* Drinfeld double

$$\mathfrak{D}_\Delta(n) = \widehat{\mathfrak{D}_\Delta(n)}/\mathscr{I}, \tag{2.1.3.2}$$

where $\mathscr{I}$ denotes the ideal generated by $1 \otimes K_\alpha - K_\alpha \otimes 1$, for all $\alpha \in \mathbb{Z}I$. By the construction, $\mathscr{I}$ is indeed a Hopf ideal of $\widehat{\mathfrak{D}_\Delta(n)}$. Thus, $\mathfrak{D}_\Delta(n)$ is again a Hopf algebra. We call $\mathfrak{D}_\Delta(n)$ the *double Ringel–Hall algebra* of the cyclic quiver $\Delta(n)$.

Let $\mathfrak{D}_\Delta(n)^+$ (resp., $\mathfrak{D}_\Delta(n)^-$) be the $\mathbb{Q}(v)$-subalgebra of $\mathfrak{D}_\Delta(n)$ generated by u_A^+ (resp., u_A^-) for all $A \in \Theta_\Delta^+(n)$. Let $\mathfrak{D}_\Delta(n)^0$ be the $\mathbb{Q}(v)$-subalgebra of $\mathfrak{D}_\Delta(n)$ generated by K_α for all $\alpha \in \mathbb{Z}I$. Then

$$\mathfrak{D}_\Delta(n)^+ \cong \mathfrak{H}_\Delta(n), \ \mathfrak{D}_\Delta(n)^- \cong \mathfrak{H}_\Delta(n)^{\mathrm{op}}, \ \text{and } \mathfrak{D}_\Delta(n)^0 \cong \mathbb{Q}(v)[K_1^{\pm 1}, \ldots, K_n^{\pm 1}]. \tag{2.1.3.3}$$

Moreover, the multiplication map

$$\mathfrak{D}_\Delta(n)^+ \otimes \mathfrak{D}_\Delta(n)^0 \otimes \mathfrak{D}_\Delta(n)^- \longrightarrow \mathfrak{D}_\Delta(n)$$

is an isomorphism of $\mathbb{Q}(v)$-vector spaces. Also, we have

$$\mathfrak{D}_\Delta(n)^{\geqslant 0} := \mathfrak{D}_\Delta(n)^+ \otimes \mathfrak{D}_\Delta(n)^0 \cong \mathfrak{H}_\Delta(n)^{\geqslant 0} \text{ and}$$
$$\mathfrak{D}_\Delta(n)^{\leqslant 0} := \mathfrak{D}_\Delta(n)^0 \otimes \mathfrak{D}_\Delta(n)^- \cong \mathfrak{H}_\Delta(n)^{\leqslant 0}.$$

We will identify $\mathfrak{D}_\Delta(n)^{\geqslant 0}$ and $\mathfrak{D}_\Delta(n)^{\leqslant 0}$ with $\mathfrak{H}_\Delta(n)^{\geqslant 0}$ and $\mathfrak{H}_\Delta(n)^{\leqslant 0}$, respectively, in the sequel. In particular, we may use the PBW type basis for $\mathfrak{H}_\Delta(n)$ to display a PBW type basis for $\mathfrak{D}_\Delta(n)$:

$$\{\widetilde{u}_A^+ K_\alpha \widetilde{u}_B^- \mid A, B \in \Theta_\Delta^+(n), \alpha \in \mathbb{Z}I\}.$$

Remark 2.1.4. By specializing v to $z \in \mathbb{C}$ which is not a root of unity, (1.2.0.3) implies that the skew-Hopf pairing ψ given in (2.1.3.1) is well-defined over $\mathbb{C}$. Hence the construction above works over $\mathbb{C}$ with $\mathfrak{H}_\Delta(n)^{\geqslant 0}$, etc., replaced by the corresponding specialization $\mathfrak{H}_\Delta(n)_{\mathbb{C}}^{\geqslant 0}$, etc. (see Remark 1.5.7). Thus, we obtain the double Ringel–Hall algebra $\mathfrak{D}_{\Delta,\mathbb{C}}(n)$ over $\mathbb{C}$. In fact, if we use the ring $\mathcal{A}$ defined in Remark 1.5.7, the same reasoning shows that the skew-Hopf pairing in (2.1.3.1) is defined over $\mathcal{A}$. Hence, we can form a double Ringel–Hall algebra $\mathfrak{D}_\Delta(n)_{\mathcal{A}}$. Then $\mathfrak{D}_{\Delta,\mathbb{C}}(n) \cong \mathfrak{D}_\Delta(n)_{\mathcal{A}} \otimes \mathbb{C}$ and $\mathfrak{D}_\Delta(n) = \mathfrak{D}_\Delta(n)_{\mathcal{A}} \otimes \mathbb{Q}(v)$.

2.2. Schiffmann–Hubery generators

In this and the following sections, we will investigate the structure of $\mathfrak{D}_\Delta(n)$ by relating it with the quantum enveloping algebra of a generalized Kac–Moody algebra based on [**67, 39**]; see also [**38, 14**].

We first construct certain primitive central elements from the central elements of $\mathfrak{H}_\Delta(n)$ defined in (1.4.0.1). Recall that an element of a Hopf algebra with comultiplication Δ is called *primitive* if

$$\Delta(x) = x \otimes 1 + 1 \otimes x.$$

For $m \geqslant 1$, let

$$c_m^{\pm} = (-1)^m v^{-2nm} \sum_A (-1)^{\dim \mathrm{End}(M(A))} \mathfrak{a}_A u_A^{\pm} \in \mathfrak{D}_\Delta(n)^{\pm}, \qquad (2.2.0.1)$$

where the sum is taken over all $A \in \Theta_\Delta^+(n)$ such that $\mathbf{d}(A) = m\delta$ and soc $M(A)$ is square-free. We also define $c_0^{\pm} = 1$ by convention. By Theorem 1.4.1, the elements c_m^+ and c_m^- are central in $\mathfrak{D}_\Delta(n)^+$ and $\mathfrak{D}_\Delta(n)^-$, respectively.

Following [**39**, §4], let $C^{\pm}(u) = 1 + \sum_{m \geqslant 1} c_m^{\pm} u^m$ be the generating functions in indeterminate u associated with the sequence $\{c_m^{\pm}\}_{m \geqslant 1}$ and define elements $x_m^{\pm}$ by

$$X^{\pm}(u) = \sum_{m \geqslant 1} x_m^{\pm} u^{m-1} = \frac{\mathrm{d}}{\mathrm{d}u} \log C^{\pm}(u) = \frac{1}{C^{\pm}(u)} \frac{\mathrm{d}}{\mathrm{d}u} C^{\pm}(u).$$

Thus, for each $m \geqslant 1$,

$$x_m^{\pm} = m c_m^{\pm} - \sum_{s=1}^{m-1} x_s^{\pm} c_{m-s}^{\pm}. \qquad (2.2.0.2)$$

Recall from (1.2.0.6) the elements

$$\widetilde{u}_A^{\pm} = v^{\dim \mathrm{End}(M(A)) - \dim M(A)} u_A^{\pm} \in \mathfrak{D}_\Delta(n)^{\pm}, \quad \text{for } A \in \Theta_\Delta^+(n).$$

In particular, for each $1 \leqslant l \leqslant n$ and $m \geqslant 1$,

$$\widetilde{u}_{E_{l,l+mn}^\Delta}^{\pm} = v^{m-nm} u_{E_{l,l+mn}^\Delta}^{\pm}.$$

The following lemma can be deduced from [**40**, Lem. 12]. However, we provide here a proof for completeness.

Lemma 2.2.1. *For each $m \geqslant 1$,*

$$x_m^{\pm} = v^{-nm}(v^m - v^{-m}) \sum_{l=1}^{n} \widetilde{u}_{E_{l,l+mn}^\Delta}^{\pm} + (v - v^{-1})^2 y_m^{\pm}, \qquad (2.2.1.1)$$

where $y_m^{\pm}$ are $\mathcal{Z}$-linear combinations of certain $\widetilde{u}_A^{\pm}$ such that $\mathbf{d}(A) = m\delta$ and $M(A)$ are decomposable.

Proof. By (1.2.0.3), for each $A \in \Theta_\Delta^+(n)$, $\mathfrak{a}_A \in \mathcal{Z}$ is divisible by $(v - v^{-1})^{\sigma(A)}$, where $\sigma(A) = \sum_{1 \leqslant i \leqslant n,\, j \in \mathbb{Z}} a_{i,j}$. Note that $\sigma(A)$ equals the number of indecomposable summands in $M(A)$. Since $\mathfrak{a}_{E_{l,l+mn}^\Delta} = v^{2(m-1)}(v^2 - 1)$, for each $1 \leqslant l \leqslant n$ and $m \geqslant 1$, it follows from (2.2.0.1) that

$$c_m^{\pm} \equiv v^{(1-n)m-1}(v - v^{-1}) \sum_{l=1}^{n} \widetilde{u}_{E_{l,l+mn}^\Delta}^{\pm} \quad \mathrm{mod}\ (v - v^{-1})^2 \mathfrak{J}^{\pm},$$

where $\mathfrak{I}^{\pm}$ denote the $\mathcal{Z}$-submodules of $\mathfrak{D}_{\Delta}(n)^{\pm}$ spanned by all the $\widetilde{u}_A^{\pm}$ satisfying that $M(A)$ are decomposable (i.e., $\sigma(A) > 1$).

Since subrepresentations and quotient representations of each indecomposable representation of $\Delta(n)$ are again indecomposable, it follows that, for each $A \in \Theta_{\Delta}^{+}(n)$,

$$\widetilde{u}_A^{\pm}\mathfrak{I}^{\pm} \subseteq \mathfrak{I}^{\pm} \text{ and } \mathfrak{I}^{\pm}\widetilde{u}_A^{\pm} \subseteq \mathfrak{I}^{\pm}.$$

It suffices to show that

$$x_m^{\pm} \equiv v^{-nm}(v^m - v^{-m}) \sum_{l=1}^{n} \widetilde{u}_{E_{l,l+mn}^{\Delta}}^{\pm} \quad \bmod (v - v^{-1})^2 \mathfrak{I}^{\pm}.$$

We proceed by induction on m. If $m = 1$, it is trivial since $x_1^{\pm} = c_1^{\pm}$. Let now $m > 1$. The inductive hypothesis together with (2.2.0.2) implies that

$$\begin{aligned} x_m^{\pm} \equiv mv^{(1-n)m-1}(v - v^{-1}) \sum_{l=1}^{n} \widetilde{u}_{E_{l,l+mn}^{\Delta}}^{\pm} - \sum_{s=1}^{m-1} \Big(v^{-ns}(v^s - v^{-s}) \sum_{l=1}^{n} \widetilde{u}_{E_{l,l+sn}^{\Delta}}^{\pm}\Big) \\ \times \Big(v^{(1-n)(m-s)-1}(v - v^{-1}) \sum_{l=1}^{n} \widetilde{u}_{E_{l,l+(m-s)n}^{\Delta}}^{\pm}\Big) \bmod (v - v^{-1})^2 \mathfrak{I}^{\pm}. \end{aligned}$$

It is clear that for $1 \leqslant l, l' \leqslant n$,

$$\widetilde{u}_{E_{l,l+sn}^{\Delta}}^{\pm} \widetilde{u}_{E_{l',l'+(m-s)n}^{\Delta}}^{\pm} \equiv \delta_{l,l'} \widetilde{u}_{E_{l,l+mn}^{\Delta}}^{\pm} \quad \bmod \mathfrak{I}^{\pm}.$$

We conclude that

$$\begin{aligned} x_m^{\pm} &\equiv v^{(1-n)m}\Big(m(1 - v^{-2}) - \sum_{s=1}^{m-1} (1 - v^{-2})(1 - v^{-2s})\Big) \sum_{l=1}^{n} \widetilde{u}_{E_{l,l+mn}^{\Delta}}^{\pm} \\ &\equiv v^{-nm}(v^m - v^{-m}) \sum_{l=1}^{n} \widetilde{u}_{E_{l,l+mn}^{\Delta}}^{\pm} \quad \bmod (v - v^{-1})^2 \mathfrak{I}^{\pm}. \end{aligned}$$

□

We further set

$$\mathsf{z}_m^{\pm} = \frac{v^{nm}}{v^m - v^{-m}} x_m^{\pm} \in \mathfrak{D}_{\Delta}(n)^{\pm}, \quad \text{for } m \geqslant 1.$$

Applying (2.2.1.1) gives that

$$\mathsf{z}_m^{\pm} = \sum_{l=1}^{n} \widetilde{u}_{E_{l,l+mn}^{\Delta}}^{\pm} + \frac{v^{nm}(v - v^{-1})}{[m]} y_m^{\pm}. \tag{2.2.1.2}$$

Hence, the elements z_m^+ (resp., z_m^-) are central in $\mathfrak{D}_\Delta(n)^+$ (resp., in $\mathfrak{D}_\Delta(n)^-$). Moreover, by Theorem 1.4.1, we have

$$\mathfrak{D}_\Delta(n)^\pm = \mathfrak{C}_\Delta(n)^\pm \otimes \mathbb{Q}(v)[c_1^\pm, c_2^\pm, \ldots] = \mathfrak{C}_\Delta(n)^\pm \otimes \mathbb{Q}(v)[\mathsf{z}_1^\pm, \mathsf{z}_2^\pm, \ldots], \tag{2.2.1.3}$$

where $\mathfrak{C}_\Delta(n)^\pm$ are the composition algebras generated by $u_i^\pm$, $1 \leqslant i \leqslant n$.

Let Δ be the comultiplication on $\mathfrak{D}_\Delta(n)$ induced by Green's comultiplication and σ be the antipode of $\mathfrak{D}_\Delta(n)$ induced by Xiao's antipode; see Proposition 1.5.4 and Corollary 1.5.6.

Proposition 2.2.2. *For each $m \geqslant 1$, the elements z_m^+ and z_m^- satisfy*

$$\Delta(\mathsf{z}_m^\pm) = \mathsf{z}_m^\pm \otimes 1 + 1 \otimes \mathsf{z}_m^\pm \text{ and } \sigma(\mathsf{z}_m^\pm) = -\mathsf{z}_m^\pm.$$

Moreover, for all $i \in I$ and $m, m' \geqslant 1$,

$$[\mathsf{z}_m^+, \mathsf{z}_{m'}^-] = 0,\ [\mathsf{z}_m^+, u_i^-] = 0,\ [u_i^+, \mathsf{z}_{m'}^-] = 0,\ \text{and}\ [\mathsf{z}_m^\pm, K_i] = 0.$$

In other words, $\mathsf{z}_m^\pm$ are central elements in $\mathfrak{D}_\Delta(n)$.

Proof. By [**39**, Prop. 9], for each $m \geqslant 1$,

$$\Delta(c_m^\pm) = \sum_{s=0}^{m} c_s^\pm \otimes c_{m-s}^\pm.$$

Applying (2.2.0.2) implies that $x_m^\pm$ are primitive; see [**39**, Cor. 10]. Hence, $\mathsf{z}_m^\pm$ are also primitive, that is,

$$\Delta(\mathsf{z}_m^\pm) = \mathsf{z}_m^\pm \otimes 1 + 1 \otimes \mathsf{z}_m^\pm.$$

Since $\mathfrak{D}_\Delta(n)$ is a Hopf algebra, we have $\mu(\sigma \otimes 1)\Delta(\mathsf{z}_m^\pm) = \eta\varepsilon(\mathsf{z}_m^\pm) = 0$ for each $m \geqslant 1$. Thus,

$$0 = \mu(\sigma \otimes 1)(\mathsf{z}_m^\pm \otimes 1 + 1 \otimes \mathsf{z}_m^\pm) = \sigma(\mathsf{z}_m^\pm) + \mathsf{z}_m^\pm,$$

i.e., $\sigma(\mathsf{z}_m^\pm) = -\mathsf{z}_m^\pm$.

Since $\Delta(\mathsf{z}_m^+) = \mathsf{z}_m^+ \otimes 1 + 1 \otimes \mathsf{z}_m^+$ and $\Delta(\mathsf{z}_{m'}^-) = \mathsf{z}_{m'}^- \otimes 1 + 1 \otimes \mathsf{z}_{m'}^-$, applying (2.1.0.1) to z_m^+ and $\mathsf{z}_{m'}^-$ gives

$$\begin{aligned} &\psi(\mathsf{z}_m^+, \mathsf{z}_{m'}^-) + \mathsf{z}_m^+\psi(1, \mathsf{z}_{m'}^-) + \mathsf{z}_{m'}^-\psi(\mathsf{z}_m^+, 1) + \mathsf{z}_{m'}^-\mathsf{z}_m^+\psi(1, 1) \\ =&\mathsf{z}_m^+\mathsf{z}_{m'}^-\psi(1, 1) + \mathsf{z}_{m'}^-\psi(\mathsf{z}_m^+, 1) + \mathsf{z}_m^+\psi(1, \mathsf{z}_{m'}^-) + \psi(\mathsf{z}_m^+, \mathsf{z}_{m'}^-). \end{aligned}$$

It follows from $\psi(1, 1) = 1$ that $\mathsf{z}_{m'}^-\mathsf{z}_m^+ = \mathsf{z}_m^+\mathsf{z}_{m'}^-$, i.e., $[\mathsf{z}_m^+, \mathsf{z}_{m'}^-] = 0$.

Similarly, we have $[\mathsf{z}_m^+, u_i^-] = [u_i^+, \mathsf{z}_{m'}^-] = [\mathsf{z}_m^\pm, K_i] = 0$. The last assertion follows from (2.2.1.3). $\square$

By Theorem 1.3.4, there are isomorphisms

$$\mathfrak{C}_\Delta(n)^+ \xrightarrow{\sim} \mathbf{U}(\widehat{\mathfrak{sl}}_n)^+, u_i^+ \longmapsto E_i, \quad \text{and } \mathfrak{C}_\Delta(n)^- \xrightarrow{\sim} \mathbf{U}(\widehat{\mathfrak{sl}}_n)^-, u_i^- \longmapsto F_i.$$

Here we have applied the anti-involution $\mathbf{U}(\widehat{\mathfrak{sl}}_n)^+ \to \mathbf{U}(\widehat{\mathfrak{sl}}_n)^-$, $E_i \mapsto F_i$.

By (2.2.1.3), there are decompositions

$$\mathfrak{D}_\Delta(n)^+ = \mathfrak{C}_\Delta(n)^+ \otimes_{\mathbb{Q}(v)} \mathbf{Z}_\Delta(n)^+ \quad \text{and} \quad \mathfrak{D}_\Delta(n)^- = \mathfrak{C}_\Delta(n)^- \otimes_{\mathbb{Q}(v)} \mathbf{Z}_\Delta(n)^-, \tag{2.2.2.1}$$

where $\mathbf{Z}_\Delta(n)^\pm := \mathbb{Q}(v)[\mathsf{z}_1^\pm, \mathsf{z}_2^\pm, \ldots]$ are the polynomial algebras in $\mathsf{z}_m^\pm$ for $m \geqslant 1$.

Remark 2.2.3. If $z \in \mathbb{C}$ is not a root of unity and $\mathfrak{H}_\Delta(n)_\mathbb{C} := \mathfrak{H}_\Delta(n) \otimes \mathbb{C}$ is the $\mathbb{C}$-algebra obtained by specializing v to z, then, by (1.2.0.3), we may use (2.2.0.2) to recursively define the central elements z_m in $\mathfrak{H}_\Delta(n)_\mathbb{C}$ and, hence, elements $\mathsf{z}_m^\pm$ in $\mathfrak{D}_{\Delta\mathbb{C}}(n)^\pm$; see Remark 2.1.4. Thus, a $\mathbb{C}$-basis similar to the one given in Corollary 1.4.2 can be constructed for $\mathfrak{D}_{\Delta\mathbb{C}}(n)^\pm$. In particular, we obtain decompositions

$$\mathfrak{D}_{\Delta\mathbb{C}}(n)^\pm = \mathfrak{C}_\Delta(n)^\pm_\mathbb{C} \otimes \mathbb{C}[\mathsf{z}_1^\pm, \mathsf{z}_2^\pm, \ldots],$$

and hence the $\mathbb{C}$-algebra $\mathfrak{D}_{\Delta\mathbb{C}}(n)$ can be presented by generators u_i^+, u_i^-, K_i, K_i^{-1}, z_s^+, z_s^-, $i \in I$, $s \in \mathbb{Z}^+$ and relations (QGL1)–(QGL8) as given in the last statement of Theorem 2.3.1 below.[1]

2.3. Presentation of $\mathfrak{D}_\Delta(n)$

Recall from (1.3.2.1) the Cartan matrix $C = (c_{i,j})$ of affine type A_{n-1} and the quantum group $\mathbf{U}(\widehat{\mathfrak{sl}}_n)$ associated with C as given in Definition 1.3.3. By identifying $\mathfrak{C}_\Delta(n)^\pm$ with $\mathbf{U}(\widehat{\mathfrak{sl}}_n)^\pm$ under the isomorphism in Theorem 1.3.4, we describe a presentation for $\mathfrak{D}_\Delta(n)$ as follows.

Theorem 2.3.1. *The double Ringel–Hall algebra $\mathfrak{D}_\Delta(n)$ of the cyclic quiver $\Delta(n)$ is the $\mathbb{Q}(v)$-algebra generated by $E_i = u_i^+$, $F_i = u_i^-$, K_i, K_i^{-1}, z_s^+, z_s^-, $i \in I$, $s \in \mathbb{Z}^+$ with relations ($i, j \in I$ and $s, t \in \mathbb{Z}^+$):*

(QGL1) $K_i K_j = K_j K_i$, $K_i K_i^{-1} = 1$;

(QGL2) $K_i E_j = v^{\delta_{i,j} - \delta_{i,j+1}} E_j K_i$, $K_i F_j = v^{-\delta_{i,j} + \delta_{i,j+1}} F_j K_i$;

(QGL3) $E_i F_j - F_j E_i = \delta_{i,j} \frac{\widetilde{K}_i - \widetilde{K}_i^{-1}}{v - v^{-1}}$, *where* $\widetilde{K}_i = K_i K_{i+1}^{-1}$;

[1] We will see in Proposition 2.4.5 that $\mathfrak{D}_{\Delta\mathbb{C}}(n)$ can be obtained as a specialization from the $\mathcal{Z}$-algebra $\mathfrak{D}_\Delta(n)$ by the base change $\mathcal{Z} \to \mathbb{C}$, $v \mapsto z$. Thus, we can use the notation $\mathfrak{D}_\Delta(n)_\mathbb{C}$ for this algebra.

(QGL4) $\sum_{a+b=1-c_{i,j}} (-1)^a \begin{bmatrix} 1-c_{i,j} \\ a \end{bmatrix} E_i^a E_j E_i^b = 0$, *for* $i \neq j$;

(QGL5) $\sum_{a+b=1-c_{i,j}} (-1)^a \begin{bmatrix} 1-c_{i,j} \\ a \end{bmatrix} F_i^a F_j F_i^b = 0$, *for* $i \neq j$;

(QGL6) $\mathsf{z}_s^+\mathsf{z}_t^+ = \mathsf{z}_t^+\mathsf{z}_s^+, \mathsf{z}_s^-\mathsf{z}_t^- = \mathsf{z}_t^-\mathsf{z}_s^-, \mathsf{z}_s^+\mathsf{z}_t^- = \mathsf{z}_t^-\mathsf{z}_s^+$;

(QGL7) $K_i\mathsf{z}_s^+ = \mathsf{z}_s^+K_i$, $K_i\mathsf{z}_s^- = \mathsf{z}_s^-K_i$;

(QGL8) $E_i\mathsf{z}_s^+ = \mathsf{z}_s^+E_i$, $E_i\mathsf{z}_s^- = \mathsf{z}_s^-E_i$, $F_i\mathsf{z}_s^- = \mathsf{z}_s^-F_i$, *and* $\mathsf{z}_s^+F_i = F_i\mathsf{z}_s^+$.

Replacing $\mathbb{Q}(v)$ *by* $\mathbb{C}$ *and* v *by a non-root-of-unity* $z \in \mathbb{C}^*$, *a similar result holds for* $\mathfrak{D}_{\triangle,\mathbb{C}}(n)$.

We will prove this result by a "standardly" defined quantum enveloping algebra associated with a Borcherds–Cartan matrix. See Proposition 2.3.4 below for the proof.

For each $m \in \mathbb{Z}^+$, let $J_m = [1, m]$ and set $J_\infty = [1, \infty) = \mathbb{Z}^+$. We then set $\widetilde{I}_m = I \cup J_m$, for $m \in \mathbb{Z}^+ \cup \{\infty\}$. Define a matrix $\widetilde{C}_m = (\widetilde{c}_{i,j})_{i,j\in\widetilde{I}_m}$ by setting

$$\widetilde{c}_{i,j} = \begin{cases} c_{i,j}, & \text{if } i, j \in I; \\ 0, & \text{otherwise.} \end{cases}$$

$\widetilde{C}_m$ is a Borcherds–Cartan matrix which defines a generalized Kac–Moody algebra; see [**5**]. Thus, with $\widetilde{C}_m$ we associate a quantum enveloping algebra $\mathbf{U}(\widetilde{C}_m)$ as follows; see [**48**].

Definition 2.3.2. Let $m \in \mathbb{Z}^+\cup\{\infty\}$ and $\widetilde{C}_m$ be defined as above. The quantum enveloping algebra $\mathbf{U}(\widetilde{C}_m)$ associated with $\widetilde{C}_m$ is the $\mathbb{Q}(v)$-algebra presented by generators

$$E_i,\ F_i,\ K_i,\ K_i^{-1},\ \mathsf{x}_s,\ \mathsf{y}_s,\ \mathsf{k}_s,\ \mathsf{k}_s^{-1},\ i \in I,\ s \in J_m,$$

and relations, for $i, j \in I$ and $s, t \in J_m$,

(R1) $K_iK_j = K_jK_i,\ K_iK_i^{-1} = 1,\ \mathsf{k}_s\mathsf{k}_t = \mathsf{k}_t\mathsf{k}_s,\ \mathsf{k}_s\mathsf{k}_s^{-1} = 1,\ K_i\mathsf{k}_s = \mathsf{k}_sK_i$;

(R2) $K_iE_j = v^{\delta_{i,j}-\delta_{i,j+1}}E_jK_i,\ K_i\mathsf{x}_s = \mathsf{x}_sK_i,\ \mathsf{k}_s\mathsf{x}_t = \mathsf{x}_t\mathsf{k}_s,\ \mathsf{k}_sE_i = E_i\mathsf{k}_s$;

(R3) $K_iF_j = v^{-\delta_{i,j}+\delta_{i,j+1}}F_jK_i,\ K_i\mathsf{y}_s = \mathsf{y}_sK_i,\ \mathsf{k}_s\mathsf{y}_t = \mathsf{y}_t\mathsf{k}_s,\ \mathsf{k}_sF_i = F_i\mathsf{k}_s$;

(R4) $E_iF_j - F_jE_i = \delta_{i,j}\frac{\widetilde{K}_i-\widetilde{K}_i^{-1}}{v-v^{-1}},\ \mathsf{x}_s\mathsf{y}_t - \mathsf{y}_t\mathsf{x}_s = \delta_{s,t}\frac{\mathsf{k}_s-\mathsf{k}_s^{-1}}{v-v^{-1}},\ E_i\mathsf{y}_s = \mathsf{y}_sE_i$, $\mathsf{x}_sF_i = F_i\mathsf{x}_s$, where $\widetilde{K}_i = K_iK_{i+1}^{-1}$;

(R5) $\sum_{a+b=1-c_{i,j}} (-1)^a \begin{bmatrix} 1-c_{i,j} \\ a \end{bmatrix} E_i^a E_j E_i^b = 0$ for $i \neq j$;

(R6) $\sum_{a+b=1-c_{i,j}} (-1)^a \begin{bmatrix} 1-c_{i,j} \\ a \end{bmatrix} F_i^a F_j F_i^b = 0$ for $i \neq j$;

(R7) $E_i \mathsf{x}_s = \mathsf{x}_s E_i, \mathsf{x}_s \mathsf{x}_t = \mathsf{x}_t \mathsf{x}_s$;
(R8) $F_i \mathsf{y}_s = \mathsf{y}_s F_i, \mathsf{y}_s \mathsf{y}_t = \mathsf{y}_t \mathsf{y}_s$.

Moreover, $\mathbf{U}(\widetilde{C}_m)$ is a Hopf algebra with comultiplication Δ, counit ε, and antipode σ defined by

$$\begin{gathered}
\Delta(E_i) = E_i \otimes \widetilde{K}_i + 1 \otimes E_i, \quad \Delta(F_i) = F_i \otimes 1 + \widetilde{K}_i^{-1} \otimes F_i, \\
\Delta(\mathsf{x}_s) = \mathsf{x}_s \otimes \mathsf{k}_s + 1 \otimes \mathsf{x}_s, \quad \Delta(\mathsf{y}_s) = \mathsf{y}_s \otimes 1 + \mathsf{k}_s^{-1} \otimes \mathsf{y}_s, \\
\Delta(K_i^{\pm 1}) = K_i^{\pm 1} \otimes K_i^{\pm 1}, \quad \Delta(\mathsf{k}_s^{\pm 1}) = \mathsf{k}_s^{\pm 1} \otimes \mathsf{k}_s^{\pm 1}; \\
\varepsilon(E_i) = \varepsilon(\mathsf{x}_s) = 0 = \varepsilon(F_i) = \varepsilon(\mathsf{y}_s), \quad \varepsilon(K_i) = \varepsilon(\mathsf{k}_s) = 1; \\
\sigma(E_i) = -E_i \widetilde{K}_i^{-1}, \quad \sigma(F_i) = -\widetilde{K}_i F_i, \quad \sigma(K_i^{\pm 1}) = K_i^{\mp 1}, \\
\sigma(\mathsf{x}_s) = -\mathsf{x}_s \mathsf{k}_s^{-1}, \quad \sigma(\mathsf{y}_s) = -\mathsf{k}_s \mathsf{y}_s, \text{ and } \sigma(\mathsf{k}_s^{\pm 1}) = \mathsf{k}_s^{\mp 1},
\end{gathered}$$

where $i \in I$ and $s \in J_m$.

Remark 2.3.3. The comultiplication here is opposite to the one given in [**54**, 3.1.10]. Consequently, the antipode is the inverse of the antipode given in [**54**, 3.3.1]. This change is necessary for its action on tensor space, commuting with the Hecke algebra action; cf. [**12**, §14.6].

Let $\widetilde{\mathbf{U}}^+ = \mathbf{U}(\widetilde{C}_m)^+$ (resp., $\widetilde{\mathbf{U}}^- = \mathbf{U}(\widetilde{C}_m)^-$, $\widetilde{\mathbf{U}}^0 = \mathbf{U}(\widetilde{C}_m)^0$) be the $\mathbb{Q}(v)$-subalgebra of $\mathbf{U}(\widetilde{C}_m)$ generated by E_i (resp., F_i, $K_i^{\pm 1}$), for all $i \in \widetilde{I}$, and x_s (resp., $\mathsf{y}_s, \mathsf{k}_s^{\pm 1}$), for all $s \in J_m$. Then, by [**48**, Th. 2.23], $\mathbf{U}(\widetilde{C}_m)$ admits a triangular decomposition

$$\mathbf{U}(\widetilde{C}_m) = \widetilde{\mathbf{U}}^+ \otimes \widetilde{\mathbf{U}}^0 \otimes \widetilde{\mathbf{U}}^-.$$

It is clear that the subalgebra of $\mathbf{U}(\widetilde{C}_m)$ generated by E_i, F_i, $\widetilde{K}_i$, $\widetilde{K}_i^{-1}$ $(i \in I)$ is the quantum enveloping algebra $\mathbf{U}(\widehat{\mathfrak{sl}}_n)$ as defined in §1.3. Also, we denote by $\mathbf{U}_\Delta(n)$ the subalgebra of $\mathbf{U}(\widetilde{C}_m)$ generated by E_i, F_i, K_i, K_i^{-1} $(i \in I)$. We will call $\mathbf{U}_\Delta(n)$ the *extended quantum affine* $\mathfrak{sl}_n$.[2]

Now Theorem 2.3.1 follows immediately from the following result which describes the structure of $\mathfrak{D}_\Delta(n)$. It is a generalization of a result for the double Ringel–Hall algebra of a finite dimensional tame hereditary algebra in [**38**] to the cyclic quiver case. Its proof is based on Theorem 1.4.1.

Proposition 2.3.4. *There is a unique surjective Hopf algebra homomorphism* $\Phi : \mathbf{U}(\widetilde{C}_\infty) \to \mathfrak{D}_\Delta(n)$ *satisfying, for each* $i \in I$ *and* $s \in J_\infty$,

$$E_i \longmapsto u_i^+, \ \mathsf{x}_s \longmapsto \mathsf{z}_s^+, \ F_i \longmapsto u_i^-, \ \mathsf{y}_s \longmapsto \mathsf{z}_s^-, \ K_i^{\pm 1} \longmapsto K_i^{\pm}, \ \mathsf{k}_s^{\pm} \longmapsto 1.$$

[2] This is called quantum affine $\mathfrak{gl}_n$ by Lusztig [**56**], which is an extension of quantum affine $\mathfrak{sl}_n$ by adding an extra generator to the 0-part. This extension is similar to the extension of quantum $\mathfrak{gl}_n$ from quantum $\mathfrak{sl}_n$. In the literature, quantum affine $\mathfrak{sl}_n$ refers also to the quantum loop algebra associated with $\mathfrak{sl}_n$; see §2.5 below.

Moreover, $\operatorname{Ker}\Phi$ *is the ideal of* $\mathbf{U}(\widetilde{C}_\infty)$ *generated by* $\mathsf{k}_s^{\pm 1}-1$, *for all* $s \in J_\infty$.

Proof. The existence and uniqueness of Φ clearly follow from Definition 2.3.2, Theorem 1.3.4, Theorem 1.4.1, and Proposition 2.2.2. Further, Φ induces surjective algebra homomorphisms

$$\Phi^\pm : \widetilde{\mathbf{U}}^\pm = \mathbf{U}(\widetilde{C}_\infty)^\pm \longrightarrow \mathfrak{D}_\Delta(n)^\pm.$$

By (R7), the multiplication map

$$\mu : \mathbf{U}(\widehat{\mathfrak{sl}}_n)^+ \otimes \mathbb{Q}(v)[\mathsf{x}_1, \mathsf{x}_2, \ldots] \longrightarrow \widetilde{\mathbf{U}}^+$$

is surjective. Hence, the composite $\Phi^+\mu$ is surjective. By Theorem 1.4.1,

$$\mathfrak{D}_\Delta(n)^+ = \mathfrak{C}_\Delta(n)^+ \otimes_{\mathbb{Q}(v)} \mathbb{Q}(v)[\mathsf{z}_1^+, \mathsf{z}_2^+, \ldots].$$

Since $\Phi^+\mu(\mathbf{U}(\widehat{\mathfrak{sl}}_n)^+) = \mathfrak{C}_\Delta(n)^+ \cong \mathbf{U}(\widehat{\mathfrak{sl}}_n)^+$, it follows that $\Phi^+\mu$ is an isomorphism. Hence, Φ^+ is an isomorphism. Similarly, Φ^- is an isomorphism, too.

Let $\mathcal{K}$ be the ideal of $\mathbf{U}(\widetilde{C}_\infty)$ generated by $\mathsf{k}_s^{\pm 1}-1$ for $s \in J_\infty$. It is clear that $\mathcal{K} \subseteq \operatorname{Ker}\Phi$. The triangular decomposition $\mathbf{U}(\widetilde{C}_\infty) = \widetilde{\mathbf{U}}^+ \otimes \widetilde{\mathbf{U}}^0 \otimes \widetilde{\mathbf{U}}^-$ implies that

$$\mathbf{U}(\widetilde{C}_\infty)/\mathcal{K} = \big(\widetilde{\mathbf{U}}^+ \otimes \mathbb{Q}(v)[K_1^{\pm 1}, \ldots, K_n^{\pm 1}] \otimes \widetilde{\mathbf{U}}^- + \mathcal{K}\big)/\mathcal{K}.$$

Since

$$\mathfrak{D}_\Delta(n) = \mathfrak{D}_\Delta(n)^+ \otimes \mathfrak{D}_\Delta(n)^0 \otimes \mathfrak{D}_\Delta(n)^- = \mathfrak{D}_\Delta(n)^+ \otimes \mathbb{Q}(v)[K_1^{\pm 1}, \ldots, K_n^{\pm 1}] \otimes \mathfrak{D}_\Delta(n)^-,$$

we conclude that $\operatorname{Ker}\Phi = \mathcal{K}$, as required.

Finally, it is straightforward to check that $\mathcal{K}$ is a Hopf ideal. □

Corollary 2.3.5. *The double Ringel–Hall algebra* $\mathfrak{D}_\Delta(n)$ *is a Hopf algebra with comultiplication* Δ, *counit* ε, *and antipode* σ *defined by*

$$\Delta(E_i) = E_i \otimes \widetilde{K}_i + 1 \otimes E_i, \quad \Delta(F_i) = F_i \otimes 1 + \widetilde{K}_i^{-1} \otimes F_i,$$
$$\Delta(K_i^{\pm 1}) = K_i^{\pm 1} \otimes K_i^{\pm 1}, \quad \Delta(\mathsf{z}_s^\pm) = \mathsf{z}_s^\pm \otimes 1 + 1 \otimes \mathsf{z}_s^\pm;$$
$$\varepsilon(E_i) = \varepsilon(F_i) = 0 = \varepsilon(\mathsf{z}_s^\pm), \quad \varepsilon(K_i) = 1;$$
$$\sigma(E_i) = -E_i\widetilde{K}_i^{-1}, \quad \sigma(F_i) = -\widetilde{K}_i F_i, \quad \sigma(K_i^{\pm 1}) = K_i^{\mp 1},$$
$$\text{and} \quad \sigma(\mathsf{z}_s^\pm) = -\mathsf{z}_s^\pm,$$

where $i \in I$ *and* $s \in J_\infty$.

Remarks 2.3.6. (1) For notational simplicity, we sometimes continue to use $u_i^\pm$ as generators of $\mathfrak{D}_\Delta(n)$. By Theorem 2.3.1, we see that there is a $\mathbb{Q}(v)$-algebra involution ς of $\mathfrak{D}_\Delta(n)$ satisfying

$$K_i^{\pm 1} \longmapsto K_i^{\mp 1}, \; u_i^\pm \longmapsto u_i^\mp, \; \mathsf{z}_s^\pm \longmapsto \mathsf{z}_s^\mp,$$

for $i \in I$ and $s \geqslant 1$. Thus, $\mathfrak{D}_\Delta(n)$ also admits a decomposition

$$\mathfrak{D}_\Delta(n) = \mathfrak{D}_\Delta(n)^- \otimes \mathfrak{D}_\Delta(n)^0 \otimes \mathfrak{D}_\Delta(n)^+. \tag{2.3.6.1}$$

Moreover, two other types of generators for Ringel–Hall algebras discussed in §1.4 result in the corresponding generators for double Ringel–Hall algebras. Thus, we may speak of semisimple generators, etc. See more discussion in §2.6.

(2) The subalgebra of $\mathfrak{D}_\Delta(n)$ generated by E_i, F_i, K_i, K_i^{-1} $(i \in I)$ is isomorphic to $\mathbf{U}_\Delta(n)$. We will always identify these two algebras and thus, view $\mathbf{U}_\Delta(n)$ as a subalgebra of $\mathfrak{D}_\Delta(n)$. In particular, we have

$$\mathbf{U}_\Delta(n) = \mathfrak{C}_\Delta(n)^- \otimes \mathfrak{D}_\Delta(n)^0 \otimes \mathfrak{C}_\Delta(n)^+.$$

Moreover, if $\mathbf{Z}_\Delta(n) = \mathbb{Q}(v)[\mathsf{z}_m^+, \mathsf{z}_m^-]_{m\geqslant 1}$ denotes the *central subalgebra* of $\mathfrak{D}_\Delta(n)$ generated by $\ldots, \mathsf{z}_2^-, \mathsf{z}_1^-, \mathsf{z}_1^+, \mathsf{z}_2^+, \ldots$, then (2.3.6.1) together with (2.2.2.1) gives

$$\mathfrak{D}_\Delta(n) \cong \mathbf{U}_\Delta(n) \otimes \mathbf{Z}_\Delta(n).$$

(3) The subalgebra $'\mathfrak{D}_\Delta(n)$ of $\mathfrak{D}_\Delta(n)$ generated by $u_A^+, u_A^-, \widetilde{K}_i, \widetilde{K}_i^{-1}$ $(A \in \Theta_\Delta^+(n), i \in I)$ is isomorphic to the reduced Drinfeld double of $'\mathfrak{H}_\Delta(n)^{\geqslant 0}$ and $'\mathfrak{H}_\Delta(n)^{\leqslant 0}$; see Remark 1.5.5(2). This subalgebra will be considered in §5.5 regarding a compatibility condition on the transfer maps.

(4) For each $m \geqslant 1$, $\mathbf{U}(\widetilde{C}_m)$ can be naturally viewed as a subalgebra of $\mathbf{U}(\widetilde{C}_\infty)$. If we let $\mathfrak{D}_\Delta(n)^{(m)}$ be the subalgebra of $\mathfrak{D}_\Delta(n)$ generated by $E_i, F_i, K_i, K_i^{-1}, \mathsf{z}_s^+, \mathsf{z}_s^-$ for $i \in I$ and $s \in J_m$, then the homomorphism Φ in Proposition 2.3.4 induces a surjective Hopf algebra homomorphism $\mathbf{U}(\widetilde{C}_m) \to \mathfrak{D}_\Delta(n)^{(m)}$. Applying Proposition 1.4.3 shows that $\mathfrak{D}_\Delta(n)^{(m)}$ is also generated by $E_i, F_i, K_i, K_i^{-1}, u_{s\delta}^+, u_{s\delta}^-$, for $i \in I$ and $s \in J_m$.

2.4. Some integral forms

We introduce an integral form for the double Ringel–Hall algebra $\mathfrak{D}_\Delta(n)$ which is similar to those considered in [**49, 42**]. This form will be used in §6.1. See Remark 3.8.7(1) for a comparison with another possible form of Lusztig type. First, we observe the following commutator formula in $\mathbf{U}(\widetilde{C}_\infty)$ about the generators x_m and y_m.

Proposition 2.4.1. *For $m, r, s \geqslant 1$,*

$$[\mathsf{x}_m^r, \mathsf{y}_m^s] = \sum_{i=1}^{\min\{r,s\}} \binom{r}{i}\binom{s}{i} i! \left(\frac{\mathsf{k}_m - \mathsf{k}_m^{-1}}{v - v^{-1}}\right)^i \mathsf{y}_m^{s-i}\mathsf{x}_m^{r-i}. \tag{2.4.1.1}$$

Proof. Since $[\mathsf{x}_m, \mathsf{y}_m] = \frac{\mathsf{k}_m - \mathsf{k}_m^{-1}}{v - v^{-1}} = \left[{\mathsf{k}_m;0 \atop 1} \right]$ and $[\mathsf{x}_m, \mathsf{k}_m^{\pm}] = 0 = [\mathsf{y}_m, \mathsf{k}_m^{\pm}]$, it is direct to check that, for $s \geqslant 1$,

$$[\mathsf{x}_m, \mathsf{y}_m^s] = s \frac{\mathsf{k}_m - \mathsf{k}_m^{-1}}{v - v^{-1}} \mathsf{y}_m^{s-1}.$$

We proceed by induction on r to prove (2.4.1.1). The case $r = 1$ is given as above. Now let $r \geqslant 1$ and suppose (2.4.1.1) holds for r. If $r < s$, then

$$\begin{aligned}
[\mathsf{x}_m^{r+1}, \mathsf{y}_m^s] &= \mathsf{x}_m[\mathsf{x}_m^r, \mathsf{y}_m^s] + [\mathsf{x}_m, \mathsf{y}_m^s]\mathsf{x}_m^r \\
&= \sum_{i=1}^{r} \binom{r}{i}\binom{s}{i} i! \left(\frac{\mathsf{k}_m - \mathsf{k}_m^{-1}}{v - v^{-1}} \right)^i \mathsf{x}_m \mathsf{y}_m^{s-i} \mathsf{x}_m^{r-i} + s \frac{\mathsf{k}_m - \mathsf{k}_m^{-1}}{v - v^{-1}} \mathsf{y}_m^{s-1} \mathsf{x}_m^r \\
&= \sum_{i=1}^{r} \binom{r}{i}\binom{s}{i} i! \left(\frac{\mathsf{k}_m - \mathsf{k}_m^{-1}}{v - v^{-1}} \right)^i \left((s-i) \frac{\mathsf{k}_m - \mathsf{k}_m^{-1}}{v - v^{-1}} \mathsf{y}_m^{s-i-1} + \mathsf{y}_m^{s-i} \mathsf{x}_m \right) \mathsf{x}_m^{r-i} \\
&\quad + s \frac{\mathsf{k}_m - \mathsf{k}_m^{-1}}{v - v^{-1}} \mathsf{y}_m^{s-1} \mathsf{x}_m^r \\
&= \sum_{i=2}^{r+1} \binom{r}{i-1}\binom{s}{i} i! \left(\frac{\mathsf{k}_m - \mathsf{k}_m^{-1}}{v - v^{-1}} \right)^i \mathsf{y}_m^{s-i} \mathsf{x}_m^{r+1-i} \\
&\quad + \sum_{i=1}^{r} \binom{r}{i}\binom{s}{i} i! \left(\frac{\mathsf{k}_m - \mathsf{k}_m^{-1}}{v - v^{-1}} \right)^i \mathsf{y}_m^{s-i} \mathsf{x}_m^{r+1-i} + s \frac{\mathsf{k}_m - \mathsf{k}_m^{-1}}{v - v^{-1}} \mathsf{y}_m^{s-1} \mathsf{x}_m^r \\
&= \sum_{i=1}^{r+1} \binom{r+1}{i}\binom{s}{i} i! \left(\frac{\mathsf{k}_m - \mathsf{k}_m^{-1}}{v - v^{-1}} \right)^i \mathsf{y}_m^{s-i} \mathsf{x}_m^{r+1-i}.
\end{aligned}$$

The case $r \geqslant s$ can be treated similarly. □

Consider the $\mathcal{Z}$-subalgebra $\widetilde{U} = \widetilde{U}_{\mathcal{Z}}$ of $\mathbf{U}(\widetilde{C}_\infty)$ generated by $K_i^{\pm 1}$, $\left[{K_i;0 \atop t} \right]$, $\mathsf{k}_s^{\pm 1}$, x_s, y_s together with the divided powers $E_i^{(m)} = E_i^m/[m]^!$ and $F_i^{(m)} = F_i^m/[m]^!$, where $i \in I$, $t \geqslant 0$, and $s, m \geqslant 1$. Further, we set

$$\widetilde{U}^+ = \widetilde{\mathbf{U}}^+ \cap \widetilde{U}, \ \widetilde{U}^- = \widetilde{\mathbf{U}}^- \cap \widetilde{U}, \ \text{ and } \widetilde{U}^0 = \widetilde{\mathbf{U}}^0 \cap \widetilde{U}.$$

Then $\widetilde{U}^+$ (resp., $\widetilde{U}^-$) is the $\mathcal{Z}$-subalgebra of $\mathbf{U}(\widetilde{C}_\infty)$ generated by x_s and $E_i^{(m)}$ (resp., y_s and $F_i^{(m)}$), for $i \in I$ and $s, m \geqslant 1$. Consequently, we obtain

$$\widetilde{U}^+ = U(\widehat{\mathfrak{sl}}_n)^+ \otimes \mathcal{Z}[\mathsf{x}_1, \mathsf{x}_2, \ldots] \ \text{ and } \ \widetilde{U}^- = U(\widehat{\mathfrak{sl}}_n)^- \otimes \mathcal{Z}[\mathsf{y}_1, \mathsf{y}_2, \ldots],$$

where $U(\widehat{\mathfrak{sl}}_n)^+$ and $U(\widehat{\mathfrak{sl}}_n)^-$ are the $\mathcal{Z}$-subalgebras of $\mathbf{U}(\widehat{\mathfrak{sl}}_n)$ generated by the divided powers $E_i^{(m)}$ and $F_i^{(m)}$, respectively. This implies in particular that both $\widetilde{U}^+$ and $\widetilde{U}^-$ are free $\mathcal{Z}$-modules.

We now look at the structure of $\widetilde{U}^0$. Let V^0 be the $\mathcal{Z}$-subalgebra generated by $K_i^{\pm 1}$, $\left[{K_i;0 \atop t} \right]$, $\mathsf{k}_s^{\pm 1}$, $\left[{\mathsf{k}_s;0 \atop 1} \right]$, for $i \in I$, $t \geqslant 0$, and $s \geqslant 1$. We first prove that V^0 is $\mathcal{Z}$-free. First, a result of Lusztig [**52**, 2.14] (see also [**22**] and [**12**, Th. 14.20]) shows that V^0 contains the $\mathcal{Z}$-subalgebra V^1 generated by $K_i^{\pm 1}$ and $\left[{K_i;0 \atop t} \right]$ which has a $\mathcal{Z}$-basis

$$\mathcal{K} = \left\{ \prod_{i=1}^{n} K_i^{a_i} \begin{bmatrix} K_i; 0 \\ t_i \end{bmatrix} \,\middle|\, a_i \in \{0,1\}, t_i \in \mathbb{N}, \forall i \in I \right\}, \text{ or}$$

$$\mathcal{K}' = \left\{ \prod_{i=1}^{n-1} \widetilde{K}_i^{a_i} \begin{bmatrix} \widetilde{K}_i; 0 \\ t_i \end{bmatrix} \cdot K_i^{a_n} \begin{bmatrix} K_n; 0 \\ t_n \end{bmatrix} \,\middle|\, a_i \in \{0,1\}, t_i \in \mathbb{N}, \forall i \in I \right\}. \tag{2.4.1.2}$$

Let V^2 be the $\mathcal{Z}$-subalgebra of V^0 generated by $\mathsf{k}_s^{\pm 1}$, $\left[{\mathsf{k}_s;0 \atop 1} \right]$, for $s \geqslant 1$, and let

$$\begin{bmatrix} \mathsf{k}_s; 0 \\ t \end{bmatrix}' = [t]^! \begin{bmatrix} \mathsf{k}_s; 0 \\ t \end{bmatrix}.$$

Then the formula [**52**, 2.3(g8)] with $t' = 1$ becomes

$$v^t \begin{bmatrix} \mathsf{k}_s; 0 \\ t \end{bmatrix}' \begin{bmatrix} \mathsf{k}_s; 0 \\ 1 \end{bmatrix} - [t]\mathsf{k}_s \begin{bmatrix} \mathsf{k}_s; 0 \\ t \end{bmatrix}' = \begin{bmatrix} \mathsf{k}_s; 0 \\ t+1 \end{bmatrix}'. \tag{2.4.1.3}$$

Lemma 2.4.2. *The set*

$$\mathcal{M} := \left\{ \mathsf{k}_1^{\delta_1} \cdots \mathsf{k}_m^{\delta_m} \begin{bmatrix} \mathsf{k}_1; 0 \\ t_1 \end{bmatrix}' \cdots \begin{bmatrix} \mathsf{k}_m; 0 \\ t_m \end{bmatrix}' \,\middle|\, m \geqslant 1,\ \delta_i \in \{0,1\},\ t_i \in \mathbb{N} \right\}$$

forms a $\mathcal{Z}$-basis of V^2.

Proof. Let $W_1 = \operatorname{span}_{\mathcal{Z}} \mathcal{M}$ and

$$W_2 = \operatorname{span}_{\mathcal{Z}} \left\{ \mathsf{k}_1^{\delta_1} \cdots \mathsf{k}_m^{\delta_m} \begin{bmatrix} \mathsf{k}_1; 0 \\ t_1 \end{bmatrix}' \cdots \begin{bmatrix} \mathsf{k}_m; 0 \\ t_m \end{bmatrix}' \,\middle|\, m \geqslant 1,\ \delta_i \in \mathbb{Z},\ t_i \in \mathbb{N} \right\}.$$

By (2.4.1.3), $\left[{\mathsf{k}_s;0 \atop t} \right]' \in V^2$, for $s \geqslant 1$ and $t \geqslant 0$. Thus $W_1 \subseteq W_2 \subseteq V^2$. Clearly, $\mathsf{k}_s^{\pm 1} W_2 \subseteq W_2$. By (2.4.1.3) again, $\left[{\mathsf{k}_s;0 \atop 1} \right] W_2 \subseteq W_2$. Thus, $V^2 W_2 \subseteq W_2$. Since $1 \in W_2$, $V^2 \subseteq W_2$ and, hence, $V^2 = W_2$. For $m \geqslant 0$, applying Lusztig's formulas

$$\mathsf{k}_s^{m+2} \begin{bmatrix} \mathsf{k}_s; 0 \\ t \end{bmatrix}' = v^t(v - v^{-1})\mathsf{k}_s^{m+1} \begin{bmatrix} \mathsf{k}_s; 0 \\ t+1 \end{bmatrix}' + v^{2t}\mathsf{k}_s^{m} \begin{bmatrix} \mathsf{k}_s; 0 \\ t \end{bmatrix}' \quad \text{and}$$

$$\mathsf{k}_s^{-m-1} \begin{bmatrix} \mathsf{k}_s; 0 \\ t \end{bmatrix}' = -v^{-t}(v - v^{-1})\mathsf{k}_s^{-m} \begin{bmatrix} \mathsf{k}_s; 0 \\ t+1 \end{bmatrix}' + v^{-2t}\mathsf{k}_s^{-m+1} \begin{bmatrix} \mathsf{k}_s; 0 \\ t \end{bmatrix}'$$

in [**52**, p. 278] yields $\mathsf{k}_s^{\pm m} \left[{\mathsf{k}_s;0 \atop t} \right]' \in W_1$, for any $m \geqslant 0$. Hence, $W_2 \subseteq W_1$ and, therefore, $W_2 = W_1 = V^2$. □

Theorem 2.4.3. *The integral* 0*-part* $\widetilde{U}^0$ *is the* $\mathcal{Z}$*-subalgebra of* $\mathbf{U}(\widetilde{C}_\infty)$ *generated by* $K_i^{\pm 1}$, $\left[{K_i;0 \atop t}\right]$, $\mathsf{k}_s^{\pm 1}$, $\left[{\mathsf{k}_s;0 \atop 1}\right]$, *for* $i \in I$ *and* $s \geqslant 1$. *Hence,* $\widetilde{U}^0$ *is a free* $\mathcal{Z}$*-module with basis* $\{xy \mid x \in \mathcal{K}, y \in \mathcal{M}\}$.

Proof. Let $U' = \widetilde{U}^+ V^0 \widetilde{U}^-$. This is clearly a subset of $\widetilde{U}$. Since, by (2.4.1.1), U' is a subalgebra,[3] it follows that $\widetilde{U} \subseteq U'$. Hence, $U' = \widetilde{U}$, and consequently, $\widetilde{U}^0 = U' \cap \mathbf{U}^0 = V^0$. The rest of the proof is clear. □

We now look at an integral form of $\boldsymbol{\mathfrak{D}}_\Delta(n)$.

Definition 2.4.4. The *integral form* $\mathfrak{D}_\Delta(n)$ of $\boldsymbol{\mathfrak{D}}_\Delta(n)$ is the $\mathcal{Z}$-subalgebra generated by $K_i^{\pm 1}$, $\left[{K_i;0 \atop t}\right]$, z_s^+, z_s^-, $(u_i^+)^{(m)}$ and $(u_i^-)^{(m)}$, for $i \in I$ and s, t, $m \geqslant 1$.

It is easy to see that the surjective $\mathbb{Q}(v)$-algebra homomorphism

$$\Phi : \mathbf{U}(\widetilde{C}_\infty) \longrightarrow \boldsymbol{\mathfrak{D}}_\Delta(n)$$

given in Proposition 2.3.4 induces a surjective $\mathcal{Z}$-algebra homomorphism $\Phi : \widetilde{U} \to \mathfrak{D}_\Delta(n)$. We also set

$$\mathfrak{D}_\Delta(n)^\epsilon = \boldsymbol{\mathfrak{D}}_\Delta(n)^\epsilon \cap \mathfrak{D}_\Delta(n) \ \text{ for } \epsilon \in \{+, -, 0\}.$$

Then $\mathfrak{D}_\Delta(n)^+$ (resp., $\mathfrak{D}_\Delta(n)^-$) is isomorphic to $\widetilde{U}^+$ (resp., $\widetilde{U}^-$). Hence,

$$\mathfrak{D}_\Delta(n)^+ = \mathfrak{C}_\Delta(n)^+ \otimes_{\mathcal{Z}} \mathcal{Z}[\mathsf{z}_1^+, \mathsf{z}_2^+, \ldots] \ \text{ and}$$
$$\mathfrak{D}_\Delta(n)^- = \mathfrak{C}_\Delta(n)^- \otimes_{\mathcal{Z}} \mathcal{Z}[\mathsf{z}_1^-, \mathsf{z}_2^-, \ldots],$$

where $\mathfrak{C}_\Delta(n)^+ \cong U(\widehat{\mathfrak{sl}}_n)^+$ (resp., $\mathfrak{C}_\Delta(n)^- \cong U(\widehat{\mathfrak{sl}}_n)^-$) is the $\mathcal{Z}$-subalgebra of $\mathfrak{D}_\Delta(n)$ generated by the $(u_i^+)^{(m)}$ (resp., $(u_i^-)^{(m)}$). Moreover, as seen above, $\mathfrak{D}_\Delta(n)^0$ is a free $\mathcal{Z}$-module with bases as given in (2.4.1.2).

Hence, the multiplication map

$$\mathfrak{D}_\Delta(n)^+ \otimes_{\mathcal{Z}} \mathfrak{D}_\Delta(n)^0 \otimes_{\mathcal{Z}} \mathfrak{D}_\Delta(n)^- \xrightarrow{\sim} \mathfrak{D}_\Delta(n) \tag{2.4.4.1}$$

is a $\mathcal{Z}$-module isomorphism; see [**12**, Cor. 6.50]. In particular, $\mathfrak{D}_\Delta(n)$ is a free $\mathcal{Z}$-module.

The $\mathcal{Z}$-algebra $\mathfrak{D}_\Delta(n)$ gives rise to a $\mathcal{Z}$-form $U_\Delta(n)$ for $\mathbf{U}_\Delta(n)$, the extended quantum affine $\mathfrak{sl}_n$:

$$U_\Delta(n) = \mathfrak{D}_\Delta(n) \cap \mathbf{U}_\Delta(n) = \mathfrak{C}_\Delta(n)^+ \mathfrak{D}_\Delta(n)^0 \mathfrak{C}_\Delta(n)^-. \tag{2.4.4.2}$$

We end this section with a description of the double Ringel–Hall algebras $\mathfrak{D}_{\Delta,\mathbb{C}}(n)$ given in Remark 2.1.4 in terms of specialization.

[3] Of course, this fact can be proved directly by the relation $[\mathsf{x}_s, \mathsf{y}_t] = \delta_{s,t}\left[{\mathsf{k}_s;0 \atop 1}\right]$.

Proposition 2.4.5. *If $z \in \mathbb{C}$ is a complex number which is not a root of unity, then the specialization $\mathfrak{D}_\Delta(n)_\mathbb{C} = \mathfrak{D}_\Delta(n) \otimes \mathbb{C}$ at $v = z$ is isomorphic to the $\mathbb{C}$-algebra $\mathfrak{D}_{\Delta,\mathbb{C}}(n)$.*

Proof. If we assign to each u_A^+ (resp., u_A^-, K_i) degree $\mathfrak{d}(A) = \dim M(A)$ (resp., $-\mathfrak{d}(A)$, 0), then $\boldsymbol{\mathfrak{D}}_\Delta(n)$ admits a $\mathbb{Z}$-grading $\boldsymbol{\mathfrak{D}}_\Delta(n) = \oplus_{m\in\mathbb{Z}}\boldsymbol{\mathfrak{D}}_\Delta(n)_m$, and $\mathfrak{D}_\Delta(n)$ inherits a $\mathbb{Z}$-grading $\mathfrak{D}_\Delta(n) = \oplus_{m\in\mathbb{Z}}\mathfrak{D}_\Delta(n)_m$, where $\mathfrak{D}_\Delta(n)_m = \mathfrak{D}_\Delta(n) \cap \boldsymbol{\mathfrak{D}}_\Delta(n)_m$. By the presentation of $\mathfrak{D}_{\Delta,\mathbb{C}}(n)$ as given in Remark 2.1.4, it is clear that there is a graded algebra homomorphism $\mathfrak{D}_{\Delta,\mathbb{C}}(n) \to \mathfrak{D}_\Delta(n)_\mathbb{C}$. This map is injective since it is so on every triangular component. Now, a dimensional comparison on homogeneous components shows that it is an isomorphism. □

Remark 2.4.6. Let $\mathfrak{H}_\Delta(n)^+$ be the $\mathcal{Z}$-submodule of $\boldsymbol{\mathfrak{D}}_\Delta(n)$ spanned by u_A^+ for $A \in \Theta_\Delta^+(n)$. Then $\mathfrak{H}_\Delta(n)^+$ is a $\mathcal{Z}$-algebra which is isomorphic to $\mathfrak{H}_\Delta(n)$. We point out that in general, $\mathfrak{D}_\Delta(n)^+$ does not coincide with $\mathfrak{H}_\Delta(n)^+$. For example, let $n = 2$. An easy calculation shows that

$$\mathsf{z}_1^+ = u_1^+u_2^+ + u_2^+u_1^+ - (v+v^{-1})u_\delta^+.$$

Since $\{u_1^+u_2^+, u_2^+u_1^+, \mathsf{z}_1^+\}$ is a $\mathcal{Z}$-basis for the homogeneous space $\mathfrak{D}_\Delta(2)_\delta^+$, we conclude that $u_\delta^+ \notin \mathfrak{D}_\Delta(2)^+$. If $n = 3$, then

$$\begin{aligned}\mathsf{z}_1^+ =& u_1^+u_2^+u_3^+ + u_2^+u_3^+u_1^+ + u_3^+u_1^+u_2^+\\ &- (v+v^{-1})(u_1^+u_3^+u_2^+ + u_2^+u_1^+u_3^+ + u_3^+u_2^+u_1^+) + (v^2+1+v^{-2})u_\delta^+.\end{aligned}$$

Analogously, we have $u_\delta^+ \notin \mathfrak{D}_\Delta(3)^+$.

It seems difficult to prove directly that all central elements z_m^+ lie in $\mathfrak{H}_\Delta(n)^+$. However, we will provide a proof of this fact in Corollary 3.7.5 via the integral quantum Schur algebras. Hence, $\mathfrak{D}_\Delta(n)^+$ is indeed a subalgebra of $\mathfrak{H}_\Delta(n)^+$.

2.5. The quantum loop algebra $\mathbf{U}(\widehat{\mathfrak{gl}}_n)$

In this section, we give an application of the presentation for $\boldsymbol{\mathfrak{D}}_\Delta(n)$ defined in §2.3. More precisely, we will modify the construction in [**40**] to extend Beck's algebra embedding of the quantum affine $\mathfrak{sl}_n$ into the quantum loop algebra $\mathbf{U}(\widehat{\mathfrak{gl}}_n)$ defined by Drinfeld's new presentation [**20**] to an explicit isomorphism from $\boldsymbol{\mathfrak{D}}_\Delta(n)$ to $\mathbf{U}(\widehat{\mathfrak{gl}}_n)$.

As discussed in §1.1, consider the loop algebra

$$\widehat{\mathfrak{gl}}_n = \mathfrak{gl}_n \otimes \mathbb{Q}[t, t^{-1}],$$

which is generated by $E_{i,i+1} \otimes t^s$, $E_{i+1,i} \otimes t^s$, and $E_{j,j} \otimes t^s$, for $s \in \mathbb{Z}$, $1 \leqslant i \leqslant n-1$, and $1 \leqslant j \leqslant n$. We have the following quantum enveloping algebra associated with $\widehat{\mathfrak{gl}}_n$; see **[20]** (and also **[28, 40]**).

Definition 2.5.1. (1) The *quantum loop algebra* $\mathbf{U}(\widehat{\mathfrak{gl}}_n)$ (or *quantum affine* $\mathfrak{gl}_n$) is the $\mathbb{Q}(v)$-algebra generated by $\mathrm{x}^{\pm}_{i,s}$ ($1 \leqslant i < n$, $s \in \mathbb{Z}$), $\mathrm{k}_i^{\pm 1}$ and $\mathrm{g}_{i,t}$ ($1 \leqslant i \leqslant n$, $t \in \mathbb{Z}\backslash\{0\}$) with the following relations:

(QLA1) $\mathrm{k}_i\mathrm{k}_i^{-1} = 1 = \mathrm{k}_i^{-1}\mathrm{k}_i,\ [\mathrm{k}_i, \mathrm{k}_j] = 0;$

(QLA2) $\mathrm{k}_i\mathrm{x}^{\pm}_{j,s} = v^{\pm(\delta_{i,j}-\delta_{i,j+1})}\mathrm{x}^{\pm}_{j,s}\mathrm{k}_i,\ [\mathrm{k}_i, \mathrm{g}_{j,s}] = 0;$

(QLA3) $[\mathrm{g}_{i,s}, \mathrm{x}^{\pm}_{j,t}] = \begin{cases} 0, & \text{if } i \neq j,\ j+1; \\ \pm v^{-js}\frac{[s]}{s}\mathrm{x}^{\pm}_{j,s+t}, & \text{if } i = j; \\ \mp v^{-js}\frac{[s]}{s}\mathrm{x}^{\pm}_{j,s+t}, & \text{if } i = j+1; \end{cases}$

(QLA4) $[\mathrm{g}_{i,s}, \mathrm{g}_{j,t}] = 0;$

(QLA5) $[\mathrm{x}^{+}_{i,s}, \mathrm{x}^{-}_{j,t}] = \delta_{i,j}\frac{\phi^{+}_{i,s+t}-\phi^{-}_{i,s+t}}{v-v^{-1}};$

(QLA6) $\mathrm{x}^{\pm}_{i,s}\mathrm{x}^{\pm}_{j,t} = \mathrm{x}^{\pm}_{j,t}\mathrm{x}^{\pm}_{i,s}$, for $|i-j| > 1$, and $[\mathrm{x}^{\pm}_{i,s+1}, \mathrm{x}^{\pm}_{j,t}]_{v^{\pm c_{ij}}} = -[\mathrm{x}^{\pm}_{j,t+1}, \mathrm{x}^{\pm}_{i,s}]_{v^{\pm c_{ij}}};$

(QLA7) $[\mathrm{x}^{\pm}_{i,s}, [\mathrm{x}^{\pm}_{j,t}, \mathrm{x}^{\pm}_{i,p}]_v]_v = -[\mathrm{x}^{\pm}_{i,p}, [\mathrm{x}^{\pm}_{j,t}, \mathrm{x}^{\pm}_{i,s}]_v]_v$ for $|i-j| = 1$,

where $[x, y]_a = xy - ayx$, and $\phi^{\pm}_{i,s}$ are defined via the generating functions in indeterminate u by

$$\Phi^{\pm}_i(u) := \widetilde{\mathrm{k}}_i^{\pm 1}\exp\Big(\pm(v-v^{-1})\sum_{m\geqslant 1}\mathrm{h}_{i,\pm m}u^{\pm m}\Big) = \sum_{s\geqslant 0}\phi^{\pm}_{i,\pm s}u^{\pm s}$$

with $\widetilde{\mathrm{k}}_i = \mathrm{k}_i/\mathrm{k}_{i+1}$ ($\mathrm{k}_{n+1} = \mathrm{k}_1$) and $\mathrm{h}_{i,\pm m} = v^{\pm(i-1)m}\mathrm{g}_{i,\pm m} - v^{\pm(i+1)m}\mathrm{g}_{i+1,\pm m}$ ($1 \leqslant i < n$).

(2) Let $\mathrm{U}_{\mathbb{C}}(\widehat{\mathfrak{gl}}_n)$ be the quantum loop algebra defined by the same generators and relations (QLA1)–(QLA7) with $\mathbb{Q}(v)$ replaced by $\mathbb{C}$ and v by $z \in \mathbb{C}^*$ with $z^m \neq 1$ for all $m \geqslant 1$.

Observe that, if we put $\Theta^{\pm}_i(u) = \mathrm{k}_i^{\pm 1}\exp\big(\pm(v-v^{-1})\sum_{m\geqslant 1}\mathrm{g}_{i,\pm m}u^{\pm m}\big)$, then

$$\Phi^{\pm}_i(u) = \frac{\Theta^{\pm}_i(uv^{i-1})}{\Theta^{\pm}_{i+1}(uv^{i+1})}.$$

Also, $\phi^{+}_{i,-s} = \phi^{-}_{i,s} = 0$, for all $s \geqslant 1$. Hence, (QLA5) becomes

$$[\mathrm{x}^{+}_{i,s}, \mathrm{x}^{-}_{i,t}] = \begin{cases} \frac{\widetilde{\mathrm{k}}_i-\widetilde{\mathrm{k}}_i^{-}}{v-v^{-1}}, & \text{if } s+t = 0; \\ \frac{\phi^{+}_{i,s+t}}{v-v^{-1}}, & \text{if } s+t > 0; \\ \frac{-\phi^{-}_{i,s+t}}{v-v^{-1}}, & \text{if } s+t < 0. \end{cases}$$

By [**2**] (see also [**44**, Th. 2.2], [**3**, Th. 1], [**28**, §3.1], and [**40**, Th. 3]), there is a monomorphism (or *Beck's embedding*) of Hopf algebras $\mathcal{E}_{\mathrm{B}} : \mathbf{U}(\widehat{\mathfrak{sl}}_n) \to \mathbf{U}(\widehat{\mathfrak{gl}}_n)$ such that

$$\widetilde{K}_i \longmapsto \widetilde{\mathrm{k}}_i,\ E_i \longmapsto \mathrm{x}^+_{i,0},\ F_i \longmapsto \mathrm{x}^-_{i,0}\ (1 \leqslant i < n),\ E_n \longmapsto \varepsilon^+_n,\ F_n \longmapsto \varepsilon^-_n,$$

where

$$\varepsilon^+_n = (-1)^n[\mathrm{x}^-_{n-1,0}, \cdots, [\mathrm{x}^-_{3,0}, [\mathrm{x}^-_{2,0}, \mathrm{x}^-_{1,1}]_{v^{-1}}]_{v^{-1}} \cdots]_{v^{-1}}\widetilde{\mathrm{k}}_n \quad \text{and}$$
$$\varepsilon^-_n = (-1)^n\widetilde{\mathrm{k}}_n^{-1}[\cdots[[\mathrm{x}^+_{1,-1}, \mathrm{x}^+_{2,0}]_v, \mathrm{x}^+_{3,0}]_v, \cdots, \mathrm{x}^+_{n-1,0}]_v.$$

The image $\mathrm{Im}(\mathcal{E}_{\mathrm{B}})$, known as the quantum loop algebra of $\mathfrak{sl}_n$, is the $\mathbb{Q}(v)$-subalgebra of $\mathbf{U}(\widehat{\mathfrak{gl}}_n)$ generated by $\widetilde{\mathrm{k}}_i^{\pm 1}$, $\mathrm{x}^\pm_{i,s}$, and $\mathrm{h}_{i,t}$ for $1 \leqslant i < n$, $s, t \in \mathbb{Z}$ with $t \neq 0$. Moreover, it can be proved (see, e.g., [**28**, §§2.1&3.1]) that the defining relations of this subalgebra are the relations (QLA5)–(QLA7) together with the relations:

(QLA8) $\widetilde{\mathrm{k}}_i\widetilde{\mathrm{k}}_j = \widetilde{\mathrm{k}}_j\widetilde{\mathrm{k}}_i,\ \widetilde{\mathrm{k}}_1\widetilde{\mathrm{k}}_2\cdots\widetilde{\mathrm{k}}_n = 1,\ [\widetilde{\mathrm{k}}_i, \mathrm{h}_{j,s}] = 0,\ [\mathrm{h}_{i,s}, \mathrm{h}_{j,t}] = 0;$
(QLA9) $\widetilde{\mathrm{k}}_i\mathrm{x}^\pm_{j,s} = v^{\pm c_{i,j}}\mathrm{x}^\pm_{j,s}\widetilde{\mathrm{k}}_i,\ [\mathrm{h}_{i,s}, \mathrm{x}^\pm_{j,t}] = \pm\frac{[s\, c_{i,j}]}{s}\mathrm{x}^\pm_{j,s+t},$

where $(c_{i,j})$ is the generalized Cartan matrix of type $\widetilde{A}_{n-1}$ as given in (1.3.2.1). This is Drinfeld's new presentation for $\mathbf{U}(\widehat{\mathfrak{sl}}_n)$. In the sequel, we will identify $\mathbf{U}(\widehat{\mathfrak{sl}}_n)$ with $\mathrm{Im}(\mathcal{E}_{\mathrm{B}})$ and, thus, view $\mathbf{U}(\widehat{\mathfrak{sl}}_n)$ as a subalgebra of $\mathbf{U}(\widehat{\mathfrak{gl}}_n)$.

In an entirely similar way we can identify $\mathrm{U}_{\mathbb{C}}(\widehat{\mathfrak{sl}}_n)$ defined in §1.3 with a subalgebra of $\mathrm{U}_{\mathbb{C}}(\widehat{\mathfrak{gl}}_n)$ under the monomorphism $\mathcal{E}_{\mathrm{B},\mathbb{C}}$.[4]

Our next aim is to extend Beck's embedding $\mathcal{E}_{\mathrm{B}}$ to a Hopf algebra isomorphism $\mathfrak{D}_\Delta(n) \to \mathbf{U}(\widehat{\mathfrak{gl}}_n)$. For each $s \geqslant 1$, define

$$\theta_{\pm s} = \mp\frac{1}{[s]}(\mathrm{g}_{1,\pm s} + \cdots + \mathrm{g}_{n,\pm s}) \in \mathbf{U}(\widehat{\mathfrak{gl}}_n). \tag{2.5.1.1}$$

It can be directly checked by the definition that the θ_s are central in $\mathbf{U}(\widehat{\mathfrak{gl}}_n)$. By [**28**, §3.3], they are all primitive, i.e., for all $s \in \mathbb{Z}\backslash\{0\}$,

$$\Delta(\theta_s) = \theta_s \otimes 1 + 1 \otimes \theta_s.$$

Remark 2.5.2. In [**28**, (4.6)], the authors introduced central elements C_s ($s \in \mathbb{Z}$) in $\mathbf{U}(\widehat{\mathfrak{gl}}_n)$ which are defined by the generating function

$$\sum_{s\geqslant 0} C_{\pm s}u^{\pm s} = \prod_{i=1}^{n} \exp\Big(\mp\sum_{m>0}\frac{\mathrm{g}_{i,\pm m}}{[m]}u^{\pm m}\Big),$$

[4] Let ϕ_a be the automorphism of $\mathrm{U}_{\mathbb{C}}(\widehat{\mathfrak{gl}}_n)$ of the form $\mathrm{x}^\pm_{i,s} \mapsto a^s\mathrm{x}^\pm_{i,s}$, $\mathrm{g}_{j,t} \mapsto a^t\mathrm{g}_{j,t}$, and $\mathrm{k}_j \mapsto \mathrm{k}_j$. Then, for $a = z^n$, $\phi_a \circ \mathcal{E}_{\mathrm{B},\mathbb{C}}$ is the isomorphism B^{-1} given in [**28**, Lem. 3.3].

that is,

$$\sum_{s\geqslant 0} C_{\pm s}u^{\pm s} = \exp\Big(\sum_{m>0}\theta_{\pm m}u^{\pm m}\Big).$$

It follows that $C_0 = 1$, $C_{\pm 1} = \theta_{\pm 1}$ and, for each $s \geqslant 1$,

$$\mathbb{Q}(v)[\theta_{\pm 1}, \ldots, \theta_{\pm s}] = \mathbb{Q}(v)[C_{\pm 1}, \ldots, C_{\pm s}] \text{ and}$$
$$\theta_{\pm(s+1)} \equiv C_{\pm(s+1)} \mod \mathbb{Q}(v)[C_{\pm 1}, \ldots, C_{\pm s}].$$

Now fix an integer $s \geqslant 1$ and define $n \times n$ matrices $X^{(\pm s)}$ over $\mathbb{Q}(v)$ by

$$X^{(\pm s)} = (X_{i,j}^{(\pm s)}) = \begin{pmatrix} 1 & -v^{\pm 2s} & 0 & \cdots & 0 & 0 \\ 0 & v^{\pm s} & -v^{\pm 3s} & \cdots & 0 & 0 \\ 0 & 0 & v^{\pm 2s} & \cdots & 0 & 0 \\ \vdots & \vdots & \vdots & \ddots & \vdots & \vdots \\ 0 & 0 & 0 & \cdots & v^{\pm(n-2)s} & -v^{\pm ns} \\ \mp\frac{1}{[s]} & \mp\frac{1}{[s]} & \mp\frac{1}{[s]} & \cdots & \mp\frac{1}{[s]} & \mp\frac{1}{[s]} \end{pmatrix}.$$

By the definition of $\mathrm{h}_{i,\pm s}$ and $\theta_{\pm s}$,

$$\mathrm{h}_{i,\pm s} = \sum_{j=1}^{n} X_{i,j}^{(\pm s)}\mathfrak{g}_{j,\pm s}, \quad \text{for } 1 \leqslant i < n, \text{ and } \theta_{\pm s} = \sum_{j=1}^{n} X_{n,j}^{(\pm s)}\mathfrak{g}_{j,\pm s}.$$

A direct calculation shows that

$$\det(X^{(\pm s)}) = \mp\frac{1}{[s]}\big(1 + v^{\pm 2s} + \cdots + v^{\pm 2(n-1)s}\big)\prod_{i=1}^{n-2} v^{\pm is} \neq 0 \quad (n \geqslant 2).$$

We denote the inverse of $X^{(\pm s)}$ by $Y^{(\pm s)} = (Y_{i,j}^{(\pm s)})$. Thus, for each $1 \leqslant i \leqslant n$,

$$\mathfrak{g}_{i,\pm s} = \sum_{j=1}^{n-1} Y_{i,j}^{(\pm s)}\mathrm{h}_{j,\pm s} + Y_{i,n}^{(\pm s)}\theta_{\pm s}.$$

Therefore, the $\mathbb{Q}(v)$-subspace of $\mathbf{U}(\widehat{\mathfrak{gl}}_n)$ spanned by $\mathfrak{g}_{1,\pm s}, \ldots, \mathfrak{g}_{n,\pm s}$ coincides with that spanned by $\mathrm{h}_{1,\pm s}, \ldots, \mathrm{h}_{n-1,\pm s}, \theta_{\pm s}$. Consequently, $\mathbf{U}(\widehat{\mathfrak{gl}}_n)$ is also generated by

$$\mathrm{k}_i^{\pm 1}\ (1 \leqslant i \leqslant n),\ \mathrm{x}_{i,t}^{\pm},\ \mathrm{h}_{i,\pm s},\ \theta_{\pm s}\ (1 \leqslant i < n, t \in \mathbb{Z}, s \geqslant 1).$$

Applying Beck's embedding $\mathcal{E}_{\mathrm{B}}$ shows that $\mathbf{U}(\widehat{\mathfrak{gl}}_n)$ can be generated by

$$\mathrm{k}_i^{\pm 1}\ (1 \leqslant i \leqslant n),\ \mathrm{x}_{j,0}^{\pm}\ (1 \leqslant j < n),\ \varepsilon_n^{\pm},\ \theta_{\pm s}\ (s \geqslant 1).$$

Moreover, these generators satisfy relations similar to (QGL1)–(QGL8) in Theorem 2.3.1 in which $K_i^{\pm 1}, u_j^{\pm}, u_n^{\pm}, \mathrm{z}_s^{\pm}$ are replaced by $\mathrm{k}_i^{\pm 1}, \mathrm{x}_{j,0}^{\pm}, \varepsilon_n^{\pm}, \theta_{\pm s}$, respectively. Thus, we conclude that there is a surjective $\mathbb{Q}(v)$-algebra homomorphism

$$\begin{aligned}&\mathcal{E}_{\mathrm{H}} : \mathfrak{D}_{\Delta}(n) \longrightarrow \mathbf{U}(\widehat{\mathfrak{gl}}_n),\\ &K_i^{\pm 1} \longmapsto \mathrm{k}_i^{\pm 1},\ u_j^{\pm} \longmapsto \mathrm{x}_{j,0}^{\pm},\ u_n^{\pm} \longmapsto \varepsilon_n^{\pm},\ \mathrm{z}_s^{\pm} \longmapsto \theta_{\pm s},\end{aligned} \tag{2.5.2.1}$$

for $1 \leqslant i \leqslant n$, $1 \leqslant j < n$, and $s \geqslant 1$. It is clear that $\mathcal{E}_{\mathrm{H}}$ is an extension of $\mathcal{E}_{\mathrm{B}}$.

The fact that the elements $\mathrm{x}_{j,t}^{\pm}$ and $\mathrm{h}_{j,\pm s}$ ($1 \leqslant i < n$, $t \in \mathbb{Z}$, $s \geqslant 1$) lie in $\mathrm{Im}(\mathcal{E}_{\mathrm{B}})$ gives rise to the elements

$$\widehat{\mathrm{x}}_{j,t}^{\pm} = (\mathcal{E}_{\mathrm{B}})^{-1}(\mathrm{x}_{j,t}^{\pm}) \text{ and } \widehat{\mathrm{h}}_{j,\pm s} = (\mathcal{E}_{\mathrm{B}})^{-1}(\mathrm{h}_{j,\pm s})$$

in $\mathfrak{D}_{\Delta}(n)$ via the induced isomorphism $\mathcal{E}_{\mathrm{B}} : \mathbf{U}(\widehat{\mathfrak{sl}}_n) \to \mathrm{Im}(\mathcal{E}_{\mathrm{B}})$.

By Remark 2.3.6(2), $\mathfrak{D}_{\Delta}(n)$ is generated by

$$\widehat{\mathrm{k}}_i^{\pm 1} := K_i^{\pm 1},\ \widehat{\mathrm{x}}_{j,t}^{\pm},\ \widehat{\mathrm{h}}_{j,\pm s},\ \mathrm{z}_s^{\pm}\ (1 \leqslant i \leqslant n, 1 \leqslant j < n, t \in \mathbb{Z}, s \geqslant 1).$$

Furthermore, for $1 \leqslant i \leqslant n$ and $s \geqslant 1$, we define the elements

$$\widehat{\mathrm{g}}_{i,\pm s} = \sum_{j=1}^{n-1} Y_{i,j}^{(\pm s)} \widehat{\mathrm{h}}_{j,\pm s} + Y_{i,n}^{(\pm s)} \mathrm{z}_s^{\pm} \in \mathfrak{D}_{\Delta}(n), \tag{2.5.2.2}$$

or equivalently,

$$\begin{aligned}&\widehat{\mathrm{h}}_{i,\pm s} = v^{\pm(i-1)s}\widehat{\mathrm{g}}_{i,\pm s} - v^{\pm(i+1)s}\widehat{\mathrm{g}}_{i+1,\pm s}\ (1 \leqslant i < n, s \geqslant 1) \text{ and}\\ &\mathrm{z}_s^{\pm} = \mp\frac{1}{[s]}(\widehat{\mathrm{g}}_{1,\pm s} + \cdots + \widehat{\mathrm{g}}_{n,\pm s}).\end{aligned}$$

This implies in particular that the set

$$\mathcal{X} := \{\widehat{\mathrm{k}}_i^{\pm 1}, \widehat{\mathrm{x}}_{j,t}^{\pm}, \widehat{\mathrm{g}}_{i,\pm s} \mid 1 \leqslant i \leqslant n, 1 \leqslant j < n, t \in \mathbb{Z}, s \geqslant 1\}$$

is also a generating set for $\mathfrak{D}_{\Delta}(n)$.

Clearly, the $\widehat{\mathrm{x}}_{j,t}^{\pm}$ satisfy the relations (QLA5)–(QLA7). Since $\widehat{\mathrm{h}}_{i,\pm s}$ together with $\widehat{\mathrm{k}}_i^{\pm 1}$ and $\widehat{\mathrm{x}}_{j,t}^{\pm}$ satisfy the relations (QLA8)–(QLA9) and the $\mathrm{z}_s^{\pm}$ are central elements, it follows from (2.5.2.2) that the $\widehat{\mathrm{g}}_{i,\pm s}$ together with $\widehat{\mathrm{k}}_i^{\pm 1}$ and $\widehat{\mathrm{x}}_{j,t}^{\pm}$ satisfy the relations (QLA2)–(QLA4). In conclusion, the generators in $\mathcal{X}$ satisfy all the relations (QLA1)–(QLA7). Therefore, there is a surjective $\mathbb{Q}(v)$-algebra homomorphism

$$\mathcal{F} : \mathbf{U}(\widehat{\mathfrak{gl}}_n) \longrightarrow \mathfrak{D}_{\Delta}(n)$$

taking $\mathrm{k}_i^{\pm 1} \mapsto \widehat{\mathrm{k}}_i^{\pm 1}$, $\mathrm{g}_{i,\pm s} \mapsto \widehat{\mathrm{g}}_{i,\pm s}$, $\mathrm{x}_{j,t}^{\pm} \mapsto \widehat{\mathrm{x}}_{j,t}^{\pm}$. Obviously, both the composites $\mathcal{E}_{\mathrm{H}}\mathcal{F}$ and $\mathcal{F}\mathcal{E}_{\mathrm{H}}$ are the identity maps. This gives the following result.

Theorem 2.5.3. *The surjective algebra homomorphism* $\mathcal{E}_{\rm H} : \mathfrak{D}_\Delta(n) \to \mathbf{U}(\widehat{\mathfrak{gl}}_n)$ *given in* (2.5.2.1) *is a Hopf algebra isomorphism. In particular, Theorem 2.3.1 gives another presentation for* $\mathbf{U}(\widehat{\mathfrak{gl}}_n)$.

Proof. Clearly, $\mathcal{E}_{\rm H}$ is an algebra isomorphism. Since $\mathcal{E}_{\rm B}$ is a Hopf algebra embedding and the elements $\mathsf{z}_s^{\pm}$ and $\theta_{\pm s}$ are primitive, it follows that $\mathcal{E}_{\rm H}$ is a bialgebra isomorphism. It is well known that if a bialgebra admits an antipode, then the antipode is unique; see, for example, [**72**, p.71]. This forces $\mathcal{E}_{\rm H}$ to be a Hopf algebra isomorphism. □

Let $\mathbf{U}(\widehat{\mathfrak{gl}}_n)^+$ (resp., $\mathbf{U}(\widehat{\mathfrak{gl}}_n)^-$) be the subalgebra of $\mathbf{U}(\widehat{\mathfrak{gl}}_n)$ generated by $\mathsf{x}_{i,0}^+$, ε_n^+, θ_s (resp., $\mathsf{x}_{i,0}^-$, ε_n^-, θ_{-s}), for $1 \leqslant i < n$, $s \geqslant 1$. Also, let $\mathbf{U}(\widehat{\mathfrak{gl}}_n)^0$ be the subalgebra of $\mathbf{U}(\widehat{\mathfrak{gl}}_n)$ generated by the $\mathsf{k}_i^{\pm 1}$. The triangular decomposition of $\mathfrak{D}_\Delta(n)$ given in (2.3.6.1) induces that of $\mathbf{U}(\widehat{\mathfrak{gl}}_n)$.

Corollary 2.5.4. *The multiplication map*

$$\mathbf{U}(\widehat{\mathfrak{gl}}_n)^+ \otimes \mathbf{U}(\widehat{\mathfrak{gl}}_n)^0 \otimes \mathbf{U}(\widehat{\mathfrak{gl}}_n)^- \longrightarrow \mathbf{U}(\widehat{\mathfrak{gl}}_n)$$

is a $\mathbb{Q}(v)$*-space isomorphism.*

Remarks 2.5.5. (1) In [**28**], the authors introduced the $\mathbb{Q}(v)$-subalgebra $\mathcal{C}$ of $\mathbf{U}(\widehat{\mathfrak{gl}}_n)$ generated by the central elements $C_{\pm s}$ as defined in Remark (2.5.2) and considered the embedding of $\mathbf{U}(\widehat{\mathfrak{sl}}_n) \otimes \mathcal{C} \to \mathbf{U}(\widehat{\mathfrak{gl}}_n)$; see [**28**, (2.12)]. Indeed, under the isomorphism $\mathcal{E}_{\rm H}$, $\mathcal{C}$ is identified with the central subalgebra $\mathbf{Z}_\Delta(n)$ of $\mathfrak{D}_\Delta(n)$.

(2) In [**40**] a Hopf algebra isomorphism $\mathfrak{H}_\Delta(n)^{\geqslant 0} \to \mathbf{U}(\widehat{\mathfrak{gl}}_n)^+ \otimes \mathbf{U}(\widehat{\mathfrak{gl}}_n)^0$ was established and, moreover, the elements $\mathcal{E}_{\rm H}^{-1}(\mathsf{x}_{j,-1}^+\mathsf{k}_j^{-1})$ and $\mathcal{E}_{\rm H}^{-1}(\mathsf{g}_{i,\pm s})$ in $\mathfrak{D}_\Delta(n)$ were explicitly described.

(3) The proof above can be easily modified to construct a $\mathbb{C}$-algebra isomorphism $\mathcal{E}_{{\rm H},\mathbb{C}} : \mathfrak{D}_{\Delta,\mathbb{C}}(n) \to \mathrm{U}_{\mathbb{C}}(\widehat{\mathfrak{gl}}_n)$, where the algebras are defined over $\mathbb{C}$ with respect to a non-root-of-unity $z \in \mathbb{C}^*$. This isomorphism will be used in linking representations of $\mathrm{U}_{\mathbb{C}}(\widehat{\mathfrak{gl}}_n)$ with those of $\mathcal{S}_\Delta(n,r)_{\mathbb{C}}$ in Chapter 4, §§4.4–4.6.

(4) With the above isomorphism, the double Ringel–Hall algebras $\mathfrak{D}_\Delta(n)$, $\mathfrak{D}_{\Delta,\mathbb{C}}(n)$ will also be called a *quantum affine* $\mathfrak{gl}_n$. Moreover, the notation $\mathfrak{D}_\Delta(n)$, $\mathfrak{D}_{\Delta,\mathbb{C}}(n)$ will not be changed to $\mathbf{U}(\widehat{\mathfrak{gl}}_n)$, $\mathrm{U}_{\mathbb{C}}(\widehat{\mathfrak{gl}}_n)$ throughout the book in order to emphasize the approach used in this book.

(5) There is another geometric realization for the +-part of the quantum loop algebra $\mathbf{U}(\widehat{\mathfrak{gl}}_n)$ in terms of the Hall algebra of a Serre subcategory of the category of coherent sheaves over a weighted projective line; see [**68**, 4.3, 5.2].

2.6. Semisimple generators and commutator formulas

By [**13**, Th. 5.2], the Ringel–Hall algebra $\mathfrak{H}_\triangle(n)$ can be generated by semisimple modules over $\mathcal{Z}$. Thus, semisimple generators would be crucial to the study of integral forms of $\mathfrak{D}_\triangle(n)$. In this section we derive commutator formulas between semisimple generators of $\mathfrak{D}_\triangle(n)$ (cf. [**78**, Prop. 5.5]).

Recall from §1.2 that for the module $M(A)$ associated with $A \in \Theta_\triangle^+(n)$, we write $\mathbf{d}(A) = \mathbf{dim}\, M(A)$ for the dimension vector of $M(A)$.

Define a subset $\Theta_\triangle^+(n)^{ss}$ of $\Theta_\triangle^+(n)$ by setting

$$\Theta_\triangle^+(n)^{ss} = \{A = (a_{i,j}) \in \Theta_\triangle^+(n) \mid a_{i,j} = 0 \text{ for all } j \neq i+1\}.$$

In other words, $A \in \Theta_\triangle^+(n)^{ss} \iff M(A)$ is semisimple. Then, by Proposition 1.4.3, $\mathfrak{D}_\triangle(n)^+$ (resp., $\mathfrak{D}_\triangle(n)^-$) is generated by u_A^+ (resp., u_A^-) for all $A \in \Theta_\triangle^+(n)^{ss}$. We sometimes identify $\Theta_\triangle^+(n)^{ss}$ with $\mathbb{N}_\triangle^n$ via the map $\mathbb{N}_\triangle^n \to \Theta_\triangle^+(n)^{ss}$ sending λ to $A = A_\lambda$ with $\lambda_i = a_{i,i+1}$ for all $i \in \mathbb{Z}$.

Lemma 2.6.1. *Let* $X^+ := \mathrm{span}\{u_A^+ \mid A \in \Theta_\triangle^+(n)^{ss}\}$ *and* $X^- := \mathrm{span}\{u_A^- \mid A \in \Theta_\triangle^+(n)^{ss}\}$. *Then*

$$\mathfrak{D}_\triangle(n) \cong \mathfrak{H}_\triangle(n)^{\geqslant 0} * \mathfrak{H}_\triangle(n)^{\leqslant 0}/\mathcal{J},$$

where $\mathcal{J}$ *is the ideal of the free product* $\mathfrak{H}_\triangle(n)^{\geqslant 0} * \mathfrak{H}_\triangle(n)^{\leqslant 0}$ *generated by*

(1) $\sum(b_2 * a_2)\psi(a_1, b_1) - \sum(a_1 * b_1)\psi(a_2, b_2)$ *for all* $a \in X^+, b \in X^-$, *and*
(2) $K_\alpha * 1 - 1 * K_\alpha$ *for all* $\alpha \in \mathbb{Z}I$,

where $\Delta(a) = \sum a_1 \otimes a_2$ *and* $\Delta(b) = \sum b_1 \otimes b_2$.

Proof. For $a \in \mathfrak{H}_\triangle(n)^{\geqslant 0}$ and $b \in \mathfrak{H}_\triangle(n)^{\leqslant 0}$, we write

$$L(a,b) = \sum(b_2 * a_2)\psi(a_1, b_1) \quad\text{and}\quad R(a,b) = \sum(a_1 * b_1)\psi(a_2, b_2),$$

where, for $x = a, b$, $\Delta(x) = \sum x_1 \otimes x_2$. Define

$$X^{\geqslant 0} := \mathrm{span}\{u_A^+ K_\alpha \mid \alpha \in \mathbb{Z}I, A \in \Theta_\triangle^+(n)^{ss}\} \text{ and}$$
$$X^{\leqslant 0} := \mathrm{span}\{K_\alpha u_A^- \mid \alpha \in \mathbb{Z}I, A \in \Theta_\triangle^+(n)^{ss}\}.$$

Then $X^{\geqslant 0}$ (resp., $X^{\leqslant 0}$) generates $\mathfrak{H}_\triangle(n)^{\geqslant 0}$ (resp., $\mathfrak{H}_\triangle(n)^{\leqslant 0}$) and satisfies

$$\Delta(X^{\geqslant 0}) \subseteq X^{\geqslant 0} \otimes X^{\geqslant 0} \quad (\text{resp., } \Delta(X^{\leqslant 0}) \subseteq X^{\leqslant 0} \otimes X^{\leqslant 0}).$$

Thus, by Lemma 2.1.1,

$$\mathfrak{D}_\triangle(n) = \mathfrak{H}_\triangle(n)^{\geqslant 0} * \mathfrak{H}_\triangle(n)^{\leqslant 0}/\widehat{\mathcal{I}},$$

where $\widehat{\mathcal{I}}$ is the ideal of $\mathfrak{H}_\triangle(n)^{\geqslant 0} * \mathfrak{H}_\triangle(n)^{\leqslant 0}$ generated by

(1′) $L(a,b)-R(a,b)$ $(a\in X^{\geqslant 0}, b\in X^{\leqslant 0})$,
(2′) $K_\alpha * 1 - 1 * K_\alpha$ $(\alpha\in\mathbb{Z}I)$.

Clearly, $\mathcal{J}\subseteq\widehat{\mathcal{I}}$. To show the reverse inclusion $\widehat{\mathcal{I}}\subseteq\mathcal{J}$, it suffices to prove that, for $u_A^+K_\alpha\in X^{\geqslant 0}$ and $K_\beta u_B^-\in X^{\leqslant 0}$,

$$L(u_A^+K_\alpha, K_\beta u_B^-)\equiv R(u_A^+K_\alpha, K_\beta u_B^-) \bmod \mathcal{J}.$$

By the definition of comultiplications in $\mathfrak{H}_\Delta(n)^{\geqslant 0}$ and $\mathfrak{H}_\Delta(n)^{\leqslant 0}$,

$$\Delta(u_A^+)=\sum_{A_1,A_2\in\Theta_\Delta^+(n)} f_{A_1,A_2}^A u_{A_2}^+\otimes u_{A_1}^+\widetilde{K}_{\mathbf{d}(A_2)} \quad\text{and}$$
$$\Delta(u_B^-)=\sum_{B_1,B_2\in\Theta_\Delta^+(n)} g_{B_1,B_2}^B \widetilde{K}_{-\mathbf{d}(B_1)}u_{B_2}^-\otimes u_{B_1}^-,$$

where $f_{A_1,A_2}^A = v^{\langle \mathbf{d}(A_1),\mathbf{d}(A_2)\rangle}\frac{\mathfrak{a}_{A_1}\mathfrak{a}_{A_2}}{\mathfrak{a}_A}\varphi_{A_1,A_2}^A$ and $g_{B_1,B_2}^B = v^{-\langle \mathbf{d}(B_2),\mathbf{d}(B_1)\rangle}\frac{\mathfrak{a}_{B_1}\mathfrak{a}_{B_2}}{\mathfrak{a}_B}\varphi_{B_1,B_2}^B$. By the definition of $\mathcal{J}$,

$$L(u_A^+,u_B^-)\equiv R(u_A^+,u_B^-) \bmod \mathcal{J}, \tag{2.6.1.1}$$

where

$$L(u_A^+,u_B^-)=\sum_{A_1,A_2,B_1,B_2} f_{A_1A_2}^A g_{B_1B_2}^B\big(u_{B_1}^- * (u_{A_1}^+\widetilde{K}_{\mathbf{d}(A_2)})\big)\times\psi(u_{A_2}^+,\widetilde{K}_{-\mathbf{d}(B_1)}u_{B_2}^-)\quad\text{and}$$
$$R(u_A^+,u_B^-)=\sum_{A_1,A_2,B_1,B_2} f_{A_1A_2}^A g_{B_1B_2}^B\big(u_{A_2}^+ * (\widetilde{K}_{-\mathbf{d}(B_1)}u_{B_2}^-)\big)\psi(u_{A_1}^+\widetilde{K}_{\mathbf{d}(A_2)},u_{B_1}^-).$$

This together with Proposition 2.1.3 implies that

$$\begin{aligned}
&L(u_A^+K_\alpha, K_\beta u_B^-)\\
&=\sum_{A_1,A_2,B_1,B_2} f_{A_1A_2}^A g_{B_1B_2}^B\big((K_\beta u_{B_1}^-) * (u_{A_1}^+\widetilde{K}_{\mathbf{d}(A_2)}K_\alpha)\big)\\
&\qquad\times\psi(u_{A_2}^+K_\alpha, K_\beta\widetilde{K}_{-\mathbf{d}(B_1)}u_{B_2}^-)\\
&\equiv v^aK_{\alpha+\beta}\sum_{A_1,A_2,B_1,B_2} f_{A_1A_2}^A g_{B_1B_2}^B\big(u_{B_1}^- * (u_{A_1}^+\widetilde{K}_{\mathbf{d}(A_2)})\big)\\
&\qquad\times\psi(u_{A_2}^+,\widetilde{K}_{-\mathbf{d}(B_1)}u_{B_2}^-)\qquad\text{(by 2.6.1(2))}\\
&\equiv v^aK_{\alpha+\beta}\sum_{A_1,A_2,B_1,B_2} f_{A_1A_2}^A g_{B_1B_2}^B\big(u_{A_2}^+ * (\widetilde{K}_{-\mathbf{d}(B_1)}u_{B_2}^-)\big)\\
&\qquad\times\psi(u_{A_1}^+\widetilde{K}_{\mathbf{d}(A_2)},u_{B_1}^-)\qquad\text{(by (2.6.1.1))}
\end{aligned}$$

$$= \sum_{A_1,A_2,B_1,B_2} f_{A_1A_2}^{A} g_{B_1B_2}^{B} \big((u_{A_2}^+ K_\alpha) * (K_\beta \widetilde{K}_{-\mathbf{d}(B_1)} u_{B_2}^-)\big)$$
$$\times\, \psi(u_{A_1}^+ \widetilde{K}_{\mathbf{d}(A_2)} K_\alpha, K_\beta u_{B_1}^-)$$
$$= R(u_A^+ K_\alpha, K_\beta u_B^-) \bmod \mathcal{J},$$

where $a = \alpha \centerdot \beta + \langle \mathbf{d}(A_1) + \mathbf{d}(A_2), \alpha\rangle = \alpha \centerdot \beta + \langle \mathbf{d}(A), \alpha\rangle$, as desired. □

Recall the order relation $\leqslant$ on $\mathbb{Z}_\Delta^n$ defined in (1.1.0.3).

Lemma 2.6.2. *For $\alpha = (\alpha_i), \beta = (\beta_i) \in \mathbb{N}_\Delta^n$, let $\gamma = \gamma(\alpha, \beta) = (\gamma_i) \in \mathbb{N}_\Delta^n$ be defined by $\gamma_i = \min\{\alpha_i, \beta_{i+1}\}$. For each $\lambda \leqslant \gamma$, define $C_\lambda \in \Theta_\Delta^+(n)$ by*

$$M(C_\lambda) = \bigoplus_{i\in I} ((\alpha_i + \beta_i - \lambda_i - \lambda_{i-1}) S_i \oplus \lambda_i S_i[2]).$$

Then

$$u_\alpha u_\beta = v^{\sum_i \alpha_i(\beta_i - \beta_{i+1})} \sum_{\lambda\leqslant\gamma} \prod_{i\in I} \left[\!\!\left[\begin{matrix} \alpha_i + \beta_i - \lambda_i - \lambda_{i-1} \\ \beta_i - \lambda_{i-1} \end{matrix} \right]\!\!\right] u_{C_\lambda}.$$

Proof. Clearly, each $M(C_\lambda)$ with $\lambda \leqslant \gamma$ is an extension of $M(A_\beta)$ by $M(A_\alpha)$, where $A_\alpha = \sum_{i=1}^n \alpha_i E_{i,i+1}^\Delta$ and $A_\beta = \sum_{i=1}^n \beta_i E_{i,i+1}^\Delta$. Conversely, each extension of $M(A_\beta)$ by $M(A_\alpha)$ is isomorphic to $M(C_\lambda)$ for some $\lambda \leqslant \gamma$. Hence,

$$u_\alpha u_\beta = v^{\langle\alpha,\beta\rangle} \sum_{\lambda\leqslant\gamma} \varphi_{A_\alpha,A_\beta}^{C_\lambda} u_{C_\lambda} = v^{\sum_i \alpha_i(\beta_i-\beta_{i+1})} \sum_{\lambda\leqslant\gamma} \varphi_{A_\alpha,A_\beta}^{C_\lambda} u_{C_\lambda}.$$

The lemma then follows from the fact that

$$\varphi_{A_\alpha,A_\beta}^{C_\lambda} = \prod_{i\in I} \left[\!\!\left[\begin{matrix} \alpha_i + \beta_i - \lambda_i - \lambda_{i-1} \\ \beta_i - \lambda_{i-1} \end{matrix} \right]\!\!\right].$$ □

Theorem 2.6.3. *The algebra $\mathfrak{D}_\Delta(n)$ has generators u_A^+, K_ν, u_A^- ($A \in \Theta_\Delta^+(n)^{ss}$, $\nu \in \mathbb{Z}I$) which satisfy the following relations: for $\nu, \nu' \in \mathbb{Z}I$, $A, B \in \Theta_\Delta^+(n)^{ss}$,*

(1) $K_0 = u_0^+ = u_0^- = 1, \quad K_\nu K_{\nu'} = K_{\nu+\nu'}$;
(2) $K_\nu u_A^+ = v^{\langle \mathbf{d}(A),\nu\rangle} u_A^+ K_\nu, \quad u_A^- K_\nu = v^{\langle \mathbf{d}(A),\nu\rangle} K_\nu u_A^-$;
(3) $u_A^+ u_B^+ = v^{\sum_i \alpha_i(\beta_i-\beta_{i+1})} \sum_{\lambda\leqslant\gamma} \prod_{i\in I} \left[\!\left[{\alpha_i+\beta_i-\lambda_i-\lambda_{i-1} \atop \beta_i-\lambda_{i-1}} \right]\!\right] u_{C_\lambda}$, *if $A = A_\alpha$, $B = B_\beta$, and $\gamma = \gamma(\alpha, \beta)$;*
(4) $u_A^- u_B^- = v^{\sum_i \beta_i(\alpha_i-\alpha_{i+1})} \sum_{\lambda\leqslant\gamma'} \prod_{i\in I} \left[\!\left[{\alpha_i+\beta_i-\lambda_i-\lambda_{i-1} \atop \alpha_i-\lambda_{i-1}} \right]\!\right] u_{C_\lambda}$, *if $A = A_\alpha$, $B = B_\beta$, and $\gamma' = \gamma(\beta, \alpha)$;*

(5) commutator relations: *for all* $A, B \in \Theta^+_\Delta(n)$,

$$v^{\langle \mathbf{d}(B),\mathbf{d}(B)\rangle} \sum_{A_1,B_1} \varphi^{A_1,B_1}_{A,B} v^{\langle \mathbf{d}(B_1),\mathbf{d}(A)+\mathbf{d}(B)-\mathbf{d}(B_1)\rangle} \widetilde{K}_{\mathbf{d}(B)-\mathbf{d}(B_1)} u^-_{B_1} u^+_{A_1}$$

$$= v^{\langle \mathbf{d}(B),\mathbf{d}(A)\rangle} \sum_{A_1,B_1} \widetilde{\varphi^{A_1,B_1}_{A,B}} v^{\langle \mathbf{d}(B)-\mathbf{d}(B_1),\mathbf{d}(A_1)\rangle + \langle \mathbf{d}(B),\mathbf{d}(B_1)\rangle} \times \widetilde{K}_{\mathbf{d}(B_1)-\mathbf{d}(B)} u^+_{A_1} u^-_{B_1},$$

where

$$\varphi^{A_1,B_1}_{A,B} = \frac{\mathfrak{a}_{A_1}\mathfrak{a}_{B_1}}{\mathfrak{a}_A\mathfrak{a}_B} \sum_{A_2\in\Theta^+_\Delta(n)} v^{2\mathfrak{d}(A_2)} \mathfrak{a}_{A_2} \varphi^A_{A_1,A_2} \varphi^B_{B_1,A_2} \quad \text{and}$$

$$\widetilde{\varphi^{A_1,B_1}_{A,B}} = \frac{\mathfrak{a}_{A_1}\mathfrak{a}_{B_1}}{\mathfrak{a}_A\mathfrak{a}_B} \sum_{A_2\in\Theta^+_\Delta(n)} v^{2\mathfrak{d}(A_2)} \mathfrak{a}_{A_2} \varphi^A_{A_2,A_1} \varphi^B_{A_2,B_1}. \tag{2.6.3.1}$$

Proof. Relations (1) and (2) follow from the definition, and (3) and (4) follow from Lemma 2.6.2. We now prove (5).

As in the proof of Lemma 2.6.1, for $A, B \in \Theta^+_\Delta(n)$,

$$L(u^+_A, u^-_B) = \sum_{A_1,A_2,B_1,B_2} v^{\langle \mathbf{d}(A_1),\mathbf{d}(A_2)\rangle} \frac{\mathfrak{a}_{A_1}\mathfrak{a}_{A_2}}{\mathfrak{a}_A} \varphi^A_{A_1,A_2} v^{-\langle \mathbf{d}(B_2),\mathbf{d}(B_1)\rangle} \frac{\mathfrak{a}_{B_1}\mathfrak{a}_{B_2}}{\mathfrak{a}_B} \varphi^B_{B_1,B_2} \times \big(u^-_{B_1} * (u^+_{A_1}\widetilde{K}_{\mathbf{d}(A_2)})\big)\psi(u^+_{A_2}, \widetilde{K}_{-\mathbf{d}(B_1)}u^-_{B_2})$$

$$= \sum_{A_1,A_2,B_1,B_2} v^{\langle \mathbf{d}(A_1),\mathbf{d}(A_2)\rangle} \frac{\mathfrak{a}_{A_1}\mathfrak{a}_{A_2}}{\mathfrak{a}_A} \varphi^A_{A_1,A_2} v^{-\langle \mathbf{d}(B_2),\mathbf{d}(B_1)\rangle} \frac{\mathfrak{a}_{B_1}\mathfrak{a}_{B_2}}{\mathfrak{a}_B} \varphi^B_{B_1,B_2} \times \big(u^-_{B_1} * (u^+_{A_1}\widetilde{K}_{\mathbf{d}(A_2)})\big) v^{-\langle \mathbf{d}(A_2),\mathbf{d}(A_2)\rangle + 2\mathfrak{d}(A_2)} \frac{1}{\mathfrak{a}_{A_2}} \delta_{A_2,B_2}.$$

Hence,

$$L(u^+_A, u^-_B) \equiv \sum_{A_1,B_1,A_2} v^{-\langle \mathbf{d}(A_2),\mathbf{d}(A)\rangle + \langle \mathbf{d}(B_1),\mathbf{d}(A_2)\rangle + 2\mathfrak{d}(A_2)} \frac{\mathfrak{a}_{A_1}\mathfrak{a}_{B_1}\mathfrak{a}_{A_2}}{\mathfrak{a}_A\mathfrak{a}_B} \varphi^A_{A_1,A_2}\varphi^B_{B_1,A_2} \times (\widetilde{K}_{\mathbf{d}(A_2)} u^-_{B_1}) * u^+_{A_1}$$

$$\equiv v^{-\langle \mathbf{d}(B),\mathbf{d}(A)\rangle} \sum_{A_1,B_1} v^{\langle \mathbf{d}(B_1),\mathbf{d}(A)+\mathbf{d}(B)-\mathbf{d}(B_1)\rangle} \varphi^{A_1,B_1}_{A,B} \times \widetilde{K}_{\mathbf{d}(B)-\mathbf{d}(B_1)} u^-_{B_1} * u^+_{A_1} \mod \mathcal{J}.$$

By interchanging the running indices A_1 and A_2, B_1 and B_2,

$$
\begin{aligned}
R(u_A^+, u_B^-) = & \sum_{A_1,A_2,B_1,B_2} v^{\langle \mathbf{d}(A_1),\mathbf{d}(A_2)\rangle} \frac{\mathfrak{a}_{A_1}\mathfrak{a}_{A_2}}{\mathfrak{a}_A} \varphi^A_{A_1,A_2} v^{-\langle \mathbf{d}(B_2),\mathbf{d}(B_1)\rangle} \frac{\mathfrak{a}_{B_1}\mathfrak{a}_{B_2}}{\mathfrak{a}_B} \varphi^B_{B_1,B_2} \\
& \times \big(u^+_{A_2} * (\widetilde{K}_{-\mathbf{d}(B_1)} u^-_{B_2})\big)\psi(u^+_{A_1}\widetilde{K}_{\mathbf{d}(A_2)}, u^-_{B_1}) \\
\equiv & \sum_{A_1,A_2,B_2} v^{\langle \mathbf{d}(A_1),\mathbf{d}(A_2)\rangle - \langle \mathbf{d}(B),\mathbf{d}(A_1)\rangle + 2\mathfrak{d}(A_1)} \frac{\mathfrak{a}_{A_1}\mathfrak{a}_{A_2}\mathfrak{a}_{B_2}}{\mathfrak{a}_A\mathfrak{a}_B} \varphi^A_{A_1,A_2}\varphi^B_{A_1,B_2} \\
& \times \widetilde{K}_{-\mathbf{d}(A_1)} u^+_{A_2} * u^-_{B_2} \\
\equiv & \, v^{-\langle \mathbf{d}(B),\mathbf{d}(B)\rangle} \sum_{A_1,B_1} v^{\langle \mathbf{d}(B)-\mathbf{d}(B_1),\mathbf{d}(A_1)\rangle + \langle \mathbf{d}(B),\mathbf{d}(B_1)\rangle} \widetilde{\varphi}^{A_1,B_1}_{A,B} \\
& \times \widetilde{K}_{-\mathbf{d}(B)+\mathbf{d}(B_1)} u^+_{A_1} * u^-_{B_1} \mod \mathcal{J}.
\end{aligned}
$$

This proves (5). □

Theorem 2.6.3 does not give a presentation for $\mathfrak{D}_\Delta(n)$ since the modules $M(C_\lambda)$ are not necessarily semisimple. It would be natural to raise the following question.

Problem 2.6.4. Find the "quantum Serre relations" associated with semisimple generators to replace relations in Theorem 2.6.3(3)–(4), and prove that the relations given in Theorem 2.6.3 are defining relations for $\mathfrak{D}_\Delta(n)$. In this way, we obtain a presentation with semisimple generators for $\mathfrak{D}_\Delta(n)$.

The relations in (5) above are usually called the *commutator relations*. We now derive a finer version of the commutator relations for semisimple generators. The next lemma follows directly from the definition of comultiplication in §1.5.

Lemma 2.6.5. *For $A \in \Theta^+_\Delta(n)$, we have*

$$
\begin{aligned}
\Delta^{(2)}(u_A^+) = & \sum_{A^{(1)},A^{(2)},A^{(3)}} v^{\sum_{i>j}\langle \mathbf{d}(A^{(i)}),\mathbf{d}(A^{(j)})\rangle} \varphi^A_{A^{(3)},A^{(2)},A^{(1)}} \frac{\mathfrak{a}_{A^{(1)}}\mathfrak{a}_{A^{(2)}}\mathfrak{a}_{A^{(3)}}}{\mathfrak{a}_A} \\
& \times u^+_{A^{(1)}} \otimes u^+_{A^{(2)}}\widetilde{K}_{\mathbf{d}(A^{(1)})} \otimes u^+_{A^{(3)}}\widetilde{K}_{\mathbf{d}(A^{(1)})+\mathbf{d}(A^{(2)})}
\end{aligned}
$$

and

$$
\begin{aligned}
\Delta^{(2)}(u_A^-) = & \sum_{A^{(1)},A^{(2)},A^{(3)}} v^{-\sum_{i<j}\langle \mathbf{d}(A^{(i)}),\mathbf{d}(A^{(j)})\rangle} \varphi^A_{A^{(3)},A^{(2)},A^{(1)}} \frac{\mathfrak{a}_{A^{(1)}}\mathfrak{a}_{A^{(2)}}\mathfrak{a}_{A^{(3)}}}{\mathfrak{a}_A} \\
& \times \widetilde{K}_{-(\mathbf{d}(A^{(2)})+\mathbf{d}(A^{(3)}))} u^-_{A^{(1)}} \otimes \widetilde{K}_{-\mathbf{d}(A^{(3)})} u^-_{A^{(2)}} \otimes u^-_{A^{(3)}}.
\end{aligned}
$$

Proposition 2.6.6. *Let $X, Y \in \Theta_\Delta^+(n)$. Then, in $\mathfrak{D}_\Delta(n)$,*

$$
\begin{aligned}
& u_Y^- u_X^+ - u_X^+ u_Y^- \\
&= \sum_{\substack{A,B,B' \in \Theta_\Delta^+(n)\\ A\neq 0}} v^{\ell_1} \varphi_{A,B}^X \varphi_{A,B'}^Y \frac{\mathfrak{a}_A \mathfrak{a}_B \mathfrak{a}_{B'}}{\mathfrak{a}_X \mathfrak{a}_Y} \widetilde{K}_{-\mathbf{d}(A)} u_B^+ u_{B'}^- \\
&\quad + \sum_{\substack{A,B,B',C,C' \in \Theta_\Delta^+(n)\\ C\neq 0\neq C'}} v^{\ell_2} \varphi_{A,B,C}^X \varphi_{A,B',C'}^Y \frac{\mathfrak{a}_A \mathfrak{a}_B \mathfrak{a}_{B'}}{\mathfrak{a}_X \mathfrak{a}_Y} \Bigg(\sum_{\substack{m\geqslant 1\\ X_1,\dots,X_m \in \Theta_\Delta^+(n)^*}} (-1)^m \\
&\quad v^{2\sum_{i<j}\langle \mathbf{d}(X_i),\mathbf{d}(X_j)\rangle} \mathfrak{a}_{X_1}\cdots \mathfrak{a}_{X_m} \varphi_{X_1,\dots,X_m}^C \varphi_{X_1,\dots,X_m}^{C'} \Bigg) \widetilde{K}_{\mathbf{d}(C)-\mathbf{d}(A)} u_B^+ u_{B'}^-,
\end{aligned}
$$

where $\ell_1 = \langle \mathbf{d}(A), \mathbf{d}(B)\rangle - \langle \mathbf{d}(Y), \mathbf{d}(A)\rangle + 2\mathfrak{d}(A)$ *and*

$\ell_2 = \langle \mathbf{d}(A), \mathbf{d}(B)\rangle + \langle \mathbf{d}(Y), \mathbf{d}(C)-\mathbf{d}(A)\rangle - \langle \mathbf{d}(C), 2\mathbf{d}(C)+\mathbf{d}(B)\rangle + 2\mathfrak{d}(A) + 2\mathfrak{d}(C)$.

Proof. By [**46**, Lem. 3.2.2(iii)], for $x \in \mathfrak{H}_\Delta(n)^{\geqslant 0}$ and $y \in \mathfrak{H}_\Delta(n)^{\leqslant 0}$, we have in $\mathfrak{D}_\Delta(n)$,

$$yx = \sum \psi(x_1, \sigma(y_1))(x_2 y_2)\psi(x_3, y_3),$$

where $\Delta^{(2)}(x) = \sum x_1 \otimes x_2 \otimes x_3$ and $\Delta^{(2)}(y) = \sum y_1 \otimes y_2 \otimes y_3$. This together with Lemma 2.6.5 gives the required equality. □

The following result is a direct consequence of the above proposition together with the fact that for $\beta = (\beta_i), \beta^{(1)} = (\beta_i^{(1)}), \dots, \beta^{(m)} = (\beta_i^{(m)}) \in \mathbb{N}_\Delta^n$,

$$\varphi_{\beta^{(1)},\dots,\beta^{(m)}}^{\beta} = \prod_{i=1}^n \left[\!\!\left[\begin{matrix} \beta_i \\ \beta_i^{(1)},\dots,\beta_i^{(m)} \end{matrix} \right]\!\!\right] =: \left[\!\!\left[\begin{matrix} \beta \\ \beta^{(1)},\dots,\beta^{(m)} \end{matrix} \right]\!\!\right], \tag{2.6.6.1}$$

where $\left[\!\!\left[\begin{matrix} \beta_i \\ \beta_i^{(1)},\dots,\beta_i^{(m)} \end{matrix} \right]\!\!\right] = \frac{[\![\beta_i]\!]^!}{[\![\beta_i^{(1)}]\!]^! \dots [\![\beta_i^{(m)}]\!]^!}$ and $\beta = \beta^{(1)} + \cdots + \beta^{(m)}$.

Corollary 2.6.7. *For $\lambda, \mu \in \mathbb{N}_\Delta^n$, we have*

$$u_\mu^- u_\lambda^+ - u_\lambda^+ u_\mu^- = \sum_{\substack{\alpha\neq 0,\, \alpha\in\mathbb{N}_\Delta^n\\ \alpha\leqslant\lambda,\, \alpha\leqslant\mu}} \Bigg(\sum_{0\leqslant\gamma\leqslant\alpha} x_{\alpha,\gamma} \widetilde{K}_{2\gamma-\alpha} \Bigg) u_{\lambda-\alpha}^+ u_{\mu-\alpha}^-,$$

where

$$x_{\alpha,\gamma} = v^{\langle\alpha,\lambda-\alpha\rangle+\langle\mu,2\gamma-\alpha\rangle+2\langle\gamma,\alpha-\gamma-\lambda\rangle+2\sigma(\alpha)}$$
$$\times \left[\!\!\left[\begin{matrix} \lambda \\ \alpha-\gamma, \lambda-\alpha, \gamma \end{matrix} \right]\!\!\right] \cdot \left[\!\!\left[\begin{matrix} \mu \\ \alpha-\gamma, \mu-\alpha, \gamma \end{matrix} \right]\!\!\right] \frac{\mathfrak{a}_{\alpha-\gamma}\mathfrak{a}_{\lambda-\alpha}\mathfrak{a}_{\mu-\alpha}}{\mathfrak{a}_\lambda \mathfrak{a}_\mu}$$
$$\times \sum_{\substack{m\geqslant 1, \gamma^{(i)}\neq 0, \forall i \\ \gamma^{(1)}+\cdots+\gamma^{(m)}=\gamma}} (-1)^m v^{2\sum_{i<j}\langle\gamma^{(i)},\gamma^{(j)}\rangle} \mathfrak{a}_{\gamma^{(1)}} \cdots \mathfrak{a}_{\gamma^{(m)}} \left[\!\!\left[\begin{matrix} \gamma \\ \gamma^{(1)}, \ldots, \gamma^{(m)} \end{matrix} \right]\!\!\right]^2.$$

This is the *commutator formula* for semisimple generators.

3

Affine quantum Schur algebras and the Schur–Weyl reciprocity

Like the quantum Schur algebra, the affine quantum Schur algebra has several equivalent definitions. We first present the geometric definition, given by Ginzburg–Vasserot and Lusztig, which uses cyclic flags and the convolution product. We then discuss the two Hecke algebra definitions given by R. Green and by Varagnolo–Vasserot. The former uses q-permutation modules, while the latter uses tensor spaces. Both versions are related by the Bernstein presentation for Hecke algebras of affine type A.

In §3.4, we review the construction of BLM type bases for affine quantum Schur algebras and the multiplication formulas between simple generators and BLM basis elements (Theorem 3.4.2) developed by the last two authors [**24**]. Through the central element presentation for $\mathfrak{D}_\triangle(n)$ as given in Theorem 2.3.1, we introduce a $\mathfrak{D}_\triangle(n)$-$\mathcal{H}_\triangle(r)$-bimodule structure on the tensor space in §3.5. This gives a homomorphism ξ_r from $\mathfrak{D}_\triangle(n)$ to $\boldsymbol{\mathcal{S}}_\triangle(n, r)$. We then prove in §3.6 that the restriction of this bimodule action coincides with the $\mathfrak{H}_\triangle(n)^{\mathrm{op}}$-$\mathcal{H}_\triangle(r)$-bimodule structure defined by Varagnolo–Vasserot in [**73**]. Thus, we obtain an explicit description of the map ξ_r (Theorem 3.6.3).

In §3.7, we develop a certain triangular relation (Proposition 3.7.3) among the structure constants relative to the BLM basis elements. With this relation, we display an integral PBW type basis and, hence, a (weak) triangular decomposition for an affine quantum Schur algebra (Theorem 3.7.7). Using the triangular decomposition, we easily establish the surjectivity of the homomorphism ξ_r from $\mathfrak{D}_\triangle(n)$ to $\boldsymbol{\mathcal{S}}_\triangle(n, r)$ in §3.8 (Theorem 3.8.1).

There are several important applications of this result which will be discussed in the next three chapters. As a first application, we end this chapter by establishing certain polynomial identities (Corollary 3.9.6) arising from the commutator formulas for semisimple generators discussed in §2.6.

3.1. Cyclic flags: the geometric definition

In this section we recall the geometric definition of affine quantum Schur algebras introduced by Ginzburg–Vasserot [**32**] and Lusztig [**56**]. Recall the notation $\Lambda_\triangle(n,r)$, $\Theta_\triangle(n,r)$, etc., introduced in §1.1.

Let $\mathbb{F}$ be a field and fix an $\mathbb{F}[\varepsilon,\varepsilon^{-1}]$-free module V of rank $r \geqslant 1$, where ε is an indeterminate. A lattice in V is, by definition, a free $\mathbb{F}[\varepsilon]$-submodule L of V satisfying $V = L \otimes_{\mathbb{F}[\varepsilon]} \mathbb{F}[\varepsilon,\varepsilon^{-1}]$. For two lattices L', L of V, $L+L'$ is again a lattice. If, in addition, $L' \subseteq L$, L/L' is a finitely generated torsion $\mathbb{F}[\varepsilon]$-module. Thus, as an $\mathbb{F}$-vector space, L/L' is finite dimensional.

Let $\mathscr{F}_\triangle = \mathscr{F}_{\triangle n}$ be the set of all *cyclic flags* $\mathbf{L} = (L_i)_{i\in\mathbb{Z}}$ of lattices of *period* n, where each L_i is a lattice in V such that $L_{i-1} \subseteq L_i$ and $L_{i-n} = \varepsilon L_i$, for all $i \in \mathbb{Z}$. The group G of automorphisms of the $\mathbb{F}[\varepsilon,\varepsilon^{-1}]$-module V acts on $\mathscr{F}_\triangle$ by $g\cdot\mathbf{L} = (g(L_i))_{i\in\mathbb{Z}}$ for $g \in G$ and $\mathbf{L} \in \mathscr{F}_\triangle$. Thus, the map

$$\phi : \mathscr{F}_\triangle \longrightarrow \Lambda_\triangle(n,r), \quad \mathbf{L} \longmapsto (\dim_{\mathbb{F}} L_i/L_{i-1})_{i\in\mathbb{Z}}$$

induces a bijection between the set $\{\mathscr{F}_{\triangle,\lambda}\}_{\lambda\in\Lambda_\triangle(n,r)}$ of G-orbits in $\mathscr{F}_\triangle$ and $\Lambda_\triangle(n,r)$.

Similarly, let $\mathscr{B}_\triangle = \mathscr{B}_{\triangle r}$ be the set of all complete cyclic flags $\mathbf{L} = (L_i)_{i\in\mathbb{Z}}$ of lattices, where each L_i is a lattice in V such that $L_{i-1} \subseteq L_i$, $L_{i-r} = \varepsilon L_i$ and $\dim_{\mathbb{F}}(L_i/L_{i-1}) = 1$, for all $i \in \mathbb{Z}$.

The group G acts on $\mathscr{F}_\triangle \times \mathscr{F}_\triangle$, $\mathscr{F}_\triangle \times \mathscr{B}_\triangle$, and $\mathscr{B}_\triangle \times \mathscr{B}_\triangle$ by $g\cdot(\mathbf{L},\mathbf{L}') = (g\cdot\mathbf{L}, g\cdot\mathbf{L}')$. For $\mathbf{L} = (L_i)_{i\in\mathbb{Z}}$ and $\mathbf{L}' = (L'_i)_{i\in\mathbb{Z}} \in \mathscr{F}_\triangle$, let $X_{i,j} := X_{i,j}(\mathbf{L},\mathbf{L}') = L_{i-1} + L_i \cap L'_j$. By lexicographically ordering the indices i, j, we obtain a filtration $(X_{i,j})$ of lattices of V. For $i, j \in \mathbb{Z}$, let

$$a_{i,j} = \dim_{\mathbb{F}}(X_{i,j}/X_{i,j-1}) = \dim_{\mathbb{F}} \frac{L_i \cap L'_j}{L_{i-1}\cap L'_j + L_i \cap L'_{j-1}}.$$

By [**56**, 1.5] there is a bijection between the set of G-orbits in $\mathscr{F}_\triangle \times \mathscr{F}_\triangle$ and the matrix set $\Theta_\triangle(n,r)$ by sending $(\mathbf{L},\mathbf{L}')$ to $A = (a_{i,j})_{i,j\in\mathbb{Z}}$. Let $\mathcal{O}_A \subseteq \mathscr{F}_\triangle \times \mathscr{F}_\triangle$ be the G-orbit corresponding to the matrix $A \in \Theta_\triangle(n,r)$. By [**56**, 1.7], for $\mathbf{L}, \mathbf{L}' \in \mathscr{F}_\triangle$,

$$(\mathbf{L},\mathbf{L}') \in \mathcal{O}_A \iff (\mathbf{L}',\mathbf{L}) \in \mathcal{O}_{{}^tA}, \tag{3.1.0.1}$$

where ${}^tA$ is the transpose of A.

Similarly, putting $\omega = (\ldots,1,1,\ldots) \in \Lambda_\triangle(r,r)$ and

$$\Theta_\triangle(r,r)_\omega = \{A \in \Theta_\triangle(r,r) \mid \mathrm{ro}(A) = \mathrm{co}(A) = \omega\}, \tag{3.1.0.2}$$

the G-orbits $\mathcal{O}_A$ on $\mathscr{B}_\Delta \times \mathscr{B}_\Delta$ are indexed by the matrices $A \in \Theta_\Delta(r,r)_\omega$, while the G-orbits $\mathcal{O}_A$ on $\mathscr{F}_\Delta \times \mathscr{B}_\Delta$ are indexed by the set[1]

$$\Theta_\Delta(\tbinom{n}{r}, r)_\omega = \{A \in \Theta_\Delta(\tbinom{n}{r}) \mid \mathrm{ro}(A) \in \Lambda_\Delta(n,r), \mathrm{co}(A) = \omega\},$$

where, like $\Theta_\Delta(n)$,

$$\Theta_\Delta(\tbinom{n}{r}) = \{(a_{i,j})_{i,j\in\mathbb{Z}} \mid a_{i,j} \in \mathbb{N}, a_{i,j} = a_{i-n,j-r}, \forall i,j \in \mathbb{Z}, \sum_{i=1}^{n}\sum_{j\in\mathbb{Z}} a_{i,j} \in \mathbb{N}\}. \tag{3.1.0.3}$$

Clearly, with this notation,

$$\Theta_\Delta(r,r)_\omega = \{A \in \Theta_\Delta(\tbinom{r}{r}) \mid \mathrm{ro}(A) = \mathrm{co}(A) = \omega\}. \tag{3.1.0.4}$$

Assume now that $\mathbb{F} = \mathbb{F}_q$ is the finite field of q elements and write $\mathscr{F}_\Delta(q)$ for $\mathscr{F}_\Delta$ and $\mathscr{B}_\Delta(q)$ for $\mathscr{B}_\Delta$, etc. By regarding $\mathbb{C}\mathscr{F}_\Delta(q)$ and $\mathbb{C}\mathscr{B}_\Delta(q)$ as permutation G-modules, the endomorphism algebra $\mathcal{S}_{\Delta,q} := \mathrm{End}_{\mathbb{C}G}(\mathbb{C}\mathscr{F}_\Delta(q))^{\mathrm{op}}$ has a basis $\{e_{A,q}\}_{A\in\Theta_\Delta(n,r)}$ while $\mathcal{H}_{\Delta,q} := \mathrm{End}_{\mathbb{C}G}(\mathbb{C}\mathscr{B}_\Delta(q))^{\mathrm{op}}$ has a basis $\{e_{A,q}\}_{A\in\Theta_\Delta(r,r)_\omega}$ with the following multiplication:

$$e_{A,q}e_{A',q} = \begin{cases} \sum_{A''\in\Theta} \mathfrak{n}_{A,A',A'';q}e_{A'',q}, & \text{if } \mathrm{co}(A) = \mathrm{ro}(A'); \\ 0, & \text{otherwise,} \end{cases} \tag{3.1.0.5}$$

where $\Theta = \Theta_\Delta(n,r)$ (resp., $\Theta = \Theta_\Delta(r,r)_\omega$) and

$$\mathfrak{n}_{A,A',A'';q} = |\{\mathbf{L}' \in \mathscr{F}_\Delta(q) \text{ (resp., } \mathscr{B}_\Delta(q)) \mid (\mathbf{L},\mathbf{L}') \in \mathcal{O}_A, (\mathbf{L}',\mathbf{L}'') \in \mathcal{O}_{A'}\}| \tag{3.1.0.6}$$

for any fixed $(\mathbf{L},\mathbf{L}'') \in \mathcal{O}_{A''}$.

By [**56**, 1.8], there exists a polynomial $p_{A,A',A''} \in \mathcal{Z}$ in v^2 such that, for each finite field $\mathbb{F}$ with q elements, $\mathfrak{n}_{A,A',A'';q} = p_{A,A',A''}|_{v^2=q}$. Thus, we have the following definition; see [**56**, 1.9].

Definition 3.1.1. The *(generic) affine quantum Schur algebra* $\mathcal{S}_\Delta(n,r)$ (resp., *affine Hecke algebra* $\mathcal{H}_\Delta(r)$) is the free $\mathcal{Z}$-module with basis $\{e_A \mid A \in \Theta_\Delta(n,r)\}$ (resp., $\{e_A \mid A \in \Theta_\Delta(r,r)_\omega\}$), and multiplication defined by

$$e_Ae_{A'} = \begin{cases} \sum_{A''\in\Theta} p_{A,A',A''}e_{A''}, & \text{if } \mathrm{co}(A) = \mathrm{ro}(A'); \\ 0, & \text{otherwise.} \end{cases} \tag{3.1.1.1}$$

Both $\mathcal{S}_\Delta(n,r)$ and $\mathcal{H}_\Delta(r)$ are associative algebras over $\mathcal{Z}$ with an anti-automorphism $e_A \mapsto e_{{}^tA}$ (see (3.1.3.4) below for a modified version).

[1] The set is denoted by $\Theta_\Delta(n,r)$ in [**24**].

Alternatively, we can interpret affine quantum Schur algebras in terms of convolution algebras defined by G-invariant functions and the convolution product. Again, assume that $\mathbb{F}$ is the finite field of q elements, and for notational simplicity, let

$$\mathscr{Y} = \mathscr{F}_\Delta(q), \quad \mathscr{X} = \mathscr{B}_\Delta(q), \quad \text{and} \quad G = G(q).$$

Define $\mathbb{C}_G(\mathscr{Y} \times \mathscr{Y})$, $\mathbb{C}_G(\mathscr{Y} \times \mathscr{X})$, and $\mathbb{C}_G(\mathscr{X} \times \mathscr{X})$ to be the $\mathbb{C}$-span of the characteristic functions $\chi_\mathcal{O}$ of the G-orbits $\mathcal{O}$ on $\mathscr{Y} \times \mathscr{Y}$, $\mathscr{Y} \times \mathscr{X}$, and $\mathscr{X} \times \mathscr{X}$, respectively. With the *convolution product*

$$(\chi_\mathcal{O} * \chi_{\mathcal{O}'})(\mathbf{L}, \mathbf{L}'') = \sum_{\mathbf{L}' \in \mathscr{F}'} \chi_\mathcal{O}(\mathbf{L}, \mathbf{L}')\chi_{\mathcal{O}'}(\mathbf{L}', \mathbf{L}''), \tag{3.1.1.2}$$

where $\mathcal{O} \subset \mathscr{F} \times \mathscr{F}'$ and $\mathcal{O}' \subset \mathscr{F}' \times \mathscr{F}''$ for various selections of $\mathscr{F}, \mathscr{F}'$, and $\mathscr{F}''$, we obtain *convolution algebras* $\mathbb{C}_G(\mathscr{Y} \times \mathscr{Y})$ and $\mathbb{C}_G(\mathscr{X} \times \mathscr{X})$, and a $\mathbb{C}_G(\mathscr{Y} \times \mathscr{Y})$-$\mathbb{C}_G(\mathscr{X} \times \mathscr{X})$-bimodule $\mathbb{C}_G(\mathscr{Y} \times \mathscr{X})$. It is clear that $\mathcal{S}_{\Delta,q} \cong \mathbb{C}_G(\mathscr{Y} \times \mathscr{Y})$ and $\mathcal{H}_{\Delta,q} \cong \mathbb{C}_G(\mathscr{X} \times \mathscr{X})$, and specializing v to $\sqrt{q}$ gives an isomorphism

$$\mathcal{S}_\Delta(n,r)_\mathbb{C} \longrightarrow \mathbb{C}_G(\mathscr{Y} \times \mathscr{Y}) \text{ (resp., } \mathcal{H}_\Delta(r)_\mathbb{C} \longrightarrow \mathbb{C}_G(\mathscr{X} \times \mathscr{X})) \tag{3.1.1.3}$$

sending $e_A \otimes 1 \mapsto \chi_A$, where χ_A denotes the characteristic function of the orbit $\mathcal{O}_A$. In the sequel, we shall identify $\mathcal{S}_\Delta(n,r)_\mathbb{C}$ with $\mathbb{C}_G(\mathscr{Y} \times \mathscr{Y})$.

Via the convolution product, $\mathbb{C}_G(\mathscr{Y} \times \mathscr{X})$ becomes a $\mathbb{C}_G(\mathscr{Y} \times \mathscr{Y})$-$\mathbb{C}_G(\mathscr{X} \times \mathscr{X})$-bimodule. Thus, if we denote by $\mathcal{T}_\Delta(n,r)$ the *generic* form of $\mathbb{C}_G(\mathscr{Y} \times \mathscr{X})$, then $\mathcal{T}_\Delta(n,r)$ becomes an $\mathcal{S}_\Delta(n,r)$-$\mathcal{H}_\Delta(r)$-bimodule with a $\mathcal{Z}$-basis $\{e_A \mid A \in \Theta_\Delta(^n_r, r)_\omega\}$.

Remark 3.1.2. It is clear from the definition that this isomorphism continues to hold if $\mathbb{C}$ is replaced by the ring $R = \mathbb{Z}[\sqrt{q}, \sqrt{q}^{-1}]$. In fact, we will frequently use the isomorphism $\mathcal{S}_\Delta(n,r)_R \cong R_G(\mathscr{Y} \times \mathscr{Y})$ to derive formulas in $\mathcal{S}_\Delta(n,r)$ by doing computations in $R_G(\mathscr{Y} \times \mathscr{Y})$; see, e.g., §§3.6–3.9 below.

Observe that, for $N \geqslant n$, $\mathscr{F}_{\Delta,n}$ is naturally a subset of $\mathscr{F}_{\Delta,N}$, since every $\mathbf{L} = (L_i) \in \mathscr{F}_{\Delta,n}$ can be regarded as $\widetilde{\mathbf{L}} = (\widetilde{L}_i) \in \mathscr{F}_{\Delta,N}$, where, for all $a \in \mathbb{Z}$, $\widetilde{L}_{i+aN} = L_{i+an}$ if $1 \leqslant i \leqslant n$, and $\widetilde{L}_{i+aN} = L_{n+an}$ if $n \leqslant i \leqslant N$. Thus, if $N = \max\{n, r\}$, then $\mathscr{F}_{\Delta,n} \times \mathscr{F}_{\Delta,n}$, $\mathscr{F}_{\Delta,n} \times \mathscr{B}_{\Delta,r}$, and $\mathscr{B}_{\Delta,r} \times \mathscr{B}_{\Delta,r}$ can always be regarded as G-stable subsets of $\mathscr{F}_{\Delta,N} \times \mathscr{F}_{\Delta,N}$, and the G-orbit $\mathcal{O}_A$ containing $(\mathbf{L}, \mathbf{L}')$ is the G-orbit $\mathcal{O}_{\widetilde{A}}$ containing $(\widetilde{\mathbf{L}}, \widetilde{\mathbf{L}}')$, where $A = (a_{i,j})$ and $\widetilde{A} = (\widetilde{a}_{i,j})$ are related by, for all $m \in \mathbb{Z}$,

$$\widetilde{a}_{k,l+mN} = \begin{cases} a_{k,l+mn}, & \text{if } 1 \leqslant k, l \leqslant n; \\ 0, & \text{if either } n < k \leqslant N \text{ or } n < l \leqslant N. \end{cases} \tag{3.1.2.1}$$

Lemma 3.1.3. *Let* $N = \max\{n, r\}$. *By sending* e_A *to* $e_{\widetilde{A}}$, *both* $\mathcal{S}_\triangle(n,r)$ *and* $\mathcal{H}_\triangle(r)$ *can be identified as (centralizer) subalgebras of* $\mathcal{S}_\triangle(N,r)$, *and* $\mathcal{T}_\triangle(n,r)$ *as a subbimodule of the* $\mathcal{S}_\triangle(n,r)$-$\mathcal{H}_\triangle(r)$-*bimodule* $\mathcal{S}_\triangle(N,r)$.

Proof. Define $\omega \in \Lambda_\triangle(N,r)$ by setting

$$\omega = \begin{cases} (\ldots, 1^r, 1^r, \ldots), & \text{if } n \leqslant r; \\ (\ldots, 1^r, 0^{n-r}, 1^r, 0^{n-r}, \ldots), & \text{if } n > r. \end{cases} \tag{3.1.3.1}$$

For $\lambda \in \Lambda_\triangle(n,r)$, let $\mathrm{diag}(\lambda) = (\delta_{i,j}\lambda_i)_{i,j\in\mathbb{Z}} \in \Theta_\triangle(n,r)$. If we embed $\Lambda_\triangle(n,r)$ into $\Lambda_\triangle(N,r)$ via the map $\mu \mapsto \widetilde{\mu}$, where

$$\widetilde{\mu} = (\ldots, \mu_1, \ldots, \mu_n, 0^{N-n}, \mu_{n+1}, \ldots, \mu_{2n}, 0^{N-n}, \ldots),$$

and put $e = \sum_{\mu\in\Lambda_\triangle(n,r)} e_{\mathrm{diag}(\widetilde{\mu})}$ and $e_\omega = e_{\mathrm{diag}(\omega)}$, then $\mathcal{S}_\triangle(n,r) \cong e\mathcal{S}_\triangle(N,r)e$ and $\mathcal{H}_\triangle(r) \cong e_\omega\mathcal{S}_\triangle(N,r)e_\omega$, and $\mathcal{T}_\triangle(n,r) \cong e\mathcal{S}_\triangle(N,r)e_\omega$ as $\mathcal{S}_\triangle(n,r)$-$\mathcal{H}_\triangle(r)$-bimodules. Here all three isomorphisms send e_A to $e_{\widetilde{A}}$. □

For $A \in \Theta_\triangle(n,r)$, let

$$[A] = v^{-d_A} e_A, \quad \text{where} \quad d_A = \sum_{\substack{1\leqslant i\leqslant n \\ i\geqslant k, j<l}} a_{i,j}a_{k,l}. \tag{3.1.3.2}$$

(See [**56**, 4.1(b),4.3] for a geometric meaning of d_A.) Then for $\lambda \in \Lambda_\triangle(n,r)$ and $A \in \Theta_\triangle(n,r)$, we have

$$[\mathrm{diag}(\lambda)] \cdot [A] = \begin{cases} [A], & \text{if } \lambda = \mathrm{ro}(A); \\ 0, & \text{otherwise,} \end{cases} \quad \text{and}$$

$$[A][\mathrm{diag}(\lambda)] = \begin{cases} [A], & \text{if } \lambda = \mathrm{co}(A); \\ 0, & \text{otherwise.} \end{cases} \tag{3.1.3.3}$$

Moreover, by [**56**, 1.11], the $\mathcal{Z}$-linear map

$$\tau_r : \mathcal{S}_\triangle(n,r) \longrightarrow \mathcal{S}_\triangle(n,r), \quad [A] \longmapsto [{}^tA] \tag{3.1.3.4}$$

is an algebra anti-involution.

We end this section with a close look at the basis $\{[A] \mid A \in \Theta_\triangle(\begin{smallmatrix} n \\ r \end{smallmatrix}, r)_\omega\}$ for $\mathcal{T}_\triangle(n,r)$ from its specialization $\mathbb{C}_G(\mathscr{Y} \times \mathscr{X})$. Let

$$I_\triangle(n,r) = \{\mathbf{i} = (i_k)_{k\in\mathbb{Z}} \mid i_k \in \mathbb{Z}, i_{k+r} = i_k + n \text{ for all } k \in \mathbb{Z}\}. \tag{3.1.3.5}$$

We may identify the elements of $I_\triangle(n,r)$ with functions $\mathbf{i} : \mathbb{Z} \to \mathbb{Z}$ satisfying $\mathbf{i}(s + r) = \mathbf{i}(s) + n$ for all $s \in \mathbb{Z}$. (Note that a periodic function $\mathbf{i}$ can be identified with $(i_1, \ldots, i_r) \in \mathbb{Z}^r$, where $i_s = \mathbf{i}(s)$ for all s. For more details, see (3.3.0.3) below.) Clearly, there is a bijection

$$I_\triangle(n,r) \longrightarrow \Theta_\triangle(\begin{smallmatrix} n \\ r \end{smallmatrix}, r)_\omega, \quad \mathbf{i} \longmapsto A^{\mathbf{i}}, \tag{3.1.3.6}$$

where $A^{\mathbf{i}} = (a_{k,l})$ with $a_{k,l} = \delta_{k,i_l}$. Thus, the orbits of the diagonal action of G on $\mathscr{Y} \times \mathscr{X}$ are labeled by the elements of $I_\Delta(n, r)$, and $\mathcal{O}_{\mathbf{i}} := \mathcal{O}_{A^{\mathbf{i}}}$ is the orbit of the pair $(\mathbf{L}_{\mathbf{i}}, \mathbf{L}_\emptyset)$, where the ith lattices of $\mathbf{L}_{\mathbf{i}}, \mathbf{L}_\emptyset$ are defined by

$$\mathbf{L}_{\mathbf{i},i} = \bigoplus_{\mathbf{i}(j)\leqslant i} \mathbb{F}\, v_j \quad \text{and} \quad \mathbf{L}_{\emptyset,i} = \bigoplus_{j\leqslant i} \mathbb{F}\, v_j. \tag{3.1.3.7}$$

Here $v_1, \ldots, v_r$ is a fixed $\mathbb{F}[\varepsilon, \varepsilon^{-1}]$-basis of V and $v_{i+rs} = \varepsilon^{-s} v_i$, for all $s \in \mathbb{Z}$. This is because the difference set

$$\{j \mid i_j \leqslant k, j \leqslant l\}\backslash(\{j \mid i_j \leqslant k-1, j \leqslant l\} \cup \{j \mid i_j \leqslant k, j \leqslant l-1\})$$
$$= \begin{cases} \{l\}, & \text{if } k = i_l; \\ \emptyset, & \text{otherwise.} \end{cases}$$

Let $d_{\mathbf{i}} = d_{A^{\mathbf{i}}}$ (see (3.1.3.2)). If we set, for each $\mathbf{L} \in \mathscr{Y}$,

$$\mathscr{X}_{\mathbf{i},\mathbf{L}} = \{\mathbf{L}' \in \mathscr{X} \mid (\mathbf{L}, \mathbf{L}') \in \mathcal{O}_{\mathbf{i}}\},$$

then, by [**56**, Lem. 4.3], $d_{\mathbf{i}}$ is the dimension of $\mathscr{X}_{\mathbf{i},\mathbf{L}}$ (in the case where $\mathbb{F}$ is an algebraically closed field). More precisely, a direct calculation gives the following result.

Lemma 3.1.4. *For each* $\mathbf{i} = (i_j) \in I_\Delta(n, r)$, *let*

$$\mathrm{Inv}(\mathbf{i}) = \{(s, t) \in \mathbb{Z}^2 \mid 1 \leqslant s \leqslant r,\ s < t,\ i_s \geqslant i_t\}.$$

Then $d_{\mathbf{i}} = |\mathrm{Inv}(\mathbf{i})|$.

Proof. Applying [**56**, Lem. 4.3] gives

$$d_{\mathbf{i}} = d_{A^{\mathbf{i}}} = \sum_{\substack{1\leqslant t\leqslant r\\ s\geqslant k,\, t<l}} \delta_{s,i_t}\delta_{k,i_l} = \sum_{1\leqslant t\leqslant r}\sum_{s\in\mathbb{Z}} \delta_{s,i_t}\Big(\sum_{s\geqslant k,\, t<l} \delta_{k,i_l}\Big)$$
$$= \sum_{1\leqslant t\leqslant r}\Big(\sum_{i_t\geqslant k,\, t<l} \delta_{k,i_l}\Big) = |\mathrm{Inv}(\mathbf{i})|,$$

as required. □

With the identification of (3.1.1.3), we have

$$[A^{\mathbf{i}}] = q^{-\frac{1}{2}d_{\mathbf{i}}}\chi_{\mathbf{i}},$$

where $\chi_{\mathbf{i}}$ is the characteristic function of the orbit $\mathcal{O}_{\mathbf{i}}$.

3.2. Affine Hecke algebras of type A: the algebraic definition

We now follow [**35, 73**] to interpret affine quantum Schur algebras as endomorphism algebras of certain tensor spaces over the affine Hecke algebras associated with affine symmetric groups.

Let $\mathfrak{S}_{\triangle,r}$ be the (extended) *affine symmetric group* consisting of all permutations $w : \mathbb{Z} \to \mathbb{Z}$ such that $w(i + r) = w(i) + r$ for $i \in \mathbb{Z}$. This is the subset of all bijections in $I_\triangle(r, r)$ which is defined in (3.1.3.5). Hence, $\mathrm{Inv}(w)$ is well-defined. More precisely, for any $a \in \mathbb{Z}$, if we set

$$\mathrm{Inv}(w, a) = \{(s, t) \in \mathbb{Z}^2 \mid a + 1 \leqslant s \leqslant a + r,\ s < t,\ w(s) > w(t)\},$$

then $\mathrm{Inv}(w) = \mathrm{Inv}(w, 0)$ and $|\mathrm{Inv}(w)| = |\mathrm{Inv}(w, a)|$ for all $a \in \mathbb{Z}$.

There are several useful subgroups of $\mathfrak{S}_{\triangle,r}$. The subgroup W of $\mathfrak{S}_{\triangle,r}$ consisting of $w \in \mathfrak{S}_{\triangle,r}$ with $\sum_{i=1}^r w(i) = \sum_{i=1}^r i$ is the affine Weyl group of type A with generators s_i $(1 \leqslant i \leqslant r)$ defined by setting

$$s_i(j) = \begin{cases} j, & \text{for } j \not\equiv i, i+1 \bmod r; \\ j-1, & \text{for } j \equiv i+1 \bmod r; \\ j+1, & \text{for } j \equiv i \bmod r \end{cases}$$

(see [**51**, 3.6]). If $S = \{s_i\}_{1 \leqslant i \leqslant r}$, then (W, S) is a Coxeter system. For convenience of writing consecutive products of the form $s_i s_{i+1} s_{i+2} \cdots$ or $s_i s_{i-1} s_{i-2} \cdots$, we set $s_{i+kr} = s_i$ for all $k \in \mathbb{Z}$.

Observe that the cyclic subgroup $\langle \rho \rangle$ of $\mathfrak{S}_{\triangle,r}$ generated by the permutation[2] ρ of $\mathbb{Z}$ sending j to $j + 1$, for all j, is in the complement of W. Observe also that $s_{j+1}\rho = \rho s_j$, for all $j \in \mathbb{Z}$.

The subgroup $\mathfrak{A}$ of $\mathfrak{S}_{\triangle,r}$ consisting of permutations y of $\mathbb{Z}$ satisfying $y(i) \equiv i (\bmod r)$ is isomorphic to $\mathbb{Z}^r$ via the map $y \mapsto \lambda = (\lambda_1, \lambda_2, \dots, \lambda_r) \in \mathbb{Z}^r$, where $y(i) = \lambda_i r + i$, for all $1 \leqslant i \leqslant r$. We will identify $\mathfrak{A}$ with $\mathbb{Z}^r$ in the sequel. In particular, $\mathfrak{A}$ is generated by $\boldsymbol{e}_i = (0, \dots, 0, \underset{(i)}{1}, 0, \dots, 0)$ for $1 \leqslant i \leqslant r$. Moreover, the subgroup of W generated by $s_1, \dots, s_{r-1}$ is isomorphic to the symmetric group $\mathfrak{S}_r$.

Recall the matrix set $\Theta_\triangle(r, r)_\omega$ defined in (3.1.0.2) which is clearly a group with matrix multiplication.

The following result is well-known; see, e.g., [**35**, Prop. 1.1.3 & 1.1.5] for the last isomorphism.

[2] The notation ρ will have a different use in §5.3.

Proposition 3.2.1. *Maintain the notation above. There are group isomorphisms*

$$\mathfrak{S}_{\triangle,r} \cong \Theta_\triangle(r,r)_\omega \cong \mathfrak{S}_r \ltimes \mathbb{Z}^r \cong \langle\rho\rangle \ltimes W.$$

Proof. The first isomorphism is the restriction of the bijection defined in (3.1.3.6) for $n = r$; see also (3.1.0.4). In particular, every $w \in \mathfrak{S}_{\triangle,r}$ is sent to the matrix $A_w = (a_{k,l})$ with $a_{k,l} = \delta_{k,w(l)}$. The second is easily seen from the fact that every $w \in \mathfrak{S}_{\triangle,r}$ can be written as $x\lambda$, where $x \in \mathfrak{S}_r$ and $\lambda \in \mathbb{Z}^r$ are defined uniquely by $w(i) = \lambda_i r + x(i)$, for all $1 \leqslant i, x(i) \leqslant r$. If $\boldsymbol{e}_k \in \mathfrak{S}_{\triangle,r}$ ($k \in [1,r]$) denotes the permutation $\boldsymbol{e}_k(i) = i$ for $i \neq k$, $i \in [1,r]$ and $\boldsymbol{e}_k(k) = r+k$, then $\boldsymbol{e}_k = \rho s_{r+k-2} \cdots s_{k+1} s_k$, proving the last isomorphism.[3] □

Let ℓ be the length function of the Coxeter system (W, S). Then $\ell(w)$ for $w \in W$ is the length m such that $w = s_{i_1} s_{i_2} \cdots s_{i_m}$ is a reduced expression. By [**41**],

$$\ell(w) = \sum_{1 \leqslant i < j \leqslant r} \left| \left[\frac{w(j) - w(i)}{r} \right] \right|.$$

Extend the length function to $\mathfrak{S}_{\triangle,r}$ by setting $\ell(\rho^i w') = \ell(w')$, for all $i \in \mathbb{Z}$ and $w' \in W$. Since $\mathrm{Inv}(\rho^i) = \emptyset$, induction on $\ell(w)$ shows that

$$\ell(w) = \mathrm{Inv}(w), \qquad \text{for all } w \in \mathfrak{S}_{\triangle,r}. \tag{3.2.1.1}$$

The group $\mathfrak{S}_{\triangle,r}$ acts on the set $I_\triangle(n,r)$ given in (3.1.3.5) by place permutation:

$$\mathbf{i}w = (i_{w(k)})_{k\in\mathbb{Z}}, \qquad \text{for } \mathbf{i} \in I_\triangle(n,r) \text{ and } w \in \mathfrak{S}_{\triangle,r}. \tag{3.2.1.2}$$

Clearly, every $\mathfrak{S}_{\triangle,r}$-orbit has a unique representative in the *fundamental set*

$$\begin{aligned} I_\triangle(n,r)_0 &= \{\mathbf{i} \in I_\triangle(n,r) \mid 1 \leqslant i_1 \leqslant i_2 \leqslant \cdots \leqslant i_r \leqslant n\} \\ &= \{\mathbf{i}_\lambda^\triangle \mid \lambda \in \Lambda_\triangle(n,r)\}, \end{aligned} \tag{3.2.1.3}$$

where $\mathbf{i}_\lambda^\triangle = (i_s)_{s\in\mathbb{Z}} \in I_\triangle(n,r)_0$ if and only if $i_s = m$, for all $s \in R_m^\lambda$ and $m \in \mathbb{Z}$, or equivalently, $\lambda_j = |\{k \in \mathbb{Z} \mid i_k = j\}|$, for all $j \in \mathbb{Z}$. Written in full, $\mathbf{i}_\lambda^\triangle$ is the sequence

$$(\ldots, \underbrace{0,\ldots,0}_{\lambda_n}, \underbrace{1,\ldots,1}_{\lambda_1}, \underbrace{2,\ldots,2}_{\lambda_2}, \ldots, \underbrace{n,\ldots,n}_{\lambda_n}, \underbrace{1+n,\ldots,1+n}_{\lambda_1}, \ldots).$$

[3] If we put $s_1, s_2, \ldots, s_r$ on a circle clockwise and break the circle by removing s_{k-1}, then $\rho^{-1}\boldsymbol{e}_k$ is obtained by flattening the circle into a line segment.

Observe that every number of the form $i+kn$, for $1 \leqslant i \leqslant n, k \in \mathbb{Z}$, determines a constant subsequence $(i_j)_{j\in R^\lambda_{i+kn}} = (i+kn, \dots, i+kn)$ of $\mathbf{i} = \mathbf{i}^\vartriangle_\lambda$ (of length λ_i) indexed by the set

$$R^\lambda_{i+kn} = \{\lambda_{k,i-1}+1, \lambda_{k,i-1}+2, \dots, \lambda_{k,i-1}+\lambda_i = \lambda_{k,i}\}, \tag{3.2.1.4}$$

where $\lambda_{k,i-1} = kr + \sum_{1\leqslant t\leqslant i-1} \lambda_t$. These sets form a partition $\bigcup_{j\in\mathbb{Z}} R^\lambda_j$ of $\mathbb{Z}$. Note that the fundamental set $I_\vartriangle(n,r)_0$ is obtained by shifting the one used in **[73]** by n.

For $\lambda \in \Lambda_\vartriangle(n,r)$, let $\mathfrak{S}_\lambda := \mathfrak{S}_{(\lambda_1,\dots,\lambda_n)}$ be the corresponding standard Young subgroup of $\mathfrak{S}_r$ (and hence, of $\mathfrak{S}_{\vartriangle,r}$), and let

$$\mathscr{D}^\vartriangle_\lambda = \{d \mid d \in \mathfrak{S}_{\vartriangle,r}, \ell(wd) = \ell(w) + \ell(d) \text{ for } w \in \mathfrak{S}_\lambda\}.$$

By (3.2.1.1), one sees easily that

$$\begin{aligned} d^{-1} \in \mathscr{D}^\vartriangle_\lambda &\iff d(\lambda_{k,i-1}+1) < d(\lambda_{k,i-1}+2) < \cdots < d(\lambda_{k,i-1}+\lambda_i), \\ &\qquad \forall 1 \leqslant i \leqslant n, k \in \mathbb{Z} \\ &\iff d(\lambda_{0,i-1}+1) < d(\lambda_{0,i-1}+2) < \cdots < d(\lambda_{0,i-1}+\lambda_i), \\ &\qquad \forall 1 \leqslant i \leqslant n. \end{aligned} \tag{3.2.1.5}$$

The following result is the affine version of a well-known result for symmetric groups (see, e.g., **[12**, Th. 4.15, (9.1.4)]. It can be deduced from **[73**, 7.4]. For a proof, see **[24**, 9.2]).

Lemma 3.2.2. *For $\lambda, \mu \in \Lambda_\vartriangle(n,r)$, let $\mathscr{D}^\vartriangle_{\lambda,\mu} = \mathscr{D}^\vartriangle_\lambda \cap (\mathscr{D}^\vartriangle_\mu)^{-1}$. There is a bijective map*

$$\jmath_\vartriangle : \{(\lambda, w, \mu) \mid w \in \mathscr{D}^\vartriangle_{\lambda,\mu}, \lambda, \mu \in \Lambda_\vartriangle(n,r)\} \longrightarrow \Theta_\vartriangle(n,r) \tag{3.2.2.1}$$

sending (λ, w, μ) to $A = (a_{k,l})$, where, if $\mathbf{i}^\vartriangle_\lambda = (i_a)_{a\in\mathbb{Z}}$ and $\mathbf{i}^\vartriangle_\mu = (j_a)_{a\in\mathbb{Z}}$, then

$$a_{k,l} = |\{t \in \mathbb{Z} \mid i_{w(t)} = k, j_t = l\}| = |R^\lambda_k \cap wR^\mu_l| \tag{3.2.2.2}$$

for all $k, l \in \mathbb{Z}$. In particular, by certain appropriate embedding, restriction gives two bijections

$$\jmath_\vartriangle : \{(\lambda, w, \omega) \mid w \in \mathscr{D}^\vartriangle_{\lambda,\mu}, \lambda \in \Lambda_\vartriangle(n,r)\} \longrightarrow \Theta_\vartriangle(\tbinom{n}{r}, r)_\omega \tag{3.2.2.3}$$

and $\jmath_\vartriangle : \{(\omega, w, \omega) \mid w \in \mathfrak{S}_{\vartriangle,r}\} \to \Theta_\vartriangle(r,r)_\omega$.

We remark that, for $w' \in \mathfrak{S}_\lambda w \mathfrak{S}_\mu$, the equality $|R^\lambda_k \cap w' R^\mu_l| = |R^\lambda_k \cap wR^\mu_l|$ holds. Hence, the matrix A is completely determined by λ, μ and the double coset $\mathfrak{S}_\lambda w \mathfrak{S}_\mu$, and is independent of the selection of the representative. Moreover, if $\jmath_\vartriangle(\lambda, d, \mu) = A$, then $\jmath_\vartriangle(\mu, d^{-1}, \lambda) = {}^tA$, the transpose of A.

Corollary 3.2.3. *For $\lambda, \mu \in \Lambda_\triangle(n,r)$ and $d \in \mathscr{D}^\triangle_{\lambda,\mu}$ with $\jmath_\triangle(\lambda, d, \mu) = A \in \Theta_\triangle(n,r)$, let $\nu^{(i)}$ be the composition of λ_i obtained by removing all zeros from row i of A. Then $\mathfrak{S}_\lambda \cap d\mathfrak{S}_\mu d^{-1} = \mathfrak{S}_\nu$, where $\nu = (\nu^{(1)}, \ldots, \nu^{(n)})$.*

Proof. Let $\mathbf{i}^\triangle_\mu = (j_s)_{s\in\mathbb{Z}} \in I_\triangle(n,r)$. Then $j_s = k$ for all $s \in R^\mu_k$ and $k \in \mathbb{Z}$. Thus, $l \in R^\lambda_i$ and $d^{-1}(l) \in R^\mu_j \iff l \in R^\lambda_i \cap dR^\mu_j$. For $1 \leqslant i \leqslant n$, if

$$I_i := (\underbrace{k, \ldots, k}_{a_{i,k}})_{k\in\mathbb{Z}} = (\ldots, \underbrace{1, \ldots, 1}_{a_{i,1}}, \underbrace{2, \ldots, 2}_{a_{i,2}}, \ldots, \underbrace{n, \ldots, n}_{a_{i,n}}, \ldots) \in \mathbb{Z}^{\lambda_i},$$

then, with the notation used in (3.2.1.4), (3.2.1.5) together with Lemma 3.2.2 implies $I_i = (j_{d^{-1}(\lambda_{0,i-1}+1)}, \ldots, j_{d^{-1}(\lambda_{0,i-1}+\lambda_i)})$. Hence, $(j_{d^{-1}(1)}, \ldots, j_{d^{-1}(r)}) = (I_1, \ldots, I_n)$. Since $\mathfrak{S}_\mu = \mathrm{Stab}_{\mathfrak{S}_{\triangle,r}}(I_\mu)$, it follows that $\mathfrak{S}_\lambda \cap d\mathfrak{S}_\mu d^{-1} = \mathfrak{S}_\lambda \cap \mathrm{Stab}_{\mathfrak{S}_{\triangle,r}}(I_\mu d^{-1}) = \mathrm{Stab}_{\mathfrak{S}_\lambda}(I_\mu d^{-1}) = \mathfrak{S}_\nu$. □

Corollary 3.2.4. *There is a bijective map*

$$\jmath^*_\triangle : I_\triangle(n,r) \longrightarrow \{(\lambda, d, \omega) \mid d \in \mathscr{D}^\triangle_\lambda, \lambda \in \Lambda_\triangle(n,r)\}, \quad \mathbf{i} \longmapsto (\lambda, d, \omega),$$

where $\mathbf{i} = \mathbf{i}^\triangle_\lambda d$. Moreover, if d^+ denotes a representative of $\mathfrak{S}_\lambda d$ with maximal length, then $d_\mathbf{i} = |\mathrm{Inv}(\mathbf{i})| = \ell(d^+)$.

Proof. Clearly, $\jmath^*_\triangle$ is the composition of the bijection given in (3.1.3.6) and the inverse of $\jmath_\triangle$ given in (3.2.2.3). We now prove the last statement.

Let $\mathcal{L} = \{(s,t) \in \mathbb{Z}^2 \mid 1 \leqslant s \leqslant r,\ s < t\}$ and $\mathbf{j} = \mathbf{i}^\triangle_\lambda$. Then

$$\mathrm{Inv}(\mathbf{i}) = \{(s,t) \in \mathcal{L} \mid j_{d(s)} \geqslant j_{d(t)}\} = X_1 \cup X_2 \quad \text{(a disjoint union)},$$

where $X_1 = \{(s,t) \in \mathcal{L} \mid j_{d(s)} > j_{d(t)}\}$ and $X_2 = \{(s,t) \in \mathcal{L} \mid j_{d(s)} = j_{d(t)}\}$. Since $(s,t) \in X_2$ if and only if $d(s), d(t) \in R^\lambda_k$ for some $k \in \mathbb{Z}$, (3.2.1.5) forces $d(s) < d(t)$. Hence, if $\mathfrak{S}_{\triangle,r}$ acts on $\mathbb{Z}^2$ diagonally, then

$$\begin{aligned} X_2 &= \{(s,t) \in \mathcal{L} \mid d(s) < d(t), d(s), d(t) \in R^\lambda_k \text{ for some } k \in \mathbb{Z}\} \\ &= d^{-1}\{(s,t) \in \mathbb{Z}^2 \mid 1 \leqslant d^{-1}(s) \leqslant r, s < t, s, t \in R^\lambda_k \text{ for some } k \in \mathbb{Z}\}. \end{aligned}$$

Thus, $|X_2|$ is the length of the longest element $w_{0,\lambda}$ in $\mathfrak{S}_\lambda$ by Lemma 3.10.1. Since, for $(s,t) \in \mathcal{L}\backslash X_2$, $j_{d(s)} > j_{d(t)} \iff d(s) > d(t)$, applying (3.2.1.5) again yields

$$\mathrm{Inv}(d) = \{(s,t) \in \mathcal{L} \mid d(s) > d(t)\} = \{(s,t) \in \mathcal{L}\backslash X_2 \mid d(s) > d(t)\} = X_1.$$

Consequently, $d_\mathbf{i} = |X_1| + |X_2| = \ell(d) + \ell(w_{0,\lambda}) = \ell(d^+)$, as required. □

We record the following generalization (to affine symmetric groups) of a well-known result for Coxeter groups.

Lemma 3.2.5. *Let $\lambda, \mu \in \Lambda_\triangle(n, r)$ and $d \in \mathscr{D}^\triangle_{\lambda,\mu}$. Then $d^{-1}\mathfrak{S}_\lambda d \cap \mathfrak{S}_\mu$ is a standard Young subgroup of $\mathfrak{S}_\mu$. Moreover, each element $w \in \mathfrak{S}_\lambda d\mathfrak{S}_\mu$ can be written uniquely as a product $w = w_1 d w_2$ with $w_1 \in \mathfrak{S}_\lambda$ and $w_2 \in \mathscr{D}^\triangle_\nu \cap \mathfrak{S}_\mu$, where $\nu \in \Lambda_\triangle(n, r)$ is defined by $\mathfrak{S}_\nu = d^{-1}\mathfrak{S}_\lambda d \cap \mathfrak{S}_\mu$, and the equality $\ell(w) = \ell(w_1) + \ell(d) + \ell(w_2)$ holds.*

Following [**45**], the (extended) affine Hecke algebra $\mathcal{H}(\mathfrak{S}_{\triangle,r})$ over $\mathcal{Z}$ is defined to be the algebra generated by T_{s_i} $(1 \leqslant i \leqslant r)$, $T_\rho^{\pm 1}$ with the following relations:

$$
\begin{aligned}
&T_{s_i}^2 = (v^2 - 1)T_{s_i} + v^2,\\
&T_{s_i}T_{s_j} = T_{s_j}T_{s_i} \quad (i - j \not\equiv \pm 1 \bmod r),\\
&T_{s_i}T_{s_j}T_{s_i} = T_{s_j}T_{s_i}T_{s_j} \quad (i - j \equiv \pm 1 \bmod r \text{ and } r \geqslant 3),\\
&T_\rho T_\rho^{-1} = T_\rho^{-1}T_\rho = 1, \quad \text{and}\\
&T_\rho T_{s_i} = T_{s_{i+1}}T_\rho,
\end{aligned}
$$

where $T_{s_{r+1}} = T_{s_1}$. This algebra has a $\mathcal{Z}$-basis $\{T_w\}_{w\in\mathfrak{S}_{\triangle,r}}$, where $T_w = T_\rho^j T_{s_{i_1}} \cdots T_{s_{i_m}}$ if $w = \rho^j s_{i_1} \cdots s_{i_m}$ is reduced.

The following result is well-known due to Iwahori–Matsumoto [**41**]. Recall the algebra $\mathcal{H}_\triangle(r)$ defined in §3.1 and the isomorphism given in (3.1.1.3).

Lemma 3.2.6. *There is a $\mathcal{Z}$-algebra isomorphism $\mathcal{H}(\mathfrak{S}_{\triangle,r}) \cong \mathcal{H}_\triangle(r)$ whose specialization of v to $\sqrt{q}$ gives a $\mathbb{C}$-algebra isomorphism*

$$\mathcal{H}(\mathfrak{S}_{\triangle,r}) \otimes \mathbb{C} \cong \mathbb{C}_G(\mathscr{X} \times \mathscr{X}).$$

Thus, *we will identify $\mathcal{H}(\mathfrak{S}_{\triangle,r})$ with $\mathcal{H}_\triangle(r)$ in the sequel.*

Let $\mathcal{H}(r) = \mathcal{H}(\mathfrak{S}_r)$ be the subalgebra of $\mathcal{H}_\triangle(r)$ generated by T_{s_i} $(1 \leqslant i < r)$. Then $\mathcal{H}(r)$ is the Hecke algebra of the symmetric group $\mathfrak{S}_r$. We finally set

$$\boldsymbol{\mathcal{H}}_\triangle(r) = \mathcal{H}_\triangle(r) \otimes_{\mathcal{Z}} \mathbb{Q}(v) \quad\text{and}\quad \boldsymbol{\mathcal{H}}(r) = \mathcal{H}(r) \otimes_{\mathcal{Z}} \mathbb{Q}(v).$$

For each $\lambda \in \Lambda_\triangle(n, r)$, let $x_\lambda = \sum_{w\in\mathfrak{S}_\lambda} T_w \in \mathcal{H}(r)$ and define

$$\mathcal{S}^{\mathcal{H}}_\triangle(n, r) := \operatorname{End}_{\mathcal{H}_\triangle(r)}\Big(\bigoplus_{\nu\in\Lambda_\triangle(n,r)} x_\nu \mathcal{H}_\triangle(r)\Big).$$

For $\lambda, \mu \in \Lambda_\triangle(n, r)$ and $d \in \mathscr{D}^\triangle_{\lambda,\mu}$, define $\phi^d_{\lambda,\mu} \in \mathcal{S}^{\mathcal{H}}_\triangle(n, r)$ as follows:

$$\phi^d_{\lambda,\mu}(x_\nu h) = \delta_{\mu\nu} \sum_{w\in\mathfrak{S}_\lambda d\mathfrak{S}_\mu} T_w h, \tag{3.2.6.1}$$

where $\nu \in \Lambda_\triangle(n, r)$ and $h \in \mathcal{H}_\triangle(r)$. Then the set $\{\phi^d_{\lambda,\mu}\}$ forms a basis for $\mathcal{S}^{\mathcal{H}}_\triangle(n, r)$.

Remarks 3.2.7. (1) We point out that, as a natural generalization of the q-Schur algebra given in [**15, 16**], the endomorphism algebra $\mathcal{S}_\Delta^{\mathcal{H}}(n,r)$ is called the *affine q-Schur algebra* and the basis $\{\phi_{\lambda,\mu}^d\}$ is the affine analogue of [**16**, 1.4].

(2) Let R be a commutative ring with 1 which is a $\mathcal{Z}$-algebra. Then, by base change to R, a similar basis can be defined for

$$\mathcal{S}_\Delta^{\mathcal{H}}(n,r;R) = \mathrm{End}_{\mathcal{H}_\Delta(r)_R}\Big(\bigoplus_{\lambda\in\Lambda_\Delta(n,r)} x_\lambda \mathcal{H}_\Delta(r)_R\Big).$$

As a result of this, the endomorphism algebra $\mathcal{S}_\Delta^{\mathcal{H}}(n,r)$ satisfies the base change property: $\mathcal{S}_\Delta^{\mathcal{H}}(n,r;R) \cong \mathcal{S}_\Delta^{\mathcal{H}}(n,r)_R$. This property has already been mentioned for $R = \mathbb{Z}[\sqrt{q}, \sqrt{q}^{-1}]$ in Remark 3.1.2.

Combining the base change property and [**73**, 7.4] gives the following result which extends the isomorphism given in Lemma 3.2.6 to affine quantum Schur algebras.

Proposition 3.2.8. *The bijection $\jmath_\Delta$ given in Lemma 3.2.2 induces a $\mathcal{Z}$-algebra isomorphism*

$$\mathfrak{h} : \mathcal{S}_\Delta(n,r) \xrightarrow{\sim} \mathcal{S}_\Delta^{\mathcal{H}}(n,r),\ e_A \longmapsto \phi_{\lambda,\mu}^d,$$

for all $A \in \Theta_\Delta(n,r)$ with $A = \jmath_\Delta(\lambda,d,\mu)$, where $\lambda,\mu \in \Lambda_\Delta(n,r)$ and $d \in \mathscr{D}_{\lambda,\mu}^\Delta$. Moreover, regarding $\bigoplus_{\lambda\in\Lambda_\Delta(n,r)} x_\lambda\mathcal{H}_\Delta(r)$ as an $\mathcal{S}_\Delta(n,r)$-module via $\mathfrak{h}$, we obtain an $\mathcal{S}_\Delta(n,r)$-$\mathcal{H}_\Delta(r)$-bimodule isomorphism

$$ev : \mathcal{T}_\Delta(n,r) \xrightarrow{\sim} \bigoplus_{\lambda\in\Lambda_\Delta(n,r)} x_\lambda\mathcal{H}_\Delta(r),\ e_A \longmapsto x_\lambda T_d,$$

for all $A \in \Theta_\Delta(^n_r, r)_\omega$ with $A = \jmath_\Delta(\lambda,d,\omega)$.

Note that if we regard $\mathcal{T}_\Delta(n,r)$ as a subset of $\mathcal{S}_\Delta(N,r)$ as in Lemma 3.1.3, the bimodule isomorphism is simply the evaluation map.

Recall that, by removing the superscript $^\Delta$, the notation $\mathscr{D}_{\lambda,\mu}$ denotes the shortest $(\mathfrak{S}_\lambda,\mathfrak{S}_\mu)$-coset representatives in $\mathfrak{S}_r$. If we identify $\Lambda_\Delta(n,r)$ with $\Lambda(n,r)$ via (1.1.0.2), we obtain the following.

Corollary 3.2.9. *The subspace spanned by all $\phi_{\lambda,\mu}^d$ with $\lambda,\mu \in \Lambda_\Delta(n,r)$ and $d \in \mathscr{D}_{\lambda,\mu}$ is a subalgebra which is isomorphic to the quantum Schur algebra $\mathcal{S}(n,r)$.*

Using the evaluation isomorphism, we now describe an explicit action of $\mathcal{H}_\Delta(r)$ on $\mathcal{T}_\Delta(n,r)$. First, for $\lambda \in \Lambda_\Delta(n,r)$, $d \in \mathscr{D}_\lambda^\Delta$, and $1 \leqslant k \leqslant r$,

$$x_\lambda T_d \cdot T_{s_k} = \begin{cases} v^2 x_\lambda T_d, & \text{if } ds_k \notin \mathscr{D}_\lambda^\Delta \text{ (then } \ell(ds_k) > \ell(d)); \\ x_\lambda T_{ds_k}, & \text{if } \ell(ds_k) > \ell(d) \text{ and } ds_k \in \mathscr{D}_\lambda^\Delta; \\ v^2 x_\lambda T_{ds_k} + (v^2-1)x_\lambda T_d, & \text{if } \ell(ds_k) < \ell(d) \text{(then } ds_k \in \mathscr{D}_\lambda^\Delta). \end{cases} \tag{3.2.9.1}$$

Second, by Corollary 3.2.4, we obtain $ev([A^{\mathbf{i}}]) = v^{-\ell(d^+)} x_\lambda T_d$ if $\jmath_\Delta^*(\mathbf{i}) = (\lambda, d, \omega)$, where d^+ is a representative of $\mathfrak{S}_\lambda d$ with maximal length. For $w \in \mathfrak{S}_{\Delta,r}$, let

$$\widetilde{T}_w = v^{-\ell(w)} T_w.$$

Thus, for $\mathbf{j} = \mathbf{i}_\lambda^\Delta$ and d as above, (3.2.9.1) becomes

$$[A^{\mathbf{j}d}]\widetilde{T}_{s_k} = \begin{cases} v[A^{\mathbf{j}d}], & \text{if } ds_k \notin \mathscr{D}_\lambda^\Delta \text{ (then } \ell(ds_k) > \ell(d)); \\ [A^{\mathbf{j}ds_k}], & \text{if } \ell(ds_k) > \ell(d) \text{ and } ds_k \in \mathscr{D}_\lambda^\Delta; \\ [A^{\mathbf{j}ds_k}] + (v - v^{-1})[A^{\mathbf{j}d}], & \text{if } \ell(ds_k) < \ell(d) \text{(then } ds_k \in \mathscr{D}_\lambda^\Delta). \end{cases}$$

This together with Lemma 3.1.4 gives the first part of the following (and part (2) is clear from the definition).

Proposition 3.2.10. *Let* $\mathbf{i} \in I_\Delta(n,r)$.

(1) *For any* $1 \leqslant k \leqslant r$, *we have*

$$[A^{\mathbf{i}}]\widetilde{T}_{s_k} = \begin{cases} v[A^{\mathbf{i}}], & \text{if } i_k = i_{k+1}; \\ [A^{\mathbf{i}s_k}], & \text{if } i_k < i_{k+1}; \\ [A^{\mathbf{i}s_k}] + (v - v^{-1})[A^{\mathbf{i}}], & \text{if } i_k > i_{k+1}. \end{cases}$$

(2) $[A^{\mathbf{i}}]T_\rho = [A^{\mathbf{i}\rho}]$, *where* $\rho \in \mathfrak{S}_{\Delta,r}$ *is the permutation sending* i *to* $i+1$ *for all* $i \in \mathbb{Z}$.

3.3. The tensor space interpretation

We now interpret the right $\mathcal{H}_\Delta(r)$-module $\mathcal{T}_\Delta(n,r)$ in terms of the tensor space, following [**73**].

The Hecke algebra $\mathcal{H}(\mathfrak{S}_{\Delta,r})$ admits the so-called *Bernstein presentation* which consists of generators

$$T_i := T_{s_i},\ X_j := \widetilde{T}_{\mathbf{e}_1+\cdots+\mathbf{e}_{j-1}} \widetilde{T}^{-1}_{\mathbf{e}_1+\cdots+\mathbf{e}_j},\ X_j^{-1}\ (1 \leqslant i \leqslant r-1, 1 \leqslant j \leqslant r),$$

and relations

$$\begin{aligned}&(T_i+1)(T_i-v^2)=0,\\&T_iT_{i+1}T_i=T_{i+1}T_iT_{i+1},\quad T_iT_j=T_jT_i\ (|i-j|>1),\\&X_iX_i^{-1}=1=X_i^{-1}X_i,\quad X_iX_j=X_jX_i,\\&T_iX_iT_i=v^2X_{i+1},\quad\text{and}\\&X_jT_i=T_iX_j\ (j\neq i,i+1).\end{aligned}$$

Note that, for any dominant $\lambda=(\lambda_i)\in\mathbb{Z}^r$ (meaning $\lambda_1\geqslant\cdots\geqslant\lambda_r$),

$$X_\lambda:=X_1^{\lambda_1}\cdots X_r^{\lambda_r}=\widetilde{T}_\lambda^{-1}.$$

In particular, $T_\rho=\widetilde{T}_\rho=X_1^{-1}\widetilde{T}_1^{-1}\cdots\widetilde{T}_{r-1}^{-1}$ since $\boldsymbol{e}_1=\rho s_{r-1}\cdots s_2s_1$ and $X_1^{-1}=\widetilde{T}_{\boldsymbol{e}_1}$.

By definition, we have, for each $1\leqslant i\leqslant r-1$,

$$\begin{aligned}&T_iX_{i+1}^{-1}=X_i^{-1}T_i+(1-v^2)X_i^{-1},\\&X_{i+1}^{-1}T_i=T_iX_i^{-1}+(1-v^2)X_i^{-1},\\&T_i^{-1}X_i^{-1}=X_{i+1}^{-1}T_i^{-1}+(1-v^{-2})X_{i+1}^{-1},\quad\text{and}\\&X_i^{-1}T_i^{-1}=T_i^{-1}X_{i+1}^{-1}+(1-v^{-2})X_{i+1}^{-1}.\end{aligned}\tag{3.3.0.1}$$

So $X_iT_i=T_iX_{i+1}+(1-v^2)X_{i+1}$ and $T_iX_i=X_{i+1}T_i+(1-v^2)X_{i+1}$, etc. Then, for each $\mathbf{a}=(a_1,\ldots,a_r)\in\mathbb{Z}^r$, an inductive argument gives the formula

$$X_{\mathbf{a}}T_i=T_iX_{\mathbf{a}s_i}+(1-v^2)\frac{X_{i+1}(X_{\mathbf{a}}-X_{\mathbf{a}s_i})}{X_i-X_{i+1}},\tag{3.3.0.2}$$

where $\mathbf{a}s_i=(a_1,\ldots,a_{i-1},a_{i+1},a_i,a_{i+2},\ldots,a_r)$.

Let Ω be the free $\mathcal{Z}$-module with basis $\{\omega_i\mid i\in\mathbb{Z}\}$. Consider the r-fold *tensor space* $\Omega^{\otimes r}$ and, for each $\mathbf{i}=(i_1,\ldots,i_r)\in\mathbb{Z}^r$, write

$$\omega_{\mathbf{i}}=\omega_{i_1}\otimes\omega_{i_2}\otimes\cdots\otimes\omega_{i_r}=\omega_{i_1}\omega_{i_2}\cdots\omega_{i_r}\in\Omega^{\otimes r}.$$

We now follow [**73**] to define a right $\mathcal{H}_\Delta(r)$-module structure on $\Omega^{\otimes r}$ and to establish an $\mathcal{H}_\Delta(r)$-module isomorphism from $\mathcal{T}_\Delta(n,r)$ to $\Omega^{\otimes r}$.

Recall the set $I_\Delta(n,r)$ defined in (3.1.3.5) and the action (3.2.1.2) of $\mathfrak{S}_{\Delta,r}$ on $I_\Delta(n,r)$. If we identify $I_\Delta(n,r)$ with $\mathbb{Z}^r$ by the following bijection

$$I_\Delta(n,r)\longrightarrow\mathbb{Z}^r,\quad \mathbf{i}\longmapsto(i_1,\ldots,i_r),\tag{3.3.0.3}$$

then the action of $\mathfrak{S}_{\Delta,r}$ on $I_\Delta(n,r)$ induces an action on $\mathbb{Z}^r$. Also, the usual action of the place permutation of $\mathfrak{S}_r$ on $I(n,r)$, where

$$I(n,r)=\{(i_1,\ldots,i_r)\in\mathbb{Z}^r\mid 1\leqslant i_k\leqslant n,\forall k\},$$

is the restriction to $\mathfrak{S}_r$ of the action of $\mathfrak{S}_{\Delta,r}$ on $\mathbb{Z}^r$ (restricted to $I(n,r)$). We often identify $I(n,r)$ as a subset of $I_\Delta(n,r)$, or $I_\Delta(n,r)_0$ as a subset of $I(n,r)$, depending on the context.

By the Bernstein presentation for $\mathcal{H}_\Delta(r)$, Varagnolo–Vasserot extended in [**73**] the action of $\mathcal{H}(r)$ on the finite tensor space $\Omega_n^{\otimes r}$, where $\Omega_n = \text{span}\{\omega_1, \ldots, \omega_n\}$, given in [**43**] (see also [**22**]) to an action on $\Omega^{\otimes r}$ via the place permutation above. In other words, $\Omega^{\otimes r}$ admits a right $\mathcal{H}_\Delta(r)$-module structure defined by

$$\begin{cases} \omega_{\mathbf{i}} \cdot X_t^{-1} = \omega_{\mathbf{i}\mathbf{e}_t} = \omega_{i_1} \cdots \omega_{i_{t-1}} \omega_{i_t + n} \omega_{i_{t+1}} \cdots \omega_{i_r}, & \text{for all } \mathbf{i} \in \mathbb{Z}^r; \\ \omega_{\mathbf{i}} \cdot T_k = \begin{cases} v^2 \omega_{\mathbf{i}}, & \text{if } i_k = i_{k+1}; \\ v\omega_{\mathbf{i}s_k}, & \text{if } i_k < i_{k+1}; \\ v\omega_{\mathbf{i}s_k} + (v^2-1)\omega_{\mathbf{i}}, & \text{if } i_k > i_{k+1}, \end{cases} & \text{for all } \mathbf{i} \in I(n,r), \end{cases} \tag{3.3.0.4}$$

where $1 \leqslant k \leqslant r-1$ and $1 \leqslant t \leqslant r$.

In general, for an arbitrary $\mathbf{i} \in \mathbb{Z}^r$, there exist $\mathbf{j} \in I(n,r)$ and $\mathbf{a} \in \mathbb{Z}^r$ satisfying $\omega_{\mathbf{j}} \cdot X_{\mathbf{a}} = \omega_{\mathbf{i}}$. Then, by applying (3.3.0.2), we define

$$\begin{aligned} \omega_{\mathbf{i}} \cdot T_k &= (\omega_{\mathbf{j}} \cdot X_{\mathbf{a}}) \cdot T_k = \omega_{\mathbf{j}} \cdot (X_{\mathbf{a}} T_k) \\ &= (\omega_{\mathbf{j}} \cdot T_k) \cdot X_{\mathbf{a}s_k} + (1-v^2)\omega_{\mathbf{j}} \cdot \frac{X_{k+1}(X_{\mathbf{a}} - X_{\mathbf{a}s_k})}{X_k - X_{k+1}}. \end{aligned}$$

Varagnolo–Vasserot have further established in [**73**, Lem. 8.3] an $\mathcal{H}_\Delta(r)$-module isomorphism between $\mathcal{T}_\Delta(n,r)$ and $\Omega^{\otimes r}$. This result justifies why the set $I_\Delta(n,r)_0$ defined in (3.2.1.3) is called a fundamental set. Recall from (3.1.3.6) the matrix $A^{\mathbf{i}}$ defined for every $\mathbf{i} \in I_\Delta(n,r)$.

Proposition 3.3.1. *There is a* unique *$\mathcal{H}_\Delta(r)$-module isomorphism*

$$g : \mathcal{T}_\Delta(n,r) \longrightarrow \Omega^{\otimes r} \text{ such that } [A^{\mathbf{i}}] \longmapsto \omega_{\mathbf{i}} \text{ for all } \mathbf{i} \in I_\Delta(n,r)_0,$$

which induces a $\mathcal{Z}$-algebra isomorphism

$$\mathsf{t} : \mathcal{S}_\Delta(n,r) \xrightarrow{\sim} \mathcal{S}_\Delta^{\mathsf{t}}(n,r) := \text{End}_{\mathcal{H}_\Delta(r)}(\Omega^{\otimes r}).$$

In particular, g is an $\mathcal{S}_\Delta(n,r)$-$\mathcal{H}_\Delta(r)$-bimodule isomorphism. Moreover, specializing v to $\sqrt{q}$ yields a $\mathbb{C}_G(\mathscr{Y} \times \mathscr{Y})$-$\mathbb{C}_G(\mathscr{X} \times \mathscr{X})$-bimodule isomorphism over $\mathbb{C}$ from $\mathbb{C}_G(\mathscr{Y} \times \mathscr{X})$ to $\Omega_{\mathbb{C}}^{\otimes r} := \Omega^{\otimes r} \otimes \mathbb{C}$ sending $[A^{\mathbf{i}}] = q^{-\frac{1}{2}d_{\mathbf{i}}}\chi_{\mathbf{i}}$ to $\omega_{\mathbf{i}}$, $\forall \mathbf{i} \in I_\Delta(n,r)_0$.

Proof. The first assertion follows from [**24**, Lem. 9.5]. By regarding $\Omega^{\otimes r}$ as an $\mathcal{S}_\Delta(n,r)$-module via t, g induces an $\mathcal{S}_\Delta(n,r)$-$\mathcal{H}_\Delta(r)$-bimodule isomorphism. The last assertion follows from the isomorphism (3.1.1.3) and the definition that $\mathcal{T}_\Delta(n,r)$ is the generic form of $\mathbb{C}_G(\mathscr{Y} \times \mathscr{X})$. □

Remark 3.3.2. As seen above, since the action of $\widetilde{T}_{s_i}$, for $1 \leqslant i \leqslant r-1$, on the basis elements $\omega_{\mathbf{i}}$ for $\mathbf{i} \in I(n,r)$ follows the same rules as the action on $[A^{\mathbf{i}}]$ given in Proposition 3.2.10(1), it follows that $g([A^{\mathbf{i}}]) = \omega_{\mathbf{i}}$ for all $\mathbf{i} \in I(n,r)$. However, the actions of $\widetilde{T}_{s_r}$ are different. Hence, if we identify $\mathcal{T}_\Delta(n,r)$ with $\Omega^{\otimes r}$ under g, then $\{[A^{\mathbf{i}}]\}_{\mathbf{i}\in I_\Delta(n,r)}$ and $\{[\omega_{\mathbf{i}}]\}_{\mathbf{i}\in I_\Delta(n,r)}$ form two different bases with the subset $\{[A^{\mathbf{i}}] = \omega_{\mathbf{i}}\}_{\mathbf{i}\in I(n,r)}$ in common.

We now identify $\mathbb{C}_G(\mathscr{Y} \times \mathscr{X})$ with $\Omega_{\mathbb{C}}^{\otimes r}$. Consequently, specializing v to $\sqrt{q}$ gives isomorphisms

$$\mathrm{End}_{\mathcal{H}_\Delta(r)_{\mathbb{C}}}(\Omega_{\mathbb{C}}^{\otimes r}) \cong \mathrm{End}_{\mathcal{H}_\Delta(r)_{\mathbb{C}}}(\mathbb{C}_G(\mathscr{Y} \times \mathscr{X})) \cong \mathbb{C}_G(\mathscr{Y} \times \mathscr{Y}) \cong \mathcal{S}_\Delta(n,r)_{\mathbb{C}}.$$

These algebras will be identified in the sequel.

Also, let $\boldsymbol{\Omega} = \Omega \otimes_{\mathbb{Z}} \mathbb{Q}(v)$, i.e., $\boldsymbol{\Omega}$ is a $\mathbb{Q}(v)$-vector space with basis $\{\omega_i \mid i \in \mathbb{Z}\}$. Then the right action of $\mathcal{H}_\Delta(r)$ on $\Omega^{\otimes r}$ extends to a right action of $\boldsymbol{\mathcal{H}}_\Delta(r) = \mathcal{H}_\Delta(r) \otimes_{\mathbb{Z}} \mathbb{Q}(v)$ on $\boldsymbol{\Omega}^{\otimes r}$. Hence, we have the $\mathbb{Q}(v)$-algebra isomorphism

$$\boldsymbol{\mathcal{S}}_\Delta(n,r) := \mathcal{S}_\Delta(n,r) \otimes_{\mathbb{Z}} \mathbb{Q}(v) \cong \mathrm{End}_{\boldsymbol{\mathcal{H}}_\Delta(r)}(\boldsymbol{\Omega}^{\otimes r}).$$

We will make the identifications $\mathcal{S}_\Delta(n,r) = \mathrm{End}_{\mathcal{H}_\Delta(r)}(\Omega^{\otimes r})$ and $\boldsymbol{\mathcal{S}}_\Delta(n,r) = \mathrm{End}_{\boldsymbol{\mathcal{H}}_\Delta(r)}(\boldsymbol{\Omega}^{\otimes r})$.

3.4. BLM bases and multiplication formulas

We now follow [4] (cf. [73]) to define BLM bases for the affine quantum Schur algebra $\mathcal{S}_\Delta(n,r)$ as discussed in [24]. Let

$$\Theta_\Delta^{\pm}(n) = \{A \in \Theta_\Delta(n) \mid a_{i,i} = 0 \text{ for all } i\} \qquad (3.4.0.1)$$

be the set of 0-diagonal matrices in $\Theta_\Delta(n)$. For $A \in \Theta_\Delta^{\pm}(n)$ and $\mathbf{j} \in \mathbb{Z}_\Delta^n$, define $A(\mathbf{j},r) \in \mathcal{S}_\Delta(n,r)$ by

$$A(\mathbf{j},r) = \begin{cases} \sum_{\lambda\in\Lambda_\Delta(n,r-\sigma(A))} v^{\lambda\cdot\mathbf{j}}[A+\mathrm{diag}(\lambda)], & \text{if } \sigma(A) \leqslant r; \\ 0, & \text{otherwise,} \end{cases} \qquad (3.4.0.2)$$

where $\lambda \cdot \mathbf{j} = \sum_{1\leqslant i\leqslant n} \lambda_i j_i$. For the convenience of later use, we extend the definition to matrices in $M_{n,\Delta}(\mathbb{Z})$ by setting $A(\mathbf{j},r) = 0$ if some off-diagonal entries of A are negative.

The elements $A(\mathbf{j},r)$ are the affine version of the elements defined in [**4**, 5.2] and have been defined in terms of characteristic functions in [**73**, 7.6].

The following result is the affine analogue of [**25**, 6.6(2)]; see [**24**, Prop. 4.1].

Proposition 3.4.1. *For a fixed $1 \leqslant i_0 \leqslant n$, the set*

$$\mathcal{B}_{\Delta,i_0,r} := \{A(\mathbf{j}, r) \mid A \in \Theta_\Delta^\pm(n), \mathbf{j} \in \mathbb{N}_\Delta^n, j_{i_0} = 0, \sigma(\mathbf{j}) + \sigma(A) \leqslant r\}$$

forms a $\mathbb{Q}(v)$-basis for $\mathcal{S}_\Delta(n, r)$. In particular, the set

$$\mathcal{B}_{\Delta,r} := \{A(\mathbf{j}, r) \mid A \in \Theta_\Delta^\pm(n), \mathbf{j} \in \mathbb{N}_\Delta^n, \sigma(A) \leqslant r\}$$

is a spanning set for $\mathcal{S}_\Delta(n, r)$.

We call $\mathcal{B}_{\Delta,i_0,r}$ a *BLM basis* of $\mathcal{S}_\Delta(n, r)$ and call $\mathcal{B}_{\Delta,r}$ the BLM spanning set. As in the finite case, one would expect that there is a basis $\mathcal{B}_\Delta$ for quantum affine $\mathfrak{gl}_n$ such that its image in $\mathcal{S}_\Delta(n, r)$ is $\mathcal{B}_{\Delta,r} \cup \{0\}$ for every $r \geqslant 0$. See §5.4 for a conjecture.

The affine analogue of the multiplication formulas given in [**4**, 5.3] has also been established in [**24**]. As seen in [**24**, Th. 4.2], these formulas are crucial to a modified approach to the realization problem for quantum affine $\mathfrak{sl}_n$. We shall also see in Chapter 5 that they are useful in finding a presentation for the affine quantum Schur algebras $\mathcal{S}_\Delta(r, r)$. Let $\boldsymbol{\alpha}_i^\Delta = \boldsymbol{e}_i^\Delta - \boldsymbol{e}_{i+1}^\Delta, \boldsymbol{\beta}_i^\Delta = -\boldsymbol{e}_i^\Delta - \boldsymbol{e}_{i+1}^\Delta \in \mathbb{Z}_\Delta^n$.

Theorem 3.4.2. *Assume $1 \leqslant h \leqslant n$. For $i \in \mathbb{Z}$, $\mathbf{j}, \mathbf{j}' \in \mathbb{Z}_\Delta^n$, and $A \in \Theta_\Delta^\pm(n)$, if we put $f(i) = f(i, A) = \sum_{j \geqslant i} a_{h,j} - \sum_{j>i} a_{h+1,j}$ and $f'(i) = f'(i, A) = \sum_{j<i} a_{h,j} - \sum_{j \leqslant i} a_{h+1,j}$, then the following identities hold in $\mathcal{S}_\Delta(n, r)$ for all $r \geqslant 0$:*

$$\begin{aligned} 0(\mathbf{j}, r)A(\mathbf{j}', r) &= v^{\mathbf{j} \centerdot \mathrm{ro}(A)} A(\mathbf{j} + \mathbf{j}', r), \\ A(\mathbf{j}', r)0(\mathbf{j}, r) &= v^{\mathbf{j} \centerdot \mathrm{co}(A)} A(\mathbf{j} + \mathbf{j}', r), \end{aligned} \tag{3.4.2.1}$$

where 0 *stands for the zero matrix,*

$$\begin{aligned} E_{h,h+1}^\Delta(\mathbf{0}, r)A(\mathbf{j}, r) = &\sum_{i<h; a_{h+1,i} \geqslant 1} v^{f(i)} \overline{\left[\!\!\left[{a_{h,i}+1 \atop 1} \right]\!\!\right]} (A + E_{h,i}^\Delta - E_{h+1,i}^\Delta)(\mathbf{j} + \boldsymbol{\alpha}_h^\Delta, r) \\ &+ \sum_{i>h+1; a_{h+1,i} \geqslant 1} v^{f(i)} \overline{\left[\!\!\left[{a_{h,i}+1 \atop 1} \right]\!\!\right]} (A + E_{h,i}^\Delta - E_{h+1,i}^\Delta)(\mathbf{j}, r) \\ &+ v^{f(h)-j_h-1} \frac{(A - E_{h+1,h}^\Delta)(\mathbf{j} + \boldsymbol{\alpha}_h^\Delta, r) - (A - E_{h+1,h}^\Delta)(\mathbf{j} + \boldsymbol{\beta}_h^\Delta, r)}{1 - v^{-2}} \\ &+ v^{f(h+1)+j_{h+1}} \overline{\left[\!\!\left[{a_{h,h+1}+1 \atop 1} \right]\!\!\right]} (A + E_{h,h+1}^\Delta)(\mathbf{j}, r), \end{aligned} \tag{3.4.2.2}$$

and

$$E^\Delta_{h+1,h}(\mathbf{0},r)A(\mathbf{j},r) = \sum_{i<h;a_{h,i}\geqslant 1} v^{f'(i)}\overline{\left[\!\!\left[{a_{h+1,i}+1 \atop 1} \right]\!\!\right]}(A - E^\Delta_{h,i} + E^\Delta_{h+1,i})(\mathbf{j},r)$$
$$+ \sum_{i>h+1;a_{h,i}\geqslant 1} v^{f'(i)}\overline{\left[\!\!\left[{a_{h+1,i}+1 \atop 1} \right]\!\!\right]}(A - E^\Delta_{h,i} + E^\Delta_{h+1,i})(\mathbf{j}-\boldsymbol{\alpha}^\Delta_h,r)$$
$$+ v^{f'(h+1)-j_{h+1}-1}\frac{(A - E^\Delta_{h,h+1})(\mathbf{j}-\boldsymbol{\alpha}^\Delta_h,r) - (A - E^\Delta_{h,h+1})(\mathbf{j}+\boldsymbol{\beta}^\Delta_h,r)}{1-v^{-2}}$$
$$+ v^{f'(h)+j_h}\overline{\left[\!\!\left[{a_{h+1,h}+1 \atop 1} \right]\!\!\right]}(A + E^\Delta_{h+1,h})(\mathbf{j},r). \qquad (3.4.2.3)$$

According to Proposition 1.4.5, the double Ringel–Hall algebra $\mathfrak{D}_\Delta(n)$ has generators corresponding to simple modules and homogeneous indecomposable modules. It would be natural to raise the following question. In fact, we will see in §5.4 that the solution to this problem is the key to prove the realization conjecture 5.4.2; cf. Problem 6.4.2.

Problem 3.4.3. Find multiplication formulas for $E^\Delta_{h,h+sn}(\mathbf{0},r)A(\mathbf{j},r)$ for all $s \neq 0$ in $\mathbb{Z}$.

3.5. The $\mathfrak{D}_\Delta(n)$-$\mathcal{H}_\Delta(r)$-bimodule structure on tensor spaces

As in §3.3, let $\boldsymbol{\Omega}$ be the $\mathbb{Q}(v)$-vector space with basis $\{\omega_s \mid s \in \mathbb{Z}\}$. Our idea is to use the presentation of $\mathfrak{D}_\Delta(n)$ to define the *natural* representation $\boldsymbol{\Omega}$ of $\mathfrak{D}_\Delta(n)$. This induces a left $\mathfrak{D}_\Delta(n)$-module structure on the tensor space $\boldsymbol{\Omega}^{\otimes r}$ which commutes with the right action of the affine Hecke algebra $\mathcal{H}_\Delta(r)$ and, hence, an algebra homomorphism $\xi_r : \mathfrak{D}_\Delta(n) \to \boldsymbol{S}_\Delta(n,r)$. We then partially establish the affine analogue of the quantum Schur–Weyl duality in §3.8.

For $i \in I$ and $m \in \mathbb{Z}^+$, we define the actions of $K_i^{\pm 1}$, E_i, F_i, and $\mathsf{z}^\pm_m$ on $\boldsymbol{\Omega}$ by

$$\begin{aligned} E_i \cdot \omega_s &= \delta_{i+1,\bar{s}}\omega_{s-1}, & F_i \cdot \omega_s &= \delta_{i,\bar{s}}\omega_{s+1}, \\ K_i^{\pm 1} \cdot \omega_s &= v^{\pm\delta_{i,\bar{s}}}\omega_s, \quad \text{and} & \mathsf{z}^\pm_m \cdot \omega_s &= \omega_{s\mp mn}. \end{aligned} \qquad (3.5.0.1)$$

Lemma 3.5.1. *With the action defined as above, $\boldsymbol{\Omega}$ becomes a left $\mathfrak{D}_\Delta(n)$-module.*

Proof. For $i \in I$ and $m \in \mathbb{Z}^+$, we denote by $\kappa_i^{\pm 1}, \phi_i^+, \phi_i^-$, and $\psi^\pm_m$ the $\mathbb{Q}(v)$-linear transformations of $\boldsymbol{\Omega}$ induced by the actions of $K_i^{\pm 1}$, E_i, F_i, and $\mathsf{z}^\pm_m$ on $\boldsymbol{\Omega}$ defined above, respectively. The assertion follows from the fact that $\kappa_i^{\pm 1}$, ϕ_i^+,

ϕ_i^-, and $\psi_m^\pm$ satisfy the relations similar to (QGL1)–(QGL8) in Theorem 2.3.1. This is because those relations involving $\psi_m^\pm$ are clear, while the others (involving only $\kappa_i^{\pm 1}$, ϕ_i^+, and ϕ_i^-) follow from the natural representation of the extended quantum affine $\mathfrak{sl}_n$ on $\mathbf{\Omega}$ (see, e.g., [**24**, (9.5.1)]). □

Let $\xi_1 : \mathfrak{D}_\triangle(n) \to \mathrm{End}_{\mathbb{Q}(v)}(\mathbf{\Omega})$ be the algebra homomorphism associated with the $\mathfrak{D}_\triangle(n)$-module $\mathbf{\Omega}$.

Proposition 3.5.2. (1) *The $\mathfrak{D}_\triangle(n)$-module $\mathbf{\Omega}$ is indecomposable.*

(2) *For each non-zero element $a \in \mathbb{Q}(v)$, the subspace $\mathbf{V}_a$ spanned by $\omega_i - a\omega_{i+n}$, for all $i \in \mathbb{Z}$, is a submodule of $\mathbf{\Omega}$. Moreover, $\mathbf{\Omega} \cong \mathbf{V}_a$, and the quotient module $\mathbf{\Omega}(a) := \mathbf{\Omega}/\mathbf{V}_\mathbf{a}$ is simple.*

Proof. (1) It is easy to check that the map $\pi : \mathbf{\Omega} \to \mathbf{\Omega}$, $\omega_i \mapsto \omega_{i+n}$ is a $\mathfrak{D}_\triangle(n)$-module homomorphism and so is π^{-1}. Take an arbitrary $f \in \mathrm{End}_{\mathfrak{D}_\triangle(n)}(\mathbf{\Omega})$ and write $f(\omega_0) = \sum_{s\in\mathbb{Z}} b_s\omega_s$, where all but finitely many $b_s \in \mathbb{Q}(v)$ are zero. Then, for each $i \in I$,

$$E_i f(\omega_0) = E_i(\sum_{s\in\mathbb{Z}} b_s\omega_s) = \sum_{s\in\mathbb{Z}} \delta_{i+1,\bar{s}} b_s\omega_{s-1} = \sum_{t\in\mathbb{Z}} b_{i+1+tn}\omega_{i+tn}.$$

On the other hand, $E_i f(\omega_0) = f(E_i\omega_0) = 0$ if $i \neq \overline{n-1}$. Hence, $b_{i+1+tn} = 0$ unless $i = \overline{n-1}$ and $f(\omega_0) = \sum_{t\in\mathbb{Z}} b_{tn}\omega_{tn}$. Thus, for each $1 \leqslant s < n$,

$$f(\omega_s) = f(F_{s-1}\cdots F_1F_n\omega_0) = F_{s-1}\cdots F_1F_nf(\omega_0) = \sum_{t\in\mathbb{Z}} b_{tn}\omega_{tn+s}.$$

For any $k = s \pm mn \in \mathbb{Z}$ with $0 \leqslant s < n$ and $m > 0$,

$$f(\omega_k) = f(\mathsf{z}_m^\mp\omega_s) = \mathsf{z}_m^\mp f(\omega_s) = \sum_{t\in\mathbb{Z}} b_{tn}\omega_{tn+k}.$$

Hence, $f = \sum_{t\in\mathbb{Z}} b_{tn}\pi^t$. We conclude that $\mathrm{End}_{\mathfrak{D}_\triangle(n)}(\mathbf{\Omega}) = \mathbb{Q}(v)[\pi, \pi^{-1}]$ is a Laurent polynomial algebra over $\mathbb{Q}(v)$, which implies that $\mathbf{\Omega}$ is indecomposable.

(2) Clearly, $\mathbf{V}_a$ is a submodule of $\mathbf{\Omega}$, and the map $\mathbf{\Omega} \to \mathbf{V}_a$, $\omega_i \mapsto \omega_i - a\omega_{i+n}$ gives the required $\mathfrak{D}_\triangle(n)$-module isomorphism. Also, the quotient module $\mathbf{\Omega}(a) = \mathbf{\Omega}/\mathbf{V}_a$ has a basis $\{\bar{\omega}_i = \omega_i + \mathbf{V}_a \mid i \in I\}$ with the $\mathfrak{D}_\triangle(n)$-module structure given by

$$E_i \cdot \bar{\omega}_j = \delta_{i+1,j}\bar{\omega}_{j-1}, \qquad F_i \cdot \bar{\omega}_j = \delta_{i,j}\bar{\omega}_{j+1},$$
$$K_i^{\pm 1} \cdot \bar{\omega}_j = v^{\pm\delta_{i,j}}\bar{\omega}_j, \quad \text{and} \quad \mathsf{z}_m^\pm \cdot \bar{\omega}_j = a^{\pm m}\bar{\omega}_j,$$

for all $i, j \in I$ and $m \geqslant 1$. It is obvious that $\mathbf{\Omega}(a)$ is a simple $\mathfrak{D}_\triangle(n)$-module. □

There will be a different construction for $\boldsymbol{\Omega}(u)$ in §4.6 through representations of the Hecke algebra $\mathcal{H}_\Delta(1)$.

As shown in §2.1, there is a PBW type basis for $\mathfrak{D}_\Delta(n)$. The action of these basis elements on the basis $\{\omega_s \mid s \in \mathbb{Z}\}$ of $\boldsymbol{\Omega}$ can be described as follows.

Proposition 3.5.3. *For each* $0 \neq A \in \Theta_\Delta^+(n)$ *and* $s \in \mathbb{Z}$, *we have*

$$\widetilde{u}_A^+ \cdot \omega_s = \sum_{t<s} \delta_{A,E_{t,s}^\Delta}\omega_t \quad \text{and} \quad \widetilde{u}_A^- \cdot \omega_s = \sum_{s<t} \delta_{A,E_{s,t}^\Delta}\omega_t,$$

where, as in (1.2.0.6), $\widetilde{u}_A^\pm = v^{d'_A}u_A^\pm$ *with* $d'_A = \dim \mathrm{End}(M(A)) - \dim M(A)$. *In particular, if* $A \neq 0$ *and* $M(A)$ *is decomposable, then* $\widetilde{u}_A^+ \cdot \omega_s = 0 = \widetilde{u}_A^- \cdot \omega_s$.

Proof. Define a $\mathbb{Q}(v)$-linear map $\phi = \phi^+ : \mathfrak{D}_\Delta(n)^+ \to \mathrm{End}_{\mathbb{Q}(v)}(\boldsymbol{\Omega})$, $\widetilde{u}_A^+ \mapsto \phi_A$ by setting $\phi_0 = \mathrm{id}$ and, for $A \neq 0$,

$$\phi_A : \boldsymbol{\Omega} \longrightarrow \boldsymbol{\Omega}, \ \omega_s \longmapsto \sum_{t<s} \delta_{A,E_{t,s}^\Delta}\omega_t.$$

Define $\phi^- : \mathfrak{D}_\Delta(n)^- \to \mathrm{End}_{\mathbb{Q}(v)}(\boldsymbol{\Omega})$ similarly.

We only prove the first equality. The second one can be proved analogously. We first show that ϕ is an algebra homomorphism. For $i < j$ in $\mathbb{Z}$, set $M^{i,j} = M(E_{i,j}^\Delta) = S_i[j-i]$ as in §1.2 and write $j - i = an + b$ for $a \geqslant 0$ and $0 \leqslant b < n$. Then it is easy to check that

$$\dim \mathrm{End}(M^{i,j}) = \begin{cases} a, & \text{if } b = 0; \\ a+1, & \text{if } b \neq 0. \end{cases} \tag{3.5.3.1}$$

We claim that, for $l < s < t$,

$$\dim \mathrm{End}(M^{l,t}) = \dim \mathrm{End}(M^{l,s}) + \dim \mathrm{End}(M^{s,t}) + \langle \mathbf{dim}\, M^{l,s}, \mathbf{dim}\, M^{s,t} \rangle \tag{3.5.3.2}$$

(cf. proof of [**13**, Lem. 8.2]). Indeed, write

$$s - l = a_1 n + b_1 \quad \text{and} \quad t - s = a_2 n + b_2,$$

for $a_1, a_2 \geqslant 0$ and $0 \leqslant b_1, b_2 < n$. Then $t - l = (a_1 + a_2)n + b_1 + b_2$. If $b_1 = 0$ or $b_2 = 0$, the equality follows from (3.5.3.1) since $\langle \mathbf{dim}\, M^{l,s}, \mathbf{dim}\, M^{s,t} \rangle = 0$ by (1.2.0.5). Suppose now $b_1 \neq 0$ and $b_2 \neq 0$. Then

$$\dim \mathrm{End}(M^{l,s}) + \dim \mathrm{End}(M^{s,t}) = a_1 + a_2 + 2 \quad \text{and}$$

$$\dim \mathrm{End}_{k\Delta}(M^{l,t}) = \begin{cases} a_1 + a_2 + 1, & \text{if } b_1 + b_2 \leqslant n; \\ a_1 + a_2 + 2, & \text{if } b_1 + b_2 > n. \end{cases}$$

Let $\mathbf{d} = (d_i) = \mathbf{dim}\, M^{l,s}$. Then

$$\langle \mathbf{dim}\, M^{l,s}, \mathbf{dim}\, M^{s,t}\rangle = d_{\overline{t-1}} - d_{\overline{s-1}} = \begin{cases} -1, & \text{if } b_1 + b_2 \leqslant n; \\ 0, & \text{if } b_1 + b_2 > n. \end{cases}$$

Therefore, the equality (3.5.3.2) holds.

For $A, B \in \Theta_\Delta^+(n)$, we have

$$\widetilde{u}_A^+ \widetilde{u}_B^+ = v^{d'_A + d'_B} u_A^+ u_B^+ = v^{\langle \mathbf{dim}\, M(A), \mathbf{dim}\, M(B)\rangle + d'_A + d'_B} \sum_C v^{-d'_C} \varphi_{A,B}^C \widetilde{u}_C^+.$$

Thus, to prove that ϕ is an algebra homomorphism, it suffices to show that for $A, B \in \Theta_\Delta^!(n)$,

$$\phi_A \phi_B = v^{\langle \mathbf{dim}\, M(A), \mathbf{dim}\, M(B)\rangle + d'_A + d'_B} \sum_C v^{-d'_C} \varphi_{A,B}^C \phi_C.$$

By definition, we have, for $s \in \mathbb{Z}$,

$$\phi_A \phi_B(\omega_s) = 0 = v^{\langle \mathbf{dim}\, M(A), \mathbf{dim}\, M(B)\rangle + d'_A + d'_B} \sum_C v^{-d'_C} \varphi_{A,B}^C \phi_C(\omega_s)$$

unless $B = E_{s,t}^\Delta$ and $A = E_{l,s}^\Delta$ for some $l < s < t$. Now, suppose that $B = E_{s,t}^\Delta$ and $A = E_{l,s}^\Delta$ for some $l < s < t$. Then

$$\phi_A \phi_B(\omega_s) = \omega_l \quad \text{and}$$

$$\sum_C v^{-d'_C} \varphi_{A,B}^C \phi_C(\omega_s) = v^{-d'_D} \omega_l,$$

where $D = E_{l,t}^\Delta$ (and $\varphi_{A,B}^D = 1$). From (3.5.3.2) it follows that

$$\phi_A \phi_B(\omega_s) = \omega_l = v^{\langle \mathbf{dim}\, M(A), \mathbf{dim}\, M(B)\rangle + d'_A + d'_B} \sum_C v^{-d'_C} \varphi_{A,B}^C \phi_C(\omega_s).$$

Hence, ϕ is an algebra homomorphism.

To complete the proof, it remains to show that ϕ coincides with the restriction of ξ_1 to $\mathfrak{D}_\Delta(n)^+$. Since $\mathfrak{D}_\Delta(n)^+$ is generated by u_i^+ and z_m^+, for $i \in I$ and $m \in \mathbb{Z}^+$, we need to check that

$$\phi(u_i^+) = \xi_1(u_i^+) \quad \text{and} \quad \phi(\mathsf{z}_m^+) = \xi_1(\mathsf{z}_m^+).$$

The equality $\phi(u_i^+) = \xi_1(u_i^+)$ is trivial. For each $s \in \mathbb{Z}$, we have by definition

$$\xi_1(\mathsf{z}_m^+)(\omega_s) = \mathsf{z}_m^+ \cdot \omega_s = \omega_{s-mn}.$$

On the other hand, by (2.2.1.2),

$$\mathsf{z}_m^+ = \sum_{l=1}^{n} \widetilde{u}_{E_{l,l+mn}^\Delta}^+ + \frac{v^{nm}(v - v^{-1})}{[m]} y_m^+,$$

where y_m^+ is a linear combination of certain $\widetilde{u}_A^+$ with $M(A)$ decomposable. Hence, $\phi(y_m^+) = 0$. Thus,

$$\phi(\mathsf{z}_m^+)(\omega_s) = \sum_{l=1}^{n} \phi(\widetilde{u}^+_{E^\Delta_{l,l+mn}})(\omega_s) = \sum_{l=1}^{n}\sum_{b<s} \delta_{E^\Delta_{l,l+mn}, E^\Delta_{b,s}}\, \omega_b = \omega_{s-mn}.$$

(Note that $E^\Delta_{a,b} = E^\Delta_{a+n,b+n}$ for $a, b \in \mathbb{Z}$.) Hence, $\phi(\mathsf{z}_m^+) = \xi_1(\mathsf{z}_m^+)$. This completes the proof. □

Remark 3.5.4. The above proposition implies that the action of $\mathfrak{D}_\Delta(n)^-$ on $\mathbf{\Omega}$ induced from ϕ^- coincides with the action given in [73, Lem. 8.1] which is defined geometrically through the algebra homomorphism $\zeta_1^- : \mathfrak{D}_\Delta(n)^- \to \mathcal{S}_\Delta(n, 1)$; see (3.6.2.1) below.

Now, for each $r \geqslant 1$, the Hopf algebra structure of $\mathfrak{D}_\Delta(n)$ induces a left $\mathfrak{D}_\Delta(n)$-module structure on the tensor space $\mathbf{\Omega}^{\otimes r}$ which has a $\mathbb{Q}(v)$-basis $\{\omega_{\mathbf{i}} \mid \mathbf{i} \in \mathbb{Z}^r\}$. Since $\mathsf{z}_m^\pm$ are primitive elements, we have by (3.5.0.1) that, for each $m \geqslant 1$ and $\omega_{\mathbf{i}} = \omega_{i_1} \otimes \cdots \otimes \omega_{i_r} \in \mathbf{\Omega}^{\otimes r}$,

$$\mathsf{z}_m^\pm \cdot \omega_{\mathbf{i}} = \sum_{s=1}^{r} \omega_{i_1} \otimes \cdots \otimes \omega_{i_{s-1}} \otimes \omega_{i_s \mp mn} \otimes \omega_{i_{s+1}} \otimes \cdots \otimes \omega_{i_r}. \tag{3.5.4.1}$$

Recall from §3.3 that $\mathcal{H}_\Delta(r)$ has a right action on $\mathbf{\Omega}^{\otimes r}$. The following result says that the left action of $\mathfrak{D}_\Delta(n)$ and the right action of $\mathcal{H}_\Delta(r)$ on $\mathbf{\Omega}^{\otimes r}$ commute.

Proposition 3.5.5. *For each $r \geqslant 1$, the actions of $\mathfrak{D}_\Delta(n)$ and $\mathcal{H}_\Delta(r)$ on $\mathbf{\Omega}^{\otimes r}$ commute. In other words, the tensor space $\mathbf{\Omega}^{\otimes r}$ is a $\mathfrak{D}_\Delta(n)$-$\mathcal{H}_\Delta(r)$-bimodule.*

Proof. Since $\mathfrak{D}_\Delta(n)$ is generated by the set

$$\mathscr{D} := \{K_i^{\pm 1}, E_i, F_i, \mathsf{z}_m^\pm \mid i \in I, m \in \mathbb{Z}^+\}$$

and $\mathcal{H}_\Delta(r)$ is generated by the set

$$\mathscr{H} := \{T_k, X_s^{\pm 1} \mid 1 \leqslant k < r, 1 \leqslant s \leqslant r\},$$

it suffices to show that

$$u \cdot (\omega_{\mathbf{i}} \cdot h) = (u \cdot \omega_{\mathbf{i}}) \cdot h \tag{3.5.5.1}$$

for all $u \in \mathscr{D}$, $h \in \mathscr{H}$, and $\mathbf{i} \in \mathbb{Z}^r$. It is easy to check from the definition that (3.5.5.1) holds for $u = K_i^{\pm 1}$ and arbitrary $h \in \mathscr{H}$ (resp., for $h = X_s^{\pm 1}$ and arbitrary $u \in \mathscr{D}$).

Furthermore, by (3.5.4.1), for $m \geqslant 1$ and $\mathbf{i} \in \mathbb{Z}^r$,

$$\mathsf{z}_m^{\pm} \cdot \omega_{\mathbf{i}} = \omega_{\mathbf{i}} \cdot \Big(\sum_{1 \leqslant s \leqslant r} X_s^{\pm m} \Big). \tag{3.5.5.2}$$

This implies that, for any $h \in \mathscr{H}$,

$$(\mathsf{z}_m^{\pm} \cdot \omega_{\mathbf{i}}) \cdot h = \Big(\omega_{\mathbf{i}} \cdot \Big(\sum_{1 \leqslant s \leqslant r} X_s^{\pm m} \Big) \Big) \cdot h = (\omega_{\mathbf{i}} \cdot h) \cdot \Big(\sum_{1 \leqslant s \leqslant r} X_s^{\pm m} \Big) = \mathsf{z}_m^{\pm} \cdot (\omega_{\mathbf{i}} \cdot h)$$

since $\sum_{1 \leqslant s \leqslant r} X_s^{\pm m}$ are central elements in $\mathcal{H}_\Delta(r)$.

Consequently, it remains to prove that (3.5.5.1) holds for $\mathbf{i} \in \mathbb{Z}^r$, $u = E_i, F_i$ ($i \in I$), and $h = T_k$ ($1 \leqslant k < r$).

Clearly, the subalgebra of $\mathfrak{D}_\Delta(n)$ generated by $K_i^{\pm 1}$, E_j, and F_j ($i \in I$ and $j \in I \backslash \{n\}$) is isomorphic to the quantum enveloping algebra $\mathbf{U}_v(\mathfrak{gl}_n)$ of $\mathfrak{gl}_n$, and the subalgebra of $\mathcal{H}_\Delta(r)$ generated by T_k ($1 \leqslant k < r$) is isomorphic to the Hecke algebra $\mathcal{H}(r)$ of $\mathfrak{S}_r$. Thus, by applying the result on quantum Schur algebras (see, e.g., [**12**, Lem. 14.23]), we obtain that (3.5.5.1) holds for $u = E_i, F_i$ ($i \in I \backslash \{n\}$), $h = T_k$ ($1 \leqslant k < r$), and $\mathbf{i} \in I(n,r) = \{(i_1, \ldots, i_r) \in \mathbb{Z}^r \mid 1 \leqslant i_s \leqslant n, \ \forall s\}$.

Suppose now $\mathbf{i} = (i_1, \ldots, i_r) \in I(n,r)$. Then, by definition,

$$E_n \cdot \omega_{\mathbf{i}} = \sum_{1 \leqslant j \leqslant r} \delta_{i_j,1} v^{g(\mathbf{i},j)} \omega_{\mathbf{i} - \boldsymbol{e}_j},$$

where $g(\mathbf{i}, j) = |\{s \mid j < s \leqslant r, \ i_s = n\}| - |\{s \mid j < s \leqslant r, \ i_s = 1\}|$ and $\boldsymbol{e}_j = (\delta_{s,j})_{1 \leqslant s \leqslant r} \in \mathbb{Z}^r$. In the following we show case by case that $(E_n \cdot \omega_{\mathbf{i}}) \cdot T_k = E_n \cdot (\omega_{\mathbf{i}} \cdot T_k)$ for $1 \leqslant k < r$.

Case $i_k = i_{k+1}$. In this case,

$$\begin{aligned}(E_n \cdot \omega_{\mathbf{i}}) \cdot T_k = v^2 \sum_{\substack{1 \leqslant j \leqslant r \\ j \neq k, k+1}} \delta_{i_j,1} v^{g(\mathbf{i},j)} \omega_{\mathbf{i} - \boldsymbol{e}_j} + \delta_{i_k,1} \big(v^{g(\mathbf{i},k)} \omega_{\mathbf{i} - \boldsymbol{e}_k} \cdot T_k \\ + v^{g(\mathbf{i},k+1)} \omega_{\mathbf{i} - \boldsymbol{e}_{k+1}} \cdot T_k \big).\end{aligned} \tag{3.5.5.3}$$

If $i_k = i_{k+1} = 1$, then $g(\mathbf{i}, k) = g(\mathbf{i}, k+1) - 1$. Hence,

$$\begin{aligned}
&\delta_{i_k,1} \big(v^{g(\mathbf{i},k)} \omega_{\mathbf{i} - \boldsymbol{e}_k} \cdot T_k + v^{g(\mathbf{i},k+1)} \omega_{\mathbf{i} - \boldsymbol{e}_{k+1}} \cdot T_k \big) \\
&= \delta_{i_k,1} \big(v^{g(\mathbf{i},k)} \omega_{\mathbf{i} + (n-1)\boldsymbol{e}_k} \cdot (X_k T_k) + v^{g(\mathbf{i},k+1)} \omega_{\mathbf{i} + (n-1)\boldsymbol{e}_{k+1}} \cdot (X_{k+1} T_k) \big) \\
&= \delta_{i_k,1} \big(v^{g(\mathbf{i},k)+2} \omega_{\mathbf{i} + (n-1)\boldsymbol{e}_k} \cdot (T_k^{-1} X_{k+1}) \\
&\qquad + v^{g(\mathbf{i},k+1)} \omega_{\mathbf{i} + (n-1)\boldsymbol{e}_{k+1}} \cdot (T_k X_k + (v^2 - 1) X_{k+1}) \big) \\
&= \delta_{i_k,1} \big(v^{g(\mathbf{i},k)+1} \omega_{\mathbf{i} - \boldsymbol{e}_{k+1}} + v^{g(\mathbf{i},k+1)} (v \omega_{\mathbf{i} - \boldsymbol{e}_k} + (v^2 - 1) \omega_{\mathbf{i} - \boldsymbol{e}_{k+1}}) \big) \\
&= \delta_{i_k,1} v^2 \big(v^{g(\mathbf{i},k)} \omega_{\mathbf{i} - \boldsymbol{e}_k} + v^{g(\mathbf{i},k+1)} \omega_{\mathbf{i} - \boldsymbol{e}_{k+1}} \big).
\end{aligned}$$

Applying (3.5.5.3) gives that

$$(E_n \cdot \omega_{\mathbf{i}}) \cdot T_k = v^2 \sum_{1\leqslant j\leqslant r} \delta_{i_j,1} v^{g(\mathbf{i},j)} \omega_{\mathbf{i}-e_j} = v^2 E_n \cdot \omega_{\mathbf{i}} = E_n \cdot (\omega_{\mathbf{i}} \cdot T_k).$$

Case $i_k < i_{k+1}$. In particular, $i_{k+1} \neq 1$. By definition, we have

$$\begin{aligned}(E_n \cdot \omega_{\mathbf{i}}) \cdot T_k &= \sum_{1\leqslant j\leqslant r,\, i_j=1} v^{g(\mathbf{i},j)} \omega_{\mathbf{i}-e_j} \cdot T_k \\
&= \sum_{1\leqslant j\leqslant r,\, i_j=1} v^{g(\mathbf{i},j)} \omega_{\mathbf{i}+(n-1)e_j} \cdot (X_j T_k) \\
&= \sum_{\substack{1\leqslant j\leqslant r,\, i_j=1\\ j\neq k,k+1}} v^{g(\mathbf{i},j)} \omega_{\mathbf{i}+(n-1)e_j} \cdot (T_k X_j) + \delta_{i_k,1} v^{g(\mathbf{i},k)} \omega_{\mathbf{i}+(n-1)e_k} \cdot (v^2 T_k^{-1} X_{k+1}) \\
&= \sum_{\substack{1\leqslant j\leqslant r,\, i_j=1\\ j\neq k,k+1}} v^{g(\mathbf{i},j)+1} \omega_{\mathbf{i}s_k - e_j} + \delta_{i_k,1} v^{g(\mathbf{i},k)} v^{1-\delta_{i_{k+1},n}} \omega_{\mathbf{i}s_k - e_{k+1}}.\end{aligned}$$

Since $g(\mathbf{i}s_k, k+1) = g(\mathbf{i}, k) - \delta_{i_{k+1},n}$ and $g(\mathbf{i}s_k, j) = g(\mathbf{i}, j)$ for $j \neq k, k+1$, we get that

$$\begin{aligned}(E_n \cdot \omega_{\mathbf{i}}) \cdot T_k &= v \sum_{\substack{1\leqslant j\leqslant r,\, i_j=1\\ j\neq k,k+1}} v^{g(\mathbf{i}s_k,j)} \omega_{\mathbf{i}s_k - e_j} + v\delta_{i_k,1} v^{g(\mathbf{i}s_k,k+1)} \omega_{\mathbf{i}s_k - e_{k+1}} \\
&= vE_n \cdot \omega_{\mathbf{i}s_k} = E_n \cdot (\omega_{\mathbf{i}} \cdot T_k).\end{aligned}$$

Case $i_k > i_{k+1}$. In this case, $i_k \neq 1$. It follows from the definition that

$$\begin{aligned}(E_n \cdot \omega_{\mathbf{i}}) \cdot T_k &= \sum_{1\leqslant j\leqslant r,\, i_j=1} v^{g(\mathbf{i},j)} \omega_{\mathbf{i}+(n-1)e_j} \cdot (X_j T_k) \\
&= \sum_{\substack{1\leqslant j\leqslant r,\, i_j=1\\ j\neq k,k+1}} v^{g(\mathbf{i},j)} \omega_{\mathbf{i}+(n-1)e_j} \cdot (T_k X_j) \\
&\quad + \delta_{i_{k+1},1} v^{g(\mathbf{i},k+1)} \omega_{\mathbf{i}+(n-1)e_{k+1}} \cdot (T_k X_k + (v^2-1) X_{k+1}) \\
&= \sum_{\substack{1\leqslant j\leqslant r,\, i_j=1\\ j\neq k,k+1}} v^{g(\mathbf{i},j)} (v\omega_{\mathbf{i}s_k - e_j} + (v^2-1)\omega_{\mathbf{i}-e_j}) \\
&\quad + \delta_{i_{k+1},1} v^{g(\mathbf{i},k+1)} (v^{1+\delta_{i_k,n}} \omega_{\mathbf{i}s_k - e_k} + (v^2-1)\omega_{\mathbf{i}-e_{k+1}}) \\
&= v\Bigg(\sum_{\substack{1\leqslant j\leqslant r,\, i_j=1\\ j\neq k,k+1}} v^{g(\mathbf{i},j)} \omega_{\mathbf{i}s_k - e_j} + \delta_{i_{k+1},1} v^{g(\mathbf{i},k+1)+\delta_{i_k,n}} \omega_{\mathbf{i}s_k - e_k} \Bigg) \\
&\quad + (v^2-1) \sum_{1\leqslant j\leqslant r,\, i_j=1} v^{g(\mathbf{i},j)} \omega_{\mathbf{i}-e_j}.\end{aligned}$$

Since $g(\mathbf{i}s_k, k) = g(\mathbf{i}, k+1) - \delta_{i_k,n}$ and $g(\mathbf{i}s_k, j) = g(\mathbf{i}, j)$ for $j \neq k, k+1$, we have

$$\begin{aligned}(E_n \cdot \omega_{\mathbf{i}}) \cdot T_k &= v \sum_{\substack{1 \leqslant j \leqslant r \\ (\mathbf{i}s_k)_j = 1}} v^{g(\mathbf{i}s_k, j)} \omega_{\mathbf{i}s_k - \mathbf{e}_j} + (v^2-1) E_n \cdot \omega_{\mathbf{i}} \\ &= v E_n \cdot \omega_{\mathbf{i}s_k} + (v^2-1) E_n \cdot \omega_{\mathbf{i}} = E_n \cdot (\omega_{\mathbf{i}} \cdot T_k).\end{aligned}$$

Similarly, we have for $\mathbf{i} \in I(n,r)$ and $1 \leqslant k < r$,

$$(F_n \cdot \omega_{\mathbf{i}}) \cdot T_k = F_n \cdot (\omega_{\mathbf{i}} \cdot T_k).$$

We conclude that (3.5.5.1) holds for $\mathbf{i} \in I(n,r)$, $u = E_i, F_i$, and $h = T_k$, where $i \in I$ and $1 \leqslant k < r$.

In general, for an arbitrary $\mathbf{j} \in \mathbb{Z}^r$, write $\omega_{\mathbf{j}} = \omega_{\mathbf{i}} \cdot (X_1^{t_1} \cdots X_r^{t_r})$ with $\mathbf{i} \in I(n,r)$ for some $t_1, \ldots, t_r \in \mathbb{Z}$. Let $\mathcal{X}$ be the subalgebra of $\mathcal{H}_\vartriangle(r)$ generated by $X_1^{\pm 1}, \ldots, X_r^{\pm 1}$. Then, by (3.3.0.2), $X_1^{t_1} \cdots X_r^{t_r} T_k = T_k x + y$ for some $x, y \in \mathcal{X}$. By the above discussion, we infer that for $u = E_i$ or F_i $(i \in I)$,

$$\begin{aligned}(u \cdot \omega_{\mathbf{j}}) \cdot T_k &= \big((u \cdot \omega_{\mathbf{i}}) \cdot (X_1^{t_1} \cdots X_r^{t_r})\big) \cdot T_k = (u \cdot \omega_{\mathbf{i}}) \cdot (T_k x + y) \\ &= \big(u \cdot (\omega_{\mathbf{i}} \cdot T_k)\big) \cdot x + u \cdot (\omega_{\mathbf{i}} \cdot y) = u \cdot \big((\omega_{\mathbf{i}} \cdot T_k) \cdot x\big) + u \cdot (\omega_{\mathbf{i}} \cdot y) \\ &= u \cdot \big(\omega_{\mathbf{i}} \cdot (T_k x + y)\big) = u \cdot (\omega_{\mathbf{j}} \cdot T_k).\end{aligned}$$

The proof is completed. □

For each $r \geqslant 1$, the $\mathfrak{D}_\vartriangle(n)$-$\mathcal{H}_\vartriangle(r)$-bimodule structure on $\mathbf{\Omega}^{\otimes r}$ induces $\mathbb{Q}(v)$-algebra homomorphisms

$$\begin{aligned}\xi_r &: \mathfrak{D}_\vartriangle(n) \longrightarrow \mathrm{End}_{\mathcal{H}_\vartriangle(r)}(\mathbf{\Omega}^{\otimes r}) \quad \text{and} \\ \xi_r^\vee &: \mathcal{H}_\vartriangle(r) \longrightarrow \mathrm{End}_{\mathfrak{D}_\vartriangle(n)}(\mathbf{\Omega}^{\otimes r})^{\mathrm{op}}.\end{aligned} \tag{3.5.5.4}$$

Remark 3.5.6. By Remark 3.5.4, the PBW type basis action for the Ringel–Hall algebra $\mathfrak{D}_\vartriangle(n)^-$ described in Proposition 3.5.3 induces via comultiplication an action on the tensor space $\mathbf{\Omega}^{\otimes r}$. Varagnolo and Vasserot [**73**, Lem. 8.2] have outlined a proof of the fact that this Ringel–Hall algebra action commutes with the action of $\mathcal{H}_\vartriangle(r)$. Proposition 3.5.3 and Remark 3.5.4 show that this Ringel–Hall algebra action coincides with the restriction of the double Ringel–Hall algebra action above, while the proof in Proposition 3.5.5 is a complete and more natural proof via the natural representation of $\mathfrak{D}_\vartriangle(n)$. See §3.6 for an explicit description of the map ξ_r and a further comparison with the work of Varagnolo and Vasserot.

We now describe the action of semisimple generators of $\mathfrak{D}_\vartriangle(n)$ on $\mathbf{\Omega}^{\otimes r}$. Recall that, for each $\mathbf{a} = (a_i) \in \mathbb{N}^n = \mathbb{N}_\vartriangle^n$,

$$S_{\mathbf{a}} = \oplus_{i \in I} a_i S_i, \quad u_{\mathbf{a}}^{\pm} = u_{[S_{\mathbf{a}}]}^{\pm}, \quad \text{and} \quad \widetilde{u}_{\mathbf{a}}^{\pm} = v^{\sum_{1 \leqslant i \leqslant n} a_i(a_i - 1)} u_{[S_{\mathbf{a}}]}^{\pm}.$$

The following result is a direct consequence of the definition of the comultiplications in Proposition 1.5.4(b) and Corollary 1.5.6(b′). See [**73**, 8.3] for the second formula.

Proposition 3.5.7. *For* $\mathbf{a} \in \mathbb{N}^n$, *we have*

$$\Delta^{(r-1)}(\widetilde{u}_{\mathbf{a}}^+) = \sum_{\mathbf{a}=\mathbf{a}^{(1)}+\cdots+\mathbf{a}^{(r)}} v^{\sum_{s>t}\langle \mathbf{a}^{(s)},\mathbf{a}^{(t)}\rangle} \widetilde{u}_{\mathbf{a}^{(1)}}^+ \otimes \widetilde{u}_{\mathbf{a}^{(2)}}^+ \widetilde{K}_{\mathbf{a}^{(1)}} \otimes \cdots \otimes \widetilde{u}_{\mathbf{a}^{(r)}}^+ \widetilde{K}_{\mathbf{a}^{(1)}+\cdots+\mathbf{a}^{(r-1)}}$$

and

$$\Delta^{(r-1)}(\widetilde{u}_{\mathbf{a}}^-) = \sum_{\mathbf{a}=\mathbf{a}^{(1)}+\cdots+\mathbf{a}^{(r)}} v^{\sum_{s>t}\langle \mathbf{a}^{(s)},\mathbf{a}^{(t)}\rangle} \widetilde{u}_{\mathbf{a}^{(1)}}^- \widetilde{K}_{-(\mathbf{a}^{(2)}+\cdots+\mathbf{a}^{(r)})} \otimes \cdots \otimes \widetilde{u}_{\mathbf{a}^{(r-1)}}^- \widetilde{K}_{-\mathbf{a}^{(r)}} \otimes \widetilde{u}_{\mathbf{a}^{(r)}}^- .$$

This proposition together with Proposition 3.5.3 gives the following corollary.

Corollary 3.5.8. *For* $\mathbf{a} = (a_i) \in \mathbb{N}_\Delta^n$ *and* $\mathbf{i} = (i_1, \ldots, i_r) \in \mathbb{Z}^r$, *we have*

$$\widetilde{u}_{\mathbf{a}}^+ \cdot \omega_{\mathbf{i}} = \sum_{\substack{m_i\in\{0,1\},\,\forall i\\ \mathbf{a}=m_1 e^\Delta_{i_1-1}+\cdots+m_r e^\Delta_{i_r-1}}} v^{\sum_{s>t} m_t(m_s-1)\langle e^\Delta_{i_s}, e^\Delta_{i_t}\rangle} \omega_{i_1-m_1} \otimes \cdots \otimes \omega_{i_r-m_r} \quad \text{and}$$

$$\widetilde{u}_{\mathbf{a}}^- \cdot \omega_{\mathbf{i}} = \sum_{\substack{m_i\in\{0,1\},\,\forall i\\ \mathbf{a}=m_1 e^\Delta_{i_1}+\cdots+m_r e^\Delta_{i_r}}} v^{\sum_{s>t} m_s(m_t-1)\langle e^\Delta_{i_s}, e^\Delta_{i_t}\rangle} \omega_{i_1+m_1} \otimes \cdots \otimes \omega_{i_r+m_r} .$$

In particular, if $\sigma(\mathbf{a}) = \sum_{i\in I} a_i > r$, *then* $\widetilde{u}_{\mathbf{a}}^+ \cdot \omega_{\mathbf{i}} = 0 = \widetilde{u}_{\mathbf{a}}^- \cdot \omega_{\mathbf{i}}$.

Corollary 3.5.9. *If* $n > r$, *then* $\xi_r(\mathfrak{D}_\Delta(n)) = \xi_r(\mathbf{U}_\Delta(n))$.

Proof. By Proposition 1.4.3, $\mathfrak{D}_\Delta(n)$ is generated by $u_i^+, u_i^-, K_i^{\pm 1}, u_{s\delta}^+, u_{s\delta}^-$ ($i \in I, s \in \mathbb{Z}^+$). If $n > r$, then $\dim S_{s\delta} = sn > r$. By Corollary 3.5.8,

$$u_{s\delta}^+ \cdot \omega_{\mathbf{i}} = 0 = u_{s\delta}^- \cdot \omega_{\mathbf{i}}, \quad \text{for all } \omega_{\mathbf{i}} \in \mathbf{\Omega}^{\otimes r}.$$

Hence, $\xi_r(u_{s\delta}^+) = 0 = \xi_r(u_{s\delta}^-)$ for all $s \geqslant 1$, and $\xi_r(\mathfrak{D}_\Delta(n))$ is generated by the images of $u_i^+, u_i^-, K_i^{\pm 1}$. This proves the equality. □

Remark 3.5.10. We remark that, if $z \in \mathbb{C}$ is not a root of unity and $\mathfrak{D}_{\Delta\mathbb{C}}(n)$ is the specialized double Ringel–Hall algebra considered in Proposition 2.4.5, then there is a $\mathfrak{D}_{\Delta\mathbb{C}}(n)$-$\mathcal{H}_\Delta(r)_\mathbb{C}$-bimodule action on the complex space $\Omega_\mathbb{C}^{\otimes r}$ which gives rise to algebra homomorphisms

$$\xi_{r,\mathbb{C}} : \mathfrak{D}_{\Delta\mathbb{C}}(n) \longrightarrow \mathcal{S}_\Delta(n,r)_\mathbb{C} \quad \text{and} \quad \xi_{r,\mathbb{C}}^\vee : \mathcal{H}_\Delta(r)_\mathbb{C} \longrightarrow \mathrm{End}_{\mathfrak{D}_{\Delta,\mathbb{C}}(n)}(\Omega_\mathbb{C}^{\otimes r})^{\mathrm{op}}. \tag{3.5.10.1}$$

3.6. A comparison with the Varagnolo–Vasserot action

In [**73**], Varagnolo–Vasserot defined a Hall algebra action on the tensor space via the action on $\mathbf{\Omega}$ (Remark 3.5.4) and the comultiplication Δ, and proved in [**73**, 8.3] that this action agrees with the affine quantum Schur algebra action via the algebra homomorphism ζ_r^- described in Proposition 3.6.1(1) below. We will define the algebra homomorphism ζ_r^+ in 3.6.1(2) opposite to ζ_r^- and show that ξ_r in (3.5.5.4) is their extension. In other words, ξ_r coincides with ζ_r^+ and ζ_r^- upon restriction (Theorem 3.6.3).

We first follow [**73**] to define an algebra homomorphism from the Ringel–Hall algebra $\mathfrak{H}_\Delta(n)$ of the cyclic quiver $\Delta(n)$ to the affine quantum Schur algebra $\mathcal{S}_\Delta(n,r)$. This definition relies on an important relation between cyclic flags and representations of cyclic quivers.

Recall from (3.1.1.3) the geometric characterization of affine quantum Schur algebras and the flag varieties $\mathscr{Y}=\mathscr{F}_\Delta(q)$, $\mathscr{X}=\mathscr{B}_\Delta(q)$, and $G=G(q)$ over the finite field $\mathbb{F}=\mathbb{F}_q$. Let $\mathbf{L}=(L_i)_{i\in\mathbb{Z}}, \mathbf{L}'=(L_i')_{i\in\mathbb{Z}}\in\mathscr{Y}$ satisfy $\mathbf{L}'\subseteq\mathbf{L}$ i.e., $L_i'\subseteq L_i$ for all $i\in\mathbb{Z}$. By [**32**, §9] and [**56**, 5.1], we can view $\mathbf{L}/\mathbf{L}'$ as a nilpotent representation $V=(V_i,f_i)$ of $\Delta(n)$ over $\mathbb{F}$ such that $V_i=L_i/L_i'$ and f_i is induced by the inclusion $L_i\subseteq L_{i+1}$ for each $i\in I$. Here we identify L_{n+1}/L_{n+1}' with L_1/L_1' via the multiplication by ε. Further, for $\mathbf{L}\subseteq\mathbf{L}'$, we define the integers

$$a(\mathbf{L},\mathbf{L}')=\sum_{1\leqslant i\leqslant n}\dim_{\mathbb{F}}(L_i'/L_i)(\dim_{\mathbb{F}}(L_{i+1}/L_i)-\dim_{\mathbb{F}}(L_i'/L_i))\quad\text{and}$$

$$c(\mathbf{L}',\mathbf{L})=\sum_{1\leqslant i\leqslant n}\dim_{\mathbb{F}}(L_{i+1}'/L_{i+1})(\dim_{\mathbb{F}}(L_{i+1}/L_i)-\dim_{\mathbb{F}}(L_i'/L_i)).\tag{3.6.0.1}$$

Recall also from (1.2.0.9) the geometric characterization $\mathtt{H}$ of the Ringel–Hall algebra $\mathfrak{H}_\Delta(n)$. Thus, by specializing v to $q^{\frac12}$, there is a $\mathbb{C}$-algebra isomorphism

$$\mathfrak{H}_\Delta(n)\otimes_{\mathcal{Z}}\mathbb{C}\longrightarrow\mathtt{H},\ \widetilde{u}_A\longmapsto\langle\mathfrak{O}_A\rangle=q^{-\frac12\dim\mathfrak{O}_A}\chi_{\mathfrak{O}_A},$$

where $\mathfrak{O}_A$ is the G_V-orbit in the representation variety E_V corresponding to the isoclass of $M(A)$.

By identifying $\mathcal{S}_\Delta(n,r)\otimes_{\mathcal{Z}}\mathbb{C}$ with $\mathbb{C}_G(\mathscr{Y}\times\mathscr{Y})$, $\mathfrak{H}_\Delta(n)\otimes_{\mathcal{Z}}\mathbb{C}$ with $\mathtt{H}$ (and hence, $\mathfrak{H}_\Delta(n)^{\mathrm{op}}\otimes_{\mathcal{Z}}\mathbb{C}$ with $\mathtt{H}^{\mathrm{op}}$), and recalling the elements $A(\mathbf{j},r)\in\mathcal{S}_\Delta(n,r)$, for each $A\in\Theta_\Delta^\pm(n)$ and $\mathbf{j}\in\mathbb{Z}_\Delta^n$, defined in (3.4.0.2), [**73**, Prop. 7.6] can now be stated as the first part of the following (cf. footnote 2 of Chapter 1).

Proposition 3.6.1. (1) *There is a $\mathcal{Z}$-algebra homomorphism*

$$\zeta_r^-:\mathfrak{H}_\Delta(n)^{\mathrm{op}}\longrightarrow\mathcal{S}_\Delta(n,r),\ \ \widetilde{u}_A\longmapsto{}^tA(\mathbf{0},r),\quad\textit{for all } A\in\Theta_\Delta^+(n)$$

such that the induced map $\zeta_r^- \otimes \mathrm{id}_{\mathbb{C}} : \mathrm{H}^{\mathrm{op}} \longrightarrow \mathbb{C}_G(\mathscr{Y} \times \mathscr{Y})$ *is given by*

$$(\zeta_r^- \otimes \mathrm{id}_{\mathbb{C}})(f)(\mathbf{L}, \mathbf{L}') = \begin{cases} q^{-\frac{1}{2}a(\mathbf{L},\mathbf{L}')} f(\mathbf{L}'/\mathbf{L}), & \textit{if } \mathbf{L} \subseteq \mathbf{L}'; \\ 0, & \textit{otherwise.} \end{cases} \tag{3.6.1.1}$$

(2) *Dually, there is a* $\mathcal{Z}$*-algebra homomorphism*

$$\zeta_r^+ : \mathfrak{H}_\Delta(n) \longrightarrow \mathcal{S}_\Delta(n,r), \quad \widetilde{u}_A \longmapsto A(\mathbf{0}, r), \quad \textit{for all } A \in \Theta_\Delta^+(n)$$

with the induced map $\zeta_r^+ \otimes \mathrm{id}_{\mathbb{C}} : \mathrm{H} \longrightarrow \mathbb{C}_G(\mathscr{Y} \times \mathscr{Y})$ *given by*

$$(\zeta_r^+ \otimes \mathrm{id}_{\mathbb{C}})(f)(\mathbf{L}, \mathbf{L}') = \begin{cases} q^{-\frac{1}{2}c(\mathbf{L},\mathbf{L}')} f(\mathbf{L}/\mathbf{L}'), & \textit{if } \mathbf{L}' \subseteq \mathbf{L}; \\ 0, & \textit{otherwise.} \end{cases} \tag{3.6.1.2}$$

Proof. Statement (1) is given in [**73**, 7.6]. We only need to prove (2). We first observe from the proof of [**56**, Lem. 1.11] that

$$d_A - d_{{}^tA} = \frac{1}{2} \sum_{1 \leqslant i \leqslant n} \Big(\sum_{j \in \mathbb{Z}} a_{i,j} \Big)^2 - \frac{1}{2} \sum_{1 \leqslant j \leqslant n} \Big(\sum_{i \in \mathbb{Z}} a_{i,j} \Big)^2.$$

Let ζ_r^+ be the composition of the algebra homomorphisms

$$\mathfrak{H}_\Delta(n) \xrightarrow{(\zeta_r^-)^{\mathrm{op}}} \mathcal{S}_\Delta(n,r)^{\mathrm{op}} \xrightarrow{\tau_r} \mathcal{S}_\Delta(n,r),$$

where τ_r is the anti-involution on $\mathcal{S}_\Delta(n,r)$ given in (3.1.3.4). Thus, τ_r induces a map $\tau_r \otimes \mathrm{id} : \mathbb{C}_G(\mathscr{Y} \times \mathscr{Y}) \to \mathbb{C}_G(\mathscr{Y} \times \mathscr{Y})$. Applying this to the characteristic function χ_A of the orbit $\mathcal{O}_A$ in $\mathscr{Y} \times \mathscr{Y}$, noting (3.1.0.1), yields

$$(\tau_r \otimes \mathrm{id})(\chi_A)(\mathbf{L}, \mathbf{L}') = q^{\frac{1}{2}(d_A - d_{{}^tA})} \chi_{{}^tA}(\mathbf{L}, \mathbf{L}') = q^{\frac{1}{2}(d_A - d_{{}^tA})} \chi_A(\mathbf{L}', \mathbf{L})$$

which is non-zero if and only if $(\mathbf{L}', \mathbf{L}) \in \mathcal{O}_A$. This implies $a_{i,j} = \dim_{\mathbb{F}} \frac{L_i' \cap L_j}{L_{i-1}' \cap L_j + L_i' \cap L_{j-1}}$. Hence, by the proof of [**56**, 1.5(a)], $\dim_{\mathbb{F}}(L_i'/L_{i-1}') = \sum_{j \in \mathbb{Z}} a_{i,j}$ and $\dim_{\mathbb{F}}(L_j/L_{j-1}) = \sum_{i \in \mathbb{Z}} a_{i,j}$. Putting

$$b(\mathbf{L}, \mathbf{L}') = \frac{1}{2}\Big(\sum_{1 \leqslant i \leqslant n} ((\dim_{\mathbb{F}}(L_i'/L_{i-1}'))^2 - \sum_{1 \leqslant i \leqslant n} (\dim_{\mathbb{F}}(L_i/L_{i-1}))^2 \Big),$$

we obtain $(\tau_r \otimes \mathrm{id})(\chi_A)(\mathbf{L}, \mathbf{L}') = q^{\frac{1}{2}b(\mathbf{L},\mathbf{L}')} \chi_A(\mathbf{L}', \mathbf{L})$. Hence,

$$(\tau_r \otimes \mathrm{id})(g)(\mathbf{L}, \mathbf{L}') = q^{\frac{1}{2}b(\mathbf{L},\mathbf{L}')} g(\mathbf{L}', \mathbf{L}), \quad \text{for } g \in \mathbb{C}_G(\mathscr{Y} \times \mathscr{Y}).$$

Taking $g = (\zeta_r^- \otimes \mathrm{id})(f)$ with $f \in \mathrm{H}$, and applying (3.6.1.1) gives

$$(\tau_r \otimes \mathrm{id})((\zeta_r^-)^{\mathrm{op}} \otimes \mathrm{id})(f)(\mathbf{L}, \mathbf{L}') = \begin{cases} q^{\frac{1}{2}(b(\mathbf{L},\mathbf{L}') - a(\mathbf{L}',\mathbf{L}))} f(\mathbf{L}/\mathbf{L}'), & \text{if } \mathbf{L}' \subseteq \mathbf{L}; \\ 0, & \text{otherwise.} \end{cases} \tag{3.6.1.3}$$

For $\mathbf{L}' \subseteq \mathbf{L}$, we have

$$\begin{aligned}
&\sum_{1\leqslant i\leqslant n} \dim_{\mathbb{F}}(L_{i+1}/L_i)^2 \\
&= \sum_{1\leqslant i\leqslant n} (\dim_{\mathbb{F}}(L_{i+1}/L'_{i+1}) + \dim_{\mathbb{F}}(L'_{i+1}/L'_i) - \dim_{\mathbb{F}}(L_i/L'_i))^2 \\
&= 2\sum_{1\leqslant i\leqslant n} (\dim_{\mathbb{F}}(L_i/L'_i))^2 + \sum_{1\leqslant i\leqslant n} (\dim_{\mathbb{F}}(L'_{i+1}/L'_i))^2 \\
&\quad + 2\sum_{1\leqslant i\leqslant n} \dim_{\mathbb{F}}(L_{i+1}/L'_{i+1}) \dim_{\mathbb{F}}(L'_{i+1}/L'_i) \\
&\quad - 2\sum_{1\leqslant i\leqslant n} \dim_{\mathbb{F}}(L_{i+1}/L'_{i+1}) \dim_{\mathbb{F}}(L_i/L'_i) \\
&\quad - 2\sum_{1\leqslant i\leqslant n} \dim_{\mathbb{F}}(L'_{i+1}/L'_i) \dim_{\mathbb{F}}(L_i/L'_i).
\end{aligned}$$

Hence, $b(\mathbf{L}, \mathbf{L}') - a(\mathbf{L}', \mathbf{L}) = -c(\mathbf{L}, \mathbf{L}')$ and (3.6.1.2) now follows from (3.6.1.3). □

For notational simplicity, we write $A(\mathbf{j}, r)$ for $A(\mathbf{j}, r) \otimes 1$ in $\mathcal{S}_\vartriangle(n, r) \otimes_{\mathcal{Z}} \mathbb{C}$. By taking f to be $q^{-\frac{1}{2}\dim \mathcal{O}_A}\chi_{\mathfrak{O}_A}$ in (3.6.1.1) and (3.6.1.2), we obtain the following.

Corollary 3.6.2. *For $A \in \Theta^+_\vartriangle(n)$ and $(\mathbf{L}, \mathbf{L}') \in \mathscr{Y} \times \mathscr{Y}$, we have*

$$A(\mathbf{0}, r)(\mathbf{L}, \mathbf{L}') = \begin{cases} q^{-\frac{1}{2}(c(\mathbf{L},\mathbf{L}')+\dim \mathfrak{O}_A)}, & \text{if } \mathbf{L}' \subseteq \mathbf{L} \text{ and } \mathbf{L}/\mathbf{L}' \in \mathfrak{O}_A; \\ 0, & \text{otherwise} \end{cases}$$

and

$${}^{t}A(\mathbf{0}, r)(\mathbf{L}, \mathbf{L}') = \begin{cases} q^{-\frac{1}{2}(a(\mathbf{L},\mathbf{L}')+\dim \mathfrak{O}_A)}, & \text{if } \mathbf{L} \subseteq \mathbf{L}' \text{ and } \mathbf{L}'/\mathbf{L} \in \mathfrak{O}_A; \\ 0, & \text{otherwise.} \end{cases}$$

We now identify $\mathfrak{H}_\vartriangle(n)$ as $\mathfrak{D}_\vartriangle(n)^+$ and $\mathfrak{H}_\vartriangle(n)^{\rm op}$ as $\mathfrak{D}_\vartriangle(n)^-$ via (2.1.3.3). Then $\zeta_r^\pm$ induce $\mathbb{Q}(v)$-algebra homomorphisms

$$\zeta_r^\pm : \mathfrak{D}_\vartriangle(n)^\pm \longrightarrow \boldsymbol{\mathcal{S}}_\vartriangle(n, r). \tag{3.6.2.1}$$

Theorem 3.6.3. *For every $r \geqslant 0$, the map $\xi_r : \mathfrak{D}_\vartriangle(n) \to \boldsymbol{\mathcal{S}}_\vartriangle(n, r)$ defined in (3.5.5.4) is the (unique) algebra homomorphism satisfying*

$$\xi_r(K_1^{j_1} \cdots K_n^{j_n}) = 0(\mathbf{j}, r),\ \xi_r(\widetilde{u}_A^+) = \zeta_r^+(\widetilde{u}_A^+) = A(\mathbf{0}, r),\ \textit{and}$$
$$\xi_r(\widetilde{u}_A^-) = \zeta_r^-(\widetilde{u}_A^-) = {}^{t}A(\mathbf{0}, r),$$

for all $\mathbf{j} = (j_1, \ldots, j_n) \in \mathbb{Z}^n$ *and* $A \in \Theta_\triangle^+(n)$. *In particular, we have* $\xi_r|_{\mathfrak{D}_\triangle(n)^\pm} = \zeta_r^\pm$.

Proof. Since $\mathfrak{D}_\triangle(n)$ is generated by $K_i^{\pm 1}$, $1 \leqslant i \leqslant n$, together with semisimple generators $u_{A_\lambda}^\pm$, $\lambda \in \mathbb{N}_\triangle^n$, where $A_\lambda = \sum_{i=1}^n \lambda_i E_{i,i+1}^\triangle$. By Proposition 3.6.1, it suffices to prove

(1) $\xi_r(K_i) = 0(\boldsymbol{e}_i^\triangle, r)$; (2) $\xi_r(\widetilde{u}_{A_\lambda}^-) = ({}^tA_\lambda)(\mathbf{0}, r)$; and (3) $\xi_r(\widetilde{u}_{A_\lambda}^+) = A_\lambda(\mathbf{0}, r)$.

To prove them, we apply Propositions 3.3.1 and 3.5.5 to compare the actions of both sides on $\omega_{\mathbf{i}} = [A^{\mathbf{i}}]$, for all $\mathbf{i} \in I_\triangle(n, r)_0$.

Suppose $\mathbf{i} = \mathbf{i}_\mu^\triangle$. By (3.5.0.1), $K_i \cdot \omega_{\mathbf{i}} = v^{\mu_i} \omega_{\mathbf{i}}$. Since $\mathrm{ro}(A^{\mathbf{i}}) = \mu$, (3.1.3.4) implies $0(\boldsymbol{e}_i^\triangle, r)[A^{\mathbf{i}}] = v^{\mu_i}[A^{\mathbf{i}}]$. Hence, $0(\boldsymbol{e}_i^\triangle, r) \cdot \omega_{\mathbf{i}} = K_i \cdot \omega_{\mathbf{i}}$, proving (1). The proof of (2) is given in [**73**, 8.3]. We now prove (3) for completeness. So we need to show that

$$A_\lambda(\mathbf{0}, r) \cdot \omega_{\mathbf{i}} = \widetilde{u}_\lambda^+ \cdot \omega_{\mathbf{i}}, \quad \text{for all } \mathbf{i} \in I_\triangle(n, r)_0 \text{ and } \lambda \in \mathbb{N}_\triangle^n. \tag{3.6.3.1}$$

The equality is trivial if $\sigma(\lambda) > r$ as both sides are 0. We now assume $\sigma(\lambda) \leqslant r$ and prove the equality by writing both sides as a linear combination of the basis $\{[A^{\mathbf{i}}]\}_{\mathbf{i} \in I_\triangle(n,r)}$; cf. Remark 3.3.2.

By Proposition 3.3.1, the left-hand side of (3.6.3.1) becomes $A_\lambda(\mathbf{0}, r) \cdot \omega_{\mathbf{i}} = A_\lambda(\mathbf{0}, r)[A^{\mathbf{i}}]$. We now compute this by regarding $A_\lambda(\mathbf{0}, r)[A^{\mathbf{i}}]$ as the convolution product $q^{-\frac{1}{2}d_{\mathbf{i}}} A_\lambda(\mathbf{0}, r) * \chi_{\mathbf{i}}$; see Remark 3.1.2 and compare [**73**, 8.3]. By the definition (3.1.1.2), for $\mathbf{j} \in I_\triangle(n, r)$ and $\lambda \in \mathbb{N}_\triangle^n$,

$$\begin{aligned}(A_\lambda(\mathbf{0}, r) * \chi_{\mathbf{i}})(\mathbf{L_j}, \mathbf{L}_\emptyset) &= \sum_{\mathbf{L} \in \mathscr{Y}} (A_\lambda(\mathbf{0}, r))(\mathbf{L_j}, \mathbf{L}) \chi_{\mathbf{i}}(\mathbf{L}, \mathbf{L}_\emptyset) \\ &= \sum_{\substack{\mathbf{L} \in \mathscr{Y} \\ (\mathbf{L}, \mathbf{L}_\emptyset) \in \mathcal{O}_{\mathbf{i}}}} (A_\lambda(\mathbf{0}, r))(\mathbf{L_j}, \mathbf{L}),\end{aligned}$$

where $\mathbf{L_j}$ is defined in (3.1.3.7), and $\mathcal{O}_{\mathbf{i}} = \mathcal{O}_{A^{\mathbf{i}}}$ is the orbit containing $(\mathbf{L_i}, \mathbf{L}_\emptyset)$ (see also (3.1.3.7) for the definition of $\mathbf{L}_\emptyset$). For $\mathbf{L} = (L_i) \in \mathscr{Y}$ and $(\mathbf{L}, \mathbf{L}_\emptyset) \in \mathcal{O}_{\mathbf{i}}$, there is $g \in G$ such that $(\mathbf{L}, \mathbf{L}_\emptyset) = g(\mathbf{L_i}, \mathbf{L}_\emptyset)$. In other words, $\mathbf{L} = g\mathbf{L_i}$ and $\mathbf{L}_\emptyset = g\mathbf{L}_\emptyset$. The fact that $\mathbf{i} \in I_\triangle(n, r)_0$ implies that for each $t \in \mathbb{Z}$, there exists $l_t \in \mathbb{Z}$ such that $\mathbf{L}_{\mathbf{i},t} = \mathbf{L}_{\emptyset, l_t}$. (More precisely, if $\mathbf{i} = \mathbf{i}_\mu^\triangle$ and $t = s + kn$ with $1 \leqslant s \leqslant n$, then $l_t = \mu_1 + \cdots + \mu_s + kr$.) Thus, $L_t = g(\mathbf{L}_{\mathbf{i},t}) = g(\mathbf{L}_{\emptyset,l_t}) = \mathbf{L}_{\emptyset,l_t} = \mathbf{L}_{\mathbf{i},t}$ for any $t \in \mathbb{Z}$. This implies that $\mathbf{L} = \mathbf{L_i}$. Hence, by Corollary 3.6.2,

$$(A_\lambda(\mathbf{0}, r) * \chi_{\mathbf{i}})(\mathbf{L_j}, \mathbf{L}_\emptyset) = (A_\lambda(\mathbf{0}, r))(\mathbf{L_j}, \mathbf{L_i}) = q^{-\frac{1}{2}(c(\mathbf{j},\mathbf{i}) + \dim \mathfrak{O}_{A_\lambda})}$$

if and only if $\mathbf{L_i} \subseteq \mathbf{L_j}$, $\mathbf{L_j}/\mathbf{L_i} \cong M(A_\lambda)$, where $c(\mathbf{j}, \mathbf{i}) = c(\mathbf{L_j}, \mathbf{L_i})$. The latter is equivalent by definition to the conditions

$$\{k \in \mathbb{Z} \mid i_k \leqslant t\} \subseteq \{k \in \mathbb{Z} \mid j_k \leqslant t\} \subseteq \{k \in \mathbb{Z} \mid i_k \leqslant t+1\} \text{ and}$$
$$\dim_{\mathbb{F}} \mathbf{L}_{\mathbf{j},t}/\mathbf{L}_{\mathbf{i},t} = \lambda_t,$$

for all $t \in \mathbb{Z}$. Hence,

$$\begin{aligned} &\mathbf{L_i} \subseteq \mathbf{L_j}, \mathbf{L_j}/\mathbf{L_i} \cong M(A_\lambda) \\ &\Longleftrightarrow i_t - 1 \leqslant j_t \leqslant i_t \text{ and } \lambda_t = |\mathbf{j}^{-1}(t) \cap \mathbf{i}^{-1}(t+1)| \text{ for all } t \in \mathbb{Z} \\ &\Longleftrightarrow \mathbf{j} = \mathbf{i} - \mathbf{m} \text{ and } \lambda = m_1 \boldsymbol{e}^{\triangle}_{i_1 - 1} + \cdots + m_r \boldsymbol{e}^{\triangle}_{i_r - 1}, \text{ for some } m_s \in \{0, 1\}, \end{aligned}$$

since, for all $s \in \mathbb{Z}$,

$$m_s = 1 \iff j_s = i_s - 1 \iff s \in \cup_{t \in \mathbb{Z}} (\mathbf{j}^{-1}(t) \cap \mathbf{i}^{-1}(t+1)).$$

Here, $\mathbf{m} \in \mathbb{Z}^r_\triangle$ is uniquely determined by $(m_1, m_2, \ldots, m_r) \in \mathbb{Z}^r$ and $\boldsymbol{e}^\triangle_i \in \mathbb{Z}^r_\triangle$ corresponds to $\boldsymbol{e}_i = (\delta_{k,i})_{1 \leqslant k \leqslant r}$ under (1.1.0.2). (Note that $\mathbf{i} - \mathbf{m} \in I_\triangle(n, r)$ if $\mathbf{i} \in I_\triangle(n, r)$.) Since $\dim \mathfrak{O}_{A_\lambda} = 0$ (see, e.g., [**12**, (1.6.2)]), we obtain, for $\mathbf{i} \in I_\triangle(n, r)_0$ and $\lambda \in \mathbb{N}^n_\triangle$,

$$A_\lambda(\mathbf{0}, r)[A^{\mathbf{i}}] = q^{-\frac{1}{2} d_{\mathbf{i}}} A_\lambda(\mathbf{0}, r) * \chi_{\mathbf{i}} = \sum_{\mathbf{m} \in \mathfrak{M}} q^{\frac{1}{2}(-c(\mathbf{i}-\mathbf{m}, \mathbf{i}) - d_{\mathbf{i}})} \chi_{\mathbf{i}-\mathbf{m}},$$

where $\mathfrak{M} = \{\mathbf{m} \in \mathbb{Z}^r_\triangle \mid m_i \in \{0,1\}, \lambda = m_1 \boldsymbol{e}^\triangle_{i_1 - 1} + \cdots + m_r \boldsymbol{e}^\triangle_{i_r - 1}\}$. Hence,

$$A_\lambda(\mathbf{0}, r) \cdot \omega_{\mathbf{i}} = \sum_{\mathbf{m} \in \mathfrak{M}} v^{-c(\mathbf{i}-\mathbf{m}, \mathbf{i}) - d_{\mathbf{i}} + d_{\mathbf{i}-\mathbf{m}}} [A^{\mathbf{i}-\mathbf{m}}]. \tag{3.6.3.2}$$

We now calculate the right-hand side of (3.6.3.1). By Corollary 3.5.8,

$$\widetilde{u}^+_\lambda \cdot \omega_{\mathbf{i}} = \widetilde{u}^+_\lambda \cdot (\omega_{i_1} \otimes \cdots \otimes \omega_{i_r}) = \sum_{\mathbf{m} \in \mathfrak{M}} v^{-c'(\mathbf{i}, \mathbf{m})} \omega_{i_1 - m_1} \otimes \cdots \otimes \omega_{i_r - m_r},$$

where

$$c'(\mathbf{i}, \mathbf{m}) := \sum_{t<s} m_t(1 - m_s) \langle \boldsymbol{e}^\triangle_{i_s}, \boldsymbol{e}^\triangle_{i_t} \rangle = \sum_{1 \leqslant t < s \leqslant r} m_t(1 - m_s) \delta_{i_t, i_s} - d(\mathbf{i}, \mathbf{m}),$$

with $d(\mathbf{i}, \mathbf{m}) = \sum_{1 \leqslant t < s \leqslant r} m_t(1 - m_s) \delta_{\overline{i_t}, \overline{i_s + 1}}$. To make a comparison with (3.6.3.2), we need to write $\omega_{\mathbf{i}-\mathbf{m}} := \omega_{i_1 - m_1} \otimes \cdots \otimes \omega_{i_r - m_r}$ as a linear combination of the basis $\{[A^{\mathbf{j}}]\}_{\mathbf{j} \in I_\triangle(n,r)}$. For $\mathbf{i} \in I_\triangle(n, r)_0$, order the set

$$\{t \in \mathbb{Z} \mid 1 \leqslant t \leqslant r, \ i_t = 1, \ m_t = 1\} = \{t_1, t_2, \ldots, t_a\}$$

by $t_1 < t_2 < \cdots < t_a$. Then, by definition, $\omega_{\mathbf{i}-\mathbf{m}} X^{-1}_{t_1} X^{-1}_{t_2} \cdots X^{-1}_{t_a} = \omega_{\mathbf{j}}$, where $\mathbf{j} = \mathbf{i} - \mathbf{m} + n(\boldsymbol{e}^\triangle_{t_1} + \boldsymbol{e}^\triangle_{t_2} + \cdots + \boldsymbol{e}^\triangle_{t_a}) \in I(n, r)$. By Proposition 3.3.1 (see also Remark 3.3.2), $\omega_{\mathbf{j}} = [A^{\mathbf{j}}]$. Thus,

$$\omega_{\mathbf{i}-\mathbf{m}} = \omega_{\mathbf{j}} X_{t_a} X_{t_{a-1}} \cdots X_{t_1} = [A^{\mathbf{j}}] X_{t_a} X_{t_{a-1}} \cdots X_{t_1}.$$

Since $\widetilde{T}_{\rho^{-1}} = \widetilde{T}_{r-1} \cdots \widetilde{T}_2 \widetilde{T}_1 X_1$, it follows that

$$X_k = \widetilde{T}_k^{-1} \cdots \widetilde{T}_{r-2}^{-1} \widetilde{T}_{r-1}^{-1} \widetilde{T}_{\rho^{-1}} \widetilde{T}_1 \widetilde{T}_2 \cdots \widetilde{T}_{k-1},$$

for $1 \leqslant k \leqslant r$. Now, by Proposition 3.2.10 and noting $\widetilde{T}_{t_a}^{-1} = \widetilde{T}_{t_a} - (v - v^{-1})$,

$$[A^{\mathbf{j}}]\widetilde{T}_{t_a}^{-1} = \begin{cases} v^{-1}[A^{\mathbf{j}s_{t_a}}], & \text{if } n = j_{t_a} = j_{t_a+1}; \\ [A^{\mathbf{j}s_{t_a}}], & \text{if } n = j_{t_a} > j_{t_a+1}. \end{cases}$$

Thus, repeated application of Proposition 3.2.10 gives

$$[A^{\mathbf{j}}]\widetilde{T}_{t_a}^{-1} \cdots \widetilde{T}_{r-2}^{-1} \widetilde{T}_{r-1}^{-1} \widetilde{T}_{\rho^{-1}} = v^{-b}[A^{\mathbf{j}s_{t_a} \cdots s_{r-1}\rho^{-1}}],$$

where $b = |\{s \in \mathbb{Z} \mid t_a < s \leqslant r,\ i_s = n,\ m_s = 0\}|$. Since $1 \leqslant i_1 \leqslant \cdots \leqslant i_r \leqslant n$, it follows that

$$b = |\{s \in \mathbb{Z} \mid 1 \leqslant s \leqslant r,\ i_s = n,\ m_s = 0\}|.$$

If $\mathbf{j}' = \mathbf{j}s_{t_a} \cdots s_{r-1}$, $\mathbf{j}'' = \mathbf{j}s_{t_a} \cdots s_{r-1}\rho^{-1}$, then $j'_r = n$ and so $j''_1 = j'_0 = 0 < j''_k$ for all $2 \leqslant k \leqslant r$. The last inequality is seen from the fact that $\{j''_k\}_{2 \leqslant k \leqslant r} = \{j'_k\}_{1 \leqslant k \leqslant r-1}$ contains only positive integers. Applying Proposition 3.2.10 again yields

$$[A^{\mathbf{j}''}]\widetilde{T}_1 \cdots \widetilde{T}_{t_a-1} = [A^{\mathbf{j}'' s_1 \cdots s_{t_a-1}}] = [A^{\mathbf{j}-ne_{t_a}}].$$

Hence, $[A^{\mathbf{j}}]X_{t_a} = v^{-b}[A^{\mathbf{j}-ne_{t_a}}]$. Continuing this argument, we obtain eventually $\omega_{\mathbf{i}-\mathbf{m}} = v^{-ab}[A^{\mathbf{i}-\mathbf{m}}]$, where

$$\begin{aligned} ab &= |\{(s,t) \in \mathbb{Z}^2 \mid 1 \leqslant s,t \leqslant r,\ i_s = n,\ m_s = 0,\ i_t = 1,\ m_t = 1\}| \\ &= \sum_{1 \leqslant s,t \leqslant r} m_t(1-m_s)\delta_{i_s,n}\delta_{i_t,1} = \sum_{1 \leqslant t < s \leqslant r} m_t(1-m_s)\delta_{i_s,n}\delta_{i_t,1} = d(\mathbf{i},\mathbf{m}), \end{aligned}$$

since $1 \leqslant t < s \leqslant r$ and $\overline{i_t} = \overline{i_s + 1}$ imply $i_s = n$ and $i_t = 1$. Therefore, we obtain

$$\widetilde{u}_\lambda^+ \cdot \omega_{\mathbf{i}} = \sum_{\mathbf{m} \in \mathfrak{M}} v^{-c'(\mathbf{i},\mathbf{m}) - d(\mathbf{i},\mathbf{m})}[A^{\mathbf{i}-\mathbf{m}}].$$

Comparing this with (3.6.3.2), it remains to prove that

$$d(\mathbf{i},\mathbf{m}) + c'(\mathbf{i},\mathbf{m}) = c(\mathbf{i}-\mathbf{m},\mathbf{i}) + d_{\mathbf{i}} - d_{\mathbf{i}-\mathbf{m}}.$$

The number $c(\mathbf{i}-\mathbf{m},\mathbf{i}) = c(\mathbf{L}_{\mathbf{i}-\mathbf{m}}, \mathbf{L}_{\mathbf{i}})$ is defined in (3.6.0.1). Since

$$\dim \mathbf{L}_{\mathbf{i},i+1}/\mathbf{L}_{\mathbf{i},i} = |\mathbf{i}^{-1}(i+1)| = \sum_{1 \leqslant s \leqslant r} \delta_{\overline{i_s},\overline{i+1}} \quad \text{and}$$

$$\lambda_i = \sum_{1 \leqslant s \leqslant r} m_s \delta_{\overline{i_s - 1},\overline{i}}, \quad \text{for } i \in \mathbb{Z},$$

it follows that

$$\begin{aligned}
c(\mathbf{i}-\mathbf{m},\mathbf{i}) &= \sum_{1\leqslant i\leqslant n} \lambda_{i+1}(|\mathbf{i}^{-1}(i+1)|-\lambda_i) = \sum_{1\leqslant s,t\leqslant r} m_t(1-m_s)\delta_{\overline{i_t-1},\overline{i_s}} \\
&= \sum_{1\leqslant s<t\leqslant r} m_t(1-m_s)\delta_{\overline{i_t-1},\overline{i_s}} + \sum_{1\leqslant t<s\leqslant r} m_t(1-m_s)\delta_{\overline{i_t-1},\overline{i_s}}. \\
&= \sum_{1\leqslant s<t\leqslant r} m_t(1-m_s)\delta_{i_t-1,i_s}(1-\delta_{i_s,n}) + d(\mathbf{i},\mathbf{m}).
\end{aligned}$$

Hence,

$$\begin{aligned}
c(\mathbf{i}-\mathbf{m},\mathbf{i}) - c'(\mathbf{i},\mathbf{m}) = 2d(\mathbf{i},\mathbf{m}) &+ \sum_{1\leqslant s<t\leqslant r} m_t(1-m_s)\delta_{i_t-1,i_s}(1-\delta_{i_s,n}) \\
&- \sum_{1\leqslant t<s\leqslant r} m_t(1-m_s)\delta_{i_s,i_t}.
\end{aligned}$$

On the other hand, for any $\mathbf{i}\in I_\Delta(n,r)$, $d_\mathbf{i} = |\mathrm{Inv}(\mathbf{i})|$ by Lemma 3.1.4. Observe that the set

$$\begin{aligned}
\mathcal{I} :=&\{(s,t)\in\mathcal{L} \mid i_s \geqslant i_t+1\} \\
&\cup \{(s,t)\in\mathcal{L} \mid i_s=i_t, m_s=m_t \text{ or } m_s=0, m_t=1\}
\end{aligned}$$

is the intersection $\mathrm{Inv}(\mathbf{i}-\mathbf{m})\cap\mathrm{Inv}(\mathbf{i})$, where

$$\mathcal{L} = \{(s,t)\in\mathbb{Z}^2 \mid 1\leqslant s\leqslant r,\ s<t\}.$$

It turns out that $\mathrm{Inv}(\mathbf{i}-\mathbf{m})\backslash\mathcal{I} = \{(s,t)\in\mathcal{L} \mid i_s+1=i_t, m_s=0, m_t=1\}$ and $\mathrm{Inv}(\mathbf{i})\backslash\mathcal{I} = \{(s,t)\in\mathcal{L} \mid i_s=i_t, m_s=1, m_t=0\}$. Hence,

$$d_{\mathbf{i}-\mathbf{m}} - d_\mathbf{i} = \sum_{\substack{1\leqslant s\leqslant r\\ t\in\mathbb{Z},\, s<t}} m_t(1-m_s)\delta_{i_t,i_s+1} - \sum_{\substack{1\leqslant s\leqslant r\\ t\in\mathbb{Z},\, s<t}} m_s(1-m_t)\delta_{i_t,i_s}.$$

Since $m_{a+r}=m_a$, $i_{a+r}=i_a+n$, for all $a\in\mathbb{Z}$, and $\mathbf{i}\in I_\Delta(n,r)_0$, it follows that $i_s\neq i_t$, for all $1\leqslant s\leqslant r, t>r$, and $i_t=n+1 \iff i_{t'}=1$ for some $1\leqslant t'\leqslant r$ with $t=t'+r$. Thus, the above sums can be rewritten as

$$\begin{aligned}
d_{\mathbf{i}-\mathbf{m}} - d_\mathbf{i} &= \sum_{1\leqslant s<t\leqslant r} m_t(1-m_s)\delta_{i_t,i_s+1}(1-\delta_{i_s,n}) \\
&\quad + \sum_{1\leqslant t<s\leqslant r} m_t(1-m_s)\delta_{i_t,1}\delta_{i_s,n} - \sum_{1\leqslant s<t\leqslant r} m_s(1-m_t)\delta_{i_t,i_s} \\
&= c(\mathbf{i}-\mathbf{m},\mathbf{i}) - c'(\mathbf{i},\mathbf{m}) - d(\mathbf{i},\mathbf{m}),
\end{aligned}$$

as required. □

3.7. Triangular decompositions of affine quantum Schur algebras

In this section we study the "triangular parts" of $\mathcal{S}_\Delta(n,r)$. We first show that $\mathcal{S}_\Delta(n,r)$ admits a triangular relation for certain structure constants relative to the BLM basis. This relation allows us to produce an integral basis from which we obtain a triangular decomposition. We will give an application of this decomposition in the next section by proving that the algebra homomorphism $\xi_r : \mathfrak{D}_\Delta(n) \to \mathcal{S}_\Delta(n,r)$ defined in §3.5 is surjective.

Keep the notation in the previous sections. We will continue to use the convolution product to derive properties or formulas via the isomorphism mentioned in Remark 3.1.2. Thus, when the convolution product $*$ is used, we automatically mean that the affine quantum Schur algebra is the algebra $\mathcal{S}_\Delta(n,r)_R$ with $R = \mathbb{Z}[\sqrt{q}, \sqrt{q}^{-1}]$ obtained by specializing v to $\sqrt{q}$, for a prime power q, and is identified with the convolution algebra $R_G(\mathscr{Y} \times \mathscr{Y})$. The following results are taken from [**56**].

Lemma 3.7.1. *Let $A = (a_{i,j}) \in \Theta_\Delta(n,r)$ and let $(\mathbf{L}, \mathbf{L}') \in \mathcal{O}_A$, where $\mathbf{L} = (L_i)_{i\in\mathbb{Z}}$ and $\mathbf{L}' = (L'_i)_{i\in\mathbb{Z}}$.*

(1) *A is upper triangular if and only if $L'_i \subseteq L_i$ for all i.*
(2) *A is lower triangular if and only if $L_i \subseteq L'_i$ for all i.*
(3) $\dim(L_i/L_{i-1}) = \sum_{k\in\mathbb{Z}} a_{i,k}$ *and* $\dim(L'_i/L'_{i-1}) = \sum_{k\in\mathbb{Z}} a_{k,i}$.
(4) $\dim\left(\frac{L_i}{L_i\cap L'_{j-1}}\right) = \sum_{s\leqslant i,t\geqslant j} a_{s,t}$ *and* $\dim\left(\frac{L'_j}{L_{i-1}\cap L'_j}\right) = \sum_{s\geqslant i,t\leqslant j} a_{s,t}$.

For $A \in M_{\Delta,n}(\mathbb{Z})$, let

$$\sigma_{i,j}(A) = \begin{cases} \sum\limits_{s\leqslant i,t\geqslant j} a_{s,t}, & \text{if } i<j;\\ \sum\limits_{s\geqslant i,t\leqslant j} a_{s,t}, & \text{if } i>j. \end{cases}$$

For any fixed $x_0 \in \mathbb{Z}$ and $i < j$, it is easy to see that there are two bijective maps

$$\begin{aligned}\{(b,s,t)\in\mathbb{Z}^3 \mid s-bn\leqslant i<j\leqslant t-bn,\\ s\in[x_0+1,x_0+n]\} &\longrightarrow \{(s,t)\in\mathbb{Z}^2 \mid s\leqslant i, t\geqslant j\},\\ (b,s,t) &\longmapsto (s-bn, t-bn)\end{aligned}$$

and

$$\{(b,s,t)\in\mathbb{Z}^3 \mid t-bn\leqslant i<j\leqslant s-bn,\\ s\in[x_0+1,x_0+n]\}\longrightarrow\{(s,t)\in\mathbb{Z}^2\mid s\geqslant j, t\leqslant i\},\\ (b,s,t)\longmapsto(s-bn,t-bn).$$

Thus, we obtain an alternative interpretation of $\sigma_{i,j}(A)$:

$$\sigma_{i,j}(A)=\begin{cases}\sum\limits_{\substack{x_0+1\leqslant s\leqslant x_0+n\\ s<t}} a_{s,t}|\{b\in\mathbb{Z}\mid s-bn\leqslant i<j\leqslant t-bn\}|, & \text{if } i<j;\\ \sum\limits_{\substack{x_0+1\leqslant s\leqslant x_0+n\\ s>t}} a_{s,t}|\{b\in\mathbb{Z}\mid t-bn\leqslant j<i\leqslant s-bn\}|, & \text{if } i>j.\end{cases}$$

In particular, $\sigma_{i,j}(A)<\infty$.

Further, for $A, B\in M_{\Delta,n}(\mathbb{Z})$, define (see [**24**, §6])

$$B\preccurlyeq A\iff\sigma_{i,j}(B)\leqslant\sigma_{i,j}(A),\quad\text{for all } i\neq j. \tag{3.7.1.1}$$

We put $B\prec A$ if $B\preccurlyeq A$ and, for some pair (i,j) with $i\neq j$, $\sigma_{i,j}(B)<\sigma_{i,j}(A)$. It is shown in [**24**, Th. 6.2] that, if $A, B\in\Theta^+_\Delta(n)$ satisfy $\mathbf{d}(A)=\mathbf{d}(B)$, then

$$B\leqslant_{\rm dg} A\iff B\preccurlyeq A.$$

In other words, the ordering $\preccurlyeq$ is an extension of the degeneration order defined in (1.2.0.2).

For $A\in\Theta_\Delta(n)$, write $A=A^++A^0+A^-=A^\pm+A^0$, where $A^+\in\Theta^+_\Delta(n)$, $A^-\in\Theta^-_\Delta(n)$, $A^\pm=A^++A^-$, and A^0 is a diagonal matrix. Here

$$\Theta^-_\Delta(n):=\{{}^tB\mid B\in\Theta^+_\Delta(n)\}.$$

Lemma 3.7.2. *Let $A\in\Theta^\pm_\Delta(n)$ with $\sigma(A)\leqslant r$.*

(1) *For $\mu\in\Lambda_\Delta(n,r-\sigma(A^+))$, $\nu\in\Lambda_\Delta(n,r-\sigma(A^-))$, if*

$$([A^++\operatorname{diag}(\mu)]*[A^-+\operatorname{diag}(\nu)])(\mathbf{L},\mathbf{L}'')\neq 0,$$

where $(\mathbf{L},\mathbf{L}'')\in\mathcal{O}_B$ for some $B\in\Theta_\Delta(n)$, then $B\preccurlyeq A$.

(2) *For $\lambda\in\Lambda_\Delta(n,r-\sigma(A))$, if $(\mathbf{L},\mathbf{L}'')\in\mathcal{O}_{A+\operatorname{diag}(\lambda)}$, then*

$$\bigcup_{\substack{\mu\in\Lambda_\Delta(n,r-\sigma(A^+))\\ \nu\in\Lambda_\Delta(n,r-\sigma(A^-))}}\{\mathbf{L}'\in\mathscr{F}_\Delta\mid(\mathbf{L},\mathbf{L}')\in\mathcal{O}_{A^++\operatorname{diag}(\mu)},(\mathbf{L}',\mathbf{L}'')\in\mathcal{O}_{A^-+\operatorname{diag}(\nu)}\}\\ =\{\mathbf{L}\cap\mathbf{L}''\}.$$

Proof. (1) Since $([A^+ + \mathrm{diag}(\mu)] * [A^- + \mathrm{diag}(\nu)])(\mathbf{L}, \mathbf{L}'') \neq 0$, there exists $\mathbf{L}' \in \mathscr{F}_\triangle$ such that $(\mathbf{L}, \mathbf{L}') \in \mathcal{O}_{A^+ + \mathrm{diag}(\lambda)}$ and $(\mathbf{L}', \mathbf{L}'') \in \mathcal{O}_{A^- + \mathrm{diag}(\mu)}$. Also, $(\mathbf{L}, \mathbf{L}'') \in \mathcal{O}_B$. Hence, by Lemma 3.7.1(1)&(2), $\mathbf{L}' \subseteq \mathbf{L}$ and $\mathbf{L}' \subseteq \mathbf{L}''$, and by Lemma 3.7.1(4),

$$\sigma_{i,j}(A) = \begin{cases} \dim\left(\dfrac{L_i}{L_i \cap L'_{j-1}}\right), & \text{if } i < j; \\ \dim\left(\dfrac{L''_j}{L'_{i-1} \cap L''_j}\right), & \text{if } i > j, \end{cases} \quad \text{and}$$

$$\sigma_{i,j}(B) = \begin{cases} \dim\left(\dfrac{L_i}{L_i \cap L''_{j-1}}\right), & \text{if } i < j; \\ \dim\left(\dfrac{L''_j}{L_{i-1} \cap L''_j}\right), & \text{if } i > j. \end{cases}$$

Therefore, $\mathbf{L}' \subseteq \mathbf{L} \cap \mathbf{L}''$ and

$$\sigma_{i,j}(A) - \sigma_{i,j}(B) = \begin{cases} \dim\left(\dfrac{L_i \cap L''_{j-1}}{L_i \cap L'_{j-1}}\right) \geqslant 0, & \text{if } i < j; \\ \dim\left(\dfrac{L_{i-1} \cap L''_j}{L'_{i-1} \cap L''_j}\right) \geqslant 0, & \text{if } i > j. \end{cases}$$

Consequently, $B \preccurlyeq A$.

(2) If $\mathbf{L}' \in \mathscr{F}_\triangle$ satisfies

$$(\mathbf{L}, \mathbf{L}') \in \mathcal{O}_{A^+ + \mathrm{diag}(\mu)} \quad \text{and} \quad (\mathbf{L}', \mathbf{L}'') \in \mathcal{O}_{A^- + \mathrm{diag}(\nu)},$$

for some $\mu \in \Lambda_\triangle(n, r - \sigma(A^+))$ and $\nu \in \Lambda_\triangle(n, r - \sigma(A^-))$, then $\mathbf{L}' \subseteq \mathbf{L} \cap \mathbf{L}''$ as seen above. Thus, for all $i \in \mathbb{Z}$,

$$\begin{aligned} \dim\left(\frac{L_i \cap L''_i}{L'_i}\right) &= \dim(L_i / L'_i) - \dim\left(\frac{L_i}{L_i \cap L''_i}\right) \\ &= \dim\left(\frac{L_i}{L_i \cap L'_i}\right) - \dim\left(\frac{L_i}{L_i \cap L''_i}\right) \\ &= \sum_{s \leqslant i < t} a_{s,t} - \sum_{s \leqslant i < t} a_{s,t} = 0, \end{aligned}$$

by Lemma 3.7.1(4) again. Hence, $\mathbf{L}' = \mathbf{L} \cap \mathbf{L}''$, proving the assertion. □

Proposition 3.7.3. *Let $A \in \Theta^\pm_\triangle(n)$. Then the following* triangular relation *relative to $\prec$ holds:*

$$\begin{aligned} A^+(\mathbf{0}, r) A^-(\mathbf{0}, r) &= A(\mathbf{0}, r) + \sum_{\substack{C \in \Theta_\triangle(n,r) \\ C \prec A}} f_{C,A}[C] \quad (\textit{in } \mathcal{S}_\triangle(n, r)) \\ &= A(\mathbf{0}, r) + \sum_{\substack{B \in \Theta^\pm_\triangle(n) \\ B \prec A, \mathbf{j} \in \mathbb{Z}^n_\triangle}} g_{B,\mathbf{j},A;r} B(\mathbf{j}, r) \quad (\textit{in } \boldsymbol{\mathcal{S}}_\triangle(n, r)), \end{aligned}$$

where $f_{C,A} \in \mathcal{Z}$, $g_{B,\mathbf{j},A;r} \in \mathbb{Q}(v)$.

Proof. Since the elements $B(\mathbf{j}, r)$ span $\boldsymbol{\mathcal{S}}_\vartriangle(n, r)$ by Proposition 3.4.1, the second equality follows from the first one. We now prove the first equality.

Let $r^\pm = \sigma(A)$, $r^+ = \sigma(A^+)$, and $r^- = \sigma(A^-)$. There is nothing to prove if $r^\pm > r$. Assume now $r^\pm \leqslant r$. By (3.1.1.1),

$$\begin{aligned} A^+(\mathbf{0}, r)A^-(\mathbf{0}, r) &= \sum_{\mu\in\Lambda(n,r-r^+)} \sum_{\nu\in\Lambda(n,r-r^-)} [A^+ + \mathrm{diag}(\mu)][A^- + \mathrm{diag}(\nu)] \\ &= \sum_{C\in\Theta_\vartriangle(n,r)} \sum_{\mu,\nu} v^{-d_{A^++\mathrm{diag}(\mu)}-d_{A^-+\mathrm{diag}(\nu)}} p_{A^++\mathrm{diag}(\mu),A^-+\mathrm{diag}(\nu),C}\, v^{d_C}[C] \\ &= \sum_{C\in\Theta_\vartriangle(n,r)} f_{C,A}[C], \end{aligned}$$

where

$$f_{C,A} = \sum_{\substack{\mu\in\Lambda_\vartriangle(n,r-r^+)\\ \nu\in\Lambda_\vartriangle(n,r-r^-)}} v^{d_C-d_{A^++\mathrm{diag}(\mu)}-d_{A^-+\mathrm{diag}(\nu)}} p_{A^++\mathrm{diag}(\mu),A^-+\mathrm{diag}(\nu),C} \in \mathcal{Z}.$$

If $f_{C,A} \neq 0$, then $p_{A^++\mathrm{diag}(\mu),A^-+\mathrm{diag}(\nu),C} \neq 0$ for some μ, ν as above. Thus, by definition, there is a finite field $\mathbb{F}$ of q elements such that

$$p_{A^++\mathrm{diag}(\mu),A^-+\mathrm{diag}(\nu),C}|_{v^2=q} = \mathfrak{n}_{A^++\mathrm{diag}(\mu),A^-+\mathrm{diag}(\nu),C;q} \neq 0,$$

where $\mathfrak{n}_{A^++\mathrm{diag}(\mu),A^-+\mathrm{diag}(\nu),C;q}$ is defined in (3.1.0.6). By Lemma 3.7.2(1), we conclude $C \preccurlyeq A$. We need to prove that $f_{C,A} = 1$, for all $C = A + \mathrm{diag}(\lambda)$ with $\lambda \in \Lambda_\vartriangle(n, r - r^\pm)$. First, Lemma 3.7.2(2) implies that there exist unique μ, ν such that $\mathfrak{n}_{A^++\mathrm{diag}(\mu),A^-+\mathrm{diag}(\nu),A+\mathrm{diag}(\lambda);q} = 1$. Thus,

$$f_{A+\mathrm{diag}(\lambda),A} = v^{d_{A+\mathrm{diag}(\lambda)}-d_{A^++\mathrm{diag}(\mu)}-d_{A^-+\mathrm{diag}(\nu)}}. \tag{3.7.3.1}$$

Second, since $d_{A+\mathrm{diag}(\lambda)} = d_{A^++\mathrm{diag}(\mu)} + d_{A^-+\mathrm{diag}(\nu)}$ by a direct computation (see Lemma 3.10.2), we obtain $f_{A+\mathrm{diag}(\lambda),A} = 1$. Finally, since $\preccurlyeq$ is a partial order on $\Theta_\vartriangle^\pm(n)$ by [**24**, Lem. 6.1], we conclude that

$$A^+(\mathbf{0}, r)A^-(\mathbf{0}, r) = A(\mathbf{0}, r) + g,$$

where g is a $\mathcal{Z}$-linear combination of $[C]$ with $C \in \Theta_\vartriangle(n, r)$ and $C \prec A$. □

We now use the triangular relation to establish a triangular decomposition for the $\mathcal{Z}$-algebra $\mathcal{S}_\vartriangle(n, r)$.

Consider the following $\mathcal{Z}$-submodules of $\mathcal{S}_\vartriangle(n, r)$

$$\begin{aligned} \mathcal{S}_\vartriangle(n, r)^+ &= \mathrm{span}_{\mathcal{Z}}\{A(\mathbf{0}, r) \mid A \in \Theta_\vartriangle^+(n)\}, \\ \mathcal{S}_\vartriangle(n, r)^- &= \mathrm{span}_{\mathcal{Z}}\{A(\mathbf{0}, r) \mid A \in \Theta_\vartriangle^-(n)\}, \quad \text{and} \\ \mathcal{S}_\vartriangle(n, r)^0 &= \mathrm{span}_{\mathcal{Z}}\{[\mathrm{diag}(\lambda)] \mid \lambda \in \Lambda_\vartriangle(n, r)\}. \end{aligned} \tag{3.7.3.2}$$

As homomorphic images of Ringel–Hall algebras (see Proposition 3.6.1), both $\mathcal{S}_\Delta(n,r)^+$ and $\mathcal{S}_\Delta(n,r)^-$ are $\mathcal{Z}$-subalgebras of $\mathcal{S}_\Delta(n,r)$. It can be directly checked that $\mathcal{S}_\Delta(n,r)^0$ is also a $\mathcal{Z}$-subalgebra of $\mathcal{S}_\Delta(n,r)$. Moreover, $\mathcal{S}_\Delta(n,r)^0$ is isomorphic to the 0-part of the (non-affine) quantum Schur algebra. We will call the subalgebras $\mathcal{S}_\Delta(n,r)^-$, $\mathcal{S}_\Delta(n,r)^0$, and $\mathcal{S}_\Delta(n,r)^+$ the *negative, zero,* and *positive parts* of $\mathcal{S}_\Delta(n,r)$, respectively. The next result shows that all three subalgebras are free as $\mathcal{Z}$-modules. Recall the notation $\left[\begin{smallmatrix} X;a \\ t \end{smallmatrix}\right]$ introduced in (1.1.0.6).

Proposition 3.7.4. *Let $\mathfrak{k}_i = 0(e_i^\Delta, r)$, $1 \leqslant i \leqslant n$.*

(1) *The elements $A(\mathbf{0},r)$ (resp., ${}^tA(\mathbf{0},r)$), for $A \in \Theta_\Delta^+(n)$ with $\sigma(A) \leqslant r$, form a $\mathcal{Z}$-basis of $\mathcal{S}_\Delta(n,r)^+$ (resp., $\mathcal{S}_\Delta(n,r)^-$). In particular, ξ_r maps $\mathfrak{H}_\Delta(n)^\pm$ onto $\mathcal{S}_\Delta(n,r)^\pm$.*

(2) *For $\lambda \in \Lambda_\Delta(n,r)$,*

$$\mathfrak{l}_\lambda := \begin{bmatrix} \mathfrak{k}_1; 0 \\ \lambda_1 \end{bmatrix}\begin{bmatrix} \mathfrak{k}_2; 0 \\ \lambda_2 \end{bmatrix}\cdots\begin{bmatrix} \mathfrak{k}_n; 0 \\ \lambda_n \end{bmatrix} = [\mathrm{diag}(\lambda)].$$

In particular, the set $\{\mathfrak{l}_\lambda \mid \lambda \in \Lambda_\Delta(n,r)\}$ forms a $\mathcal{Z}$-basis of $\mathcal{S}_\Delta(n,r)^0$.

Proof. Assertion (1) follows from the definition of $\mathcal{S}_\Delta(n,r)^\pm$ and Proposition 3.4.1. To see (2), since $\mathfrak{k}_i[\mathrm{diag}(\mu)] = \sum_{\nu\in\Lambda_\Delta(n,r)} v^{\nu_i}[\mathrm{diag}(\nu)][\mathrm{diag}(\mu)] = v^{\mu_i}[\mathrm{diag}(\mu)]$, by (3.1.3.4), it follows that $\left[\begin{smallmatrix} \mathfrak{k}_i;0 \\ t \end{smallmatrix}\right][\mathrm{diag}(\mu)] = \left[\begin{smallmatrix} \mu_i \\ t \end{smallmatrix}\right][\mathrm{diag}(\mu)]$. Hence, for $\lambda \in \Lambda_\Delta(n,r)$,

$$\mathfrak{l}_\lambda = \sum_{\mu\in\Lambda_\Delta(n,r)} \mathfrak{l}_\lambda[\mathrm{diag}(\mu)] = \sum_{\mu\in\Lambda_\Delta(n,r)} \begin{bmatrix} \mu_1 \\ \lambda_1 \end{bmatrix}\cdots\begin{bmatrix} \mu_n \\ \lambda_n \end{bmatrix}[\mathrm{diag}(\mu)] = [\mathrm{diag}(\lambda)],$$

as desired.[4] □

As in §2.3, let $\mathfrak{D}_\Delta(n)^\pm$ (resp., $\mathfrak{H}_\Delta(n)^\pm$) be the $\mathbb{Q}(v)$-submodules (resp., $\mathcal{Z}$-submodules) of $\mathfrak{D}_\Delta(n)$ spanned by $u_A^\pm$, for $A \in \Theta_\Delta^+(n)$. They are respectively the $\mathbb{Q}(v)$-subalgebras and $\mathcal{Z}$-subalgebras of $\mathfrak{D}_\Delta(n)$. The above proposition together with the results in §3.6 gives the following result; see Remark 2.4.6.

Corollary 3.7.5. *For each $s \geqslant 1$, we have $\mathsf{z}_s^+ \in \mathfrak{H}_\Delta(n)^+$ and $\mathsf{z}_s^- \in \mathfrak{H}_\Delta(n)^-$.*

Proof. We only prove $\mathsf{z}_s^+ \in \mathfrak{H}_\Delta(n)^+$. The proof for $\mathsf{z}_s^- \in \mathfrak{H}_\Delta(n)^-$ is similar.

By Proposition 3.6.1 and Theorem 3.6.3, the restriction of $\xi_r : \mathfrak{D}_\Delta(n) \to \mathcal{S}_\Delta(n,r)$ gives an algebra homomorphism

$$\zeta_r^+ : \mathfrak{D}_\Delta(n)^+ \longrightarrow \boldsymbol{\mathcal{S}}_\Delta(n,r) = \mathrm{End}_{\mathcal{H}_\Delta(r)}(\boldsymbol{\Omega}^{\otimes r})$$

[4] In the literature, $\mathfrak{l}_\lambda$ is denoted by k_λ or 1_λ. We modified the notation in order to introduce its preimage $\mathfrak{L}_\lambda$ in $\mathfrak{D}_\Delta(n)$; see (5.1.1.1).

taking $\widetilde{u}_A^+ \mapsto A(\mathbf{0}, r)$ for $A \in \Theta_\Delta^+(n)$. Write

$$\mathsf{z}_s^+ = \sum_{A \in \Theta_\Delta^+(n)} f_A \widetilde{u}_A^+,$$

where all but finitely many $f_A \in \mathbb{Q}(v)$ are zero. By (3.5.4.1), $\mathsf{z}_s^+(\Omega^{\otimes r}) \subseteq \Omega^{\otimes r}$. Hence, $\zeta_r^+(\mathsf{z}_s^+) \in \mathrm{End}_{\mathcal{H}_\Delta(r)}(\Omega^{\otimes r}) = \mathcal{S}_\Delta(n, r)$; see Proposition 3.3.1. In other words,

$$\zeta_r^+(\mathsf{z}_s^+) = \sum_{A \in \Theta_\Delta^+(n)} f_A A(\mathbf{0}, r) = \sum_{A \in \Theta_\Delta^+(n),\, \sigma(A) \leqslant r} f_A A(\mathbf{0}, r) \in \mathcal{S}_\Delta(n, r)^+.$$

By Proposition 3.7.4(1), $\{A(\mathbf{0}, r) \mid A \in \Theta_\Delta^+(n), \sigma(A) \leqslant r\}$ is a $\mathcal{Z}$-basis for $\mathcal{S}_\Delta(n, r)^+$. Hence, $f_A \in \mathcal{Z}$, for all $A \in \Theta_\Delta^+(n)$ with $\sigma(A) \leqslant r$. Since r can be chosen to be an arbitrary positive integer, it follows that $f_A \in \mathcal{Z}$, for all $A \in \Theta_\Delta^+(n)$. We conclude that $\mathsf{z}_s^+ \in \mathfrak{H}_\Delta(n)^+$. □

We now patch the three bases to obtain a basis for $\mathcal{S}_\Delta(n, r)$. For $A \in \Theta_\Delta(n)$, define (cf. [**4**])

$$\|A\| = \sum_{\substack{1 \leqslant i \leqslant n \\ i < j}} \frac{(j-i)(j-i+1)}{2} a_{i,j} + \sum_{\substack{1 \leqslant i \leqslant n \\ i > j}} \frac{(i-j)(i-j+1)}{2} a_{i,j}.$$

Lemma 3.7.6. *For $A \in \Theta_\Delta(n)$, the equality*

$$\|A\| = \sum_{\substack{1 \leqslant i \leqslant n \\ i < j}} \sigma_{i,j}(A) + \sum_{\substack{1 \leqslant i \leqslant n \\ i > j}} \sigma_{i,j}(A)$$

holds. In particular, if $A, B \in \Theta_\Delta(n)$ satisfy $A \prec B$, then $\|A\| < \|B\|$.

Proof. By definition, we have

$$\begin{aligned}\sum_{\substack{1 \leqslant i \leqslant n \\ i < j}} \sigma_{i,j}(A) &= \sum_{\substack{1 \leqslant i \leqslant n \\ s \leqslant i < j \leqslant t}} a_{s,t} = \sum_{\substack{1 \leqslant s \leqslant n \\ s < t}} a_{s,t} |\{(i, j) \mid s \leqslant i < j \leqslant t\}| \\ &= \sum_{\substack{1 \leqslant s \leqslant n \\ s < t}} \frac{(t-s)(t-s+1)}{2} a_{s,t} \quad \text{and}\end{aligned}$$

$$\begin{aligned}\sum_{\substack{1 \leqslant i \leqslant n \\ i > j}} \sigma_{i,j}(A) &= \sum_{\substack{1 \leqslant i \leqslant n \\ t \leqslant j < i \leqslant s}} a_{s,t} = \sum_{\substack{1 \leqslant s \leqslant n \\ s > t}} a_{s,t} |\{(i, j) \mid t \leqslant j < i \leqslant s\}| \\ &= \sum_{\substack{1 \leqslant s \leqslant n \\ s > t}} \frac{(s-t)(s-t+1)}{2} a_{s,t}.\end{aligned}$$

The assertion follows from the definition of $\|A\|$. □

For $A \in \Theta_\triangle(n)$ and $i \in \mathbb{Z}$, define the "hook sum"

$$\sigma_i(A) = a_{i,i} + \sum_{j<i}(a_{i,j} + a_{j,i}).$$

It is easy to see that $\sigma(A) = \sum_{1\leqslant i\leqslant n} \sigma_i(A)$. Let

$$\boldsymbol{\sigma}(A) = (\sigma_i(A))_{i\in\mathbb{Z}} \in \mathbb{N}_\triangle^n \quad \text{and} \quad \mathfrak{p}_A = A^+(\mathbf{0},r)\mathfrak{l}_{\boldsymbol{\sigma}(A)}A^-(\mathbf{0},r). \quad (3.7.6.1)$$

We now describe a PBW type basis for $\mathcal{S}_\triangle(n,r)$.

Theorem 3.7.7. *Keep the notation introduced above. There exist $g_{B,A} \in \mathcal{Z}$ such that*

$$\mathfrak{p}_A = [A] + \sum_{B\in\Theta_\triangle(n,r), B\prec A} g_{B,A}[B].$$

Moreover, the set $\mathcal{P}_r := \{\mathfrak{p}_A \mid A \in \Theta_\triangle(n,r)\}$ forms a $\mathcal{Z}$-basis for $\mathcal{S}_\triangle(n,r)$. In particular, we obtain a (weak) triangular decomposition:

$$\mathcal{S}_\triangle(n,r) = \mathcal{S}_\triangle(n,r)^+\mathcal{S}_\triangle(n,r)^0\mathcal{S}_\triangle(n,r)^-.$$

Proof. By the notational convention right above Lemma 3.7.2, if $A \in \Theta_\triangle(n,r)$, then $\sigma_i(A^\pm)$ is the ith component of $\mathrm{co}(A^+) + \mathrm{ro}(A^-)$ and

$$\begin{aligned}\mathfrak{l}_{\boldsymbol{\sigma}(A)}A^-(\mathbf{0},r) &= [\mathrm{diag}(\boldsymbol{\sigma}(A))]A^-(\mathbf{0},r)\\ &= A^-(\mathbf{0},r)[\mathrm{diag}(\boldsymbol{\sigma}(A) + \mathrm{co}(A^-) - \mathrm{ro}(A^-))].\end{aligned}$$

On the other hand, $A^+(\mathbf{0},r)A^-(\mathbf{0},r) = A^\pm(\mathbf{0},r) + g$, where g is a $\mathcal{Z}$-linear combination of $[B]$ with $B \in \Theta_\triangle(n,r)$ and $B \prec A$, by Proposition 3.7.3. Thus,

$$\begin{aligned}\mathfrak{p}_A &= A^+(\mathbf{0},r)\mathfrak{l}_{\boldsymbol{\sigma}(A)}A^-(\mathbf{0},r)\\ &= A^+(\mathbf{0},r)A^-(\mathbf{0},r)[\mathrm{diag}(\boldsymbol{\sigma}(A) + \mathrm{co}(A^-) - \mathrm{ro}(A^-))]\\ &= A^\pm(\mathbf{0},r)[\mathrm{diag}(\boldsymbol{\sigma}(A) + \mathrm{co}(A^-) - \mathrm{ro}(A^-))] + g'\\ &= [A^\pm + \mathrm{diag}(\boldsymbol{\sigma}(A) - (\mathrm{co}(A^+) + \mathrm{ro}(A^-))] + g'\\ &= [A] + g',\end{aligned}$$

where g' is the $\mathcal{Z}$-linear combination of $[B]$ with $B \in \Theta_\triangle(n,r)$ and $B \prec A$. Thus, the set $\mathcal{P}_r$ is linearly independent. To see that it spans, we can apply Lemma 3.7.6 and an induction on $\|A\|$ to show that $[A]$ is a $\mathcal{Z}$-linear combination of $\mathfrak{p}_B$ with $B \in \Theta_\triangle(n,r)$. Hence, $\mathcal{P}_r$ forms a $\mathcal{Z}$-basis for $\mathcal{S}_\triangle(n,r)$. The last assertion follows from Proposition 3.7.4. □

3.8. Affine quantum Schur–Weyl duality, I

We now use the triangular decomposition given in Theorem 3.7.7 to partially establish an affine analogue of the quantum Schur–Weyl reciprocity.

As in Remark 2.3.6(4), for each $m \geqslant 0$, let $\mathfrak{D}_\Delta(n)^{(m)}$ (resp., $\mathfrak{D}_\Delta(n)^{(0)}$) denote the subalgebra of $\mathfrak{D}_\Delta(n)$ generated by $u_i^+, u_i^-, K_i^{\pm 1}, \mathsf{z}_s^+, \mathsf{z}_s^-$ (resp., $u_i^+, u_i^-, K_i^{\pm 1}$) for $i \in I$, $1 \leqslant s \leqslant m$ (resp., for $i \in I$, $m = 0$). Thus, $\mathfrak{D}_\Delta(n)^{(0)} = \mathbf{U}_\Delta(n)$ as defined in Remark 2.3.6(2).

Theorem 3.8.1. *Let n, r be two positive integers with $n \geqslant 2$.*

(1) *The algebra homomorphism $\xi_r : \mathfrak{D}_\Delta(n) \to \boldsymbol{\mathcal{S}}_\Delta(n,r)$ is surjective.*
(2) *If we write $r = mn + m_0$ with $m \geqslant 0$ and $0 \leqslant m_0 < n$, then ξ_r induces a surjective algebra homomorphism $\mathfrak{D}_\Delta(n)^{(m)} \to \boldsymbol{\mathcal{S}}_\Delta(n,r)$. In particular, if $n > r$, then ξ_r induces a surjective algebra homomorphism $\mathbf{U}_\Delta(n) \to \boldsymbol{\mathcal{S}}_\Delta(n,r)$.*
(3) *If $n \geqslant r$, then the map $\xi_r^\vee : \boldsymbol{\mathcal{H}}_\Delta(r) \to \mathrm{End}_{\mathfrak{D}_\Delta(n)}(\boldsymbol{\Omega}^{\otimes r})^{\mathrm{op}}$ given in (3.5.5.4) is an isomorphism.*

Proof. (1) As in Remark 2.4.6, let $\mathfrak{H}_\Delta(n)^+$ (resp., $\mathfrak{H}_\Delta(n)^-$) be the $\mathcal{Z}$-subalgebra of $\mathfrak{D}_\Delta(n)$ generated by u_A^+ (resp., u_A^-) for all $A \in \Theta_\Delta^+(n)$, and let $\mathfrak{D}_\Delta(n)^0$ be the $\mathcal{Z}$-subalgebra of $\mathfrak{D}_\Delta(n)$ generated by $K_i^{\pm 1}$ and $\left[{K_i;0 \atop t} \right]$, for $i \in I$ and $t > 0$ (see (2.4.1.2) for bases of $\mathfrak{D}_\Delta(n)^0$), and set

$$\widetilde{\mathfrak{D}}_\Delta(n) := \mathfrak{H}_\Delta(n)^+ \mathfrak{D}_\Delta(n)^0 \mathfrak{H}_\Delta(n)^- \cong \mathfrak{H}_\Delta(n)^+ \otimes \mathfrak{D}_\Delta(n)^0 \otimes \mathfrak{H}_\Delta(n)^-. \quad (3.8.1.1)$$

By Theorem 3.6.3 and Proposition 3.7.4, ξ_r maps $\mathfrak{H}_\Delta(n)^\varepsilon$ onto $\mathcal{S}_\Delta(n,r)^\varepsilon$ for $\varepsilon = +, -$ and $\mathfrak{D}_\Delta(n)^0$ onto $\mathcal{S}_\Delta(n,r)^0$. Hence, ξ_r maps $\widetilde{\mathfrak{D}}_\Delta(n)$ onto $\mathcal{S}_\Delta(n,r)$ by Theorem 3.7.7. Taking a base change to $\mathbb{Q}(v)$ gives the required surjectivity.

(2) By Proposition 1.4.3, $\mathfrak{D}_\Delta(n)$ is generated by $u_i^+, u_i^-, K_i^{\pm 1}, u_{s\delta}^+, u_{s\delta}^-$ ($i \in I$, $s \in \mathbb{Z}^+$). For each $s \geqslant 1$, the semisimple module $S_{s\delta}$ has dimension sn. Thus, if $s > m \geqslant 1$, then $\dim S_{s\delta} > r$. Moreover, in this case, we have by Corollary 3.5.8 that, for each $\omega_{\mathbf{i}} \in \boldsymbol{\Omega}^{\otimes r}$,

$$u_{s\delta}^+ \cdot \omega_{\mathbf{i}} = 0 = u_{s\delta}^- \cdot \omega_{\mathbf{i}}.$$

In other words, $\xi_r(u_{s\delta}^+) = 0 = \xi_r(u_{s\delta}^-)$ whenever $s > m$ and $m \geqslant 1$. The assertion follows from the fact that $\mathfrak{D}_\Delta(n)^{(m)}$ is generated by $u_i^+, u_i^-, K_i^{\pm 1}, u_{s\delta}^+, u_{s\delta}^-$ ($i \in I$, $0 \leqslant s \leqslant m$); see Remark 2.3.6(4). The last assertion follows from Corollary 3.5.9.

(3) By (1), we have $\mathrm{End}_{\mathfrak{D}_\Delta(n)}(\boldsymbol{\Omega}^{\otimes r}) = \mathrm{End}_{\boldsymbol{\mathcal{S}}_\Delta(n,r)}(\boldsymbol{\Omega}^{\otimes r})$. On the other hand, by Lemma 3.1.3 and Proposition 3.3.1, $\boldsymbol{\mathcal{H}}_\Delta(r) \cong e_\omega \boldsymbol{\mathcal{S}}_\Delta(n,r) e_\omega$ and $\boldsymbol{\Omega}^{\otimes r} \cong$

$\mathcal{S}_\Delta(n,r)e_\omega$ (see the proof of Lemma 3.1.3). Hence,

$$\mathrm{End}_{\mathcal{S}_\Delta(n,r)}(\Omega^{\otimes r}) \cong \mathrm{End}_{\mathcal{S}_\Delta(n,r)}(\mathcal{S}_\Delta(n,r)e_\omega) \cong (e_\omega \mathcal{S}_\Delta(n,r)e_\omega)^{\mathrm{op}}.$$

This implies that $\xi_r^\vee$ is an isomorphism. □

Combining the above theorem with Corollary 3.5.8 yields the following result which can also be derived from [**56**, §4.1, Th 8.2].[5]

Corollary 3.8.2. *Suppose $n > r$. Then $\xi_r : \mathfrak{D}_\Delta(n) \to \mathcal{S}_\Delta(n,r)$ induces a surjective $\mathcal{Z}$-algebra homomorphism*

$$\theta_r : U_\Delta(n) \longrightarrow \mathcal{S}_\Delta(n,r),$$

where $U_\Delta(n)$ is the $\mathcal{Z}$-subalgebra of $\mathfrak{D}_\Delta(n)$ generated by $K_i^{\pm 1}$, $\left[{K_i;0 \atop t} \right]$, $(u_i^+)^{(m)}$ and $(u_i^-)^{(m)}$ for $i \in I$ and $t, m \geqslant 1$ (see §2.3).

Proof. By Proposition 3.7.4, $\xi_r : \mathfrak{D}_\Delta(n) \to \mathcal{S}_\Delta(n,r)$ induces surjective $\mathcal{Z}$-algebra homomorphisms

$$\xi_{r,\mathcal{Z}}^+ : \mathfrak{H}_\Delta(n)^+ \longrightarrow \mathcal{S}_\Delta(n,r)^+ \quad \text{and} \quad \xi_{r,\mathcal{Z}}^- : \mathfrak{H}_\Delta(n)^- \longrightarrow \mathcal{S}_\Delta(n,r)^-.$$

By [**13**, Th. 5.2], $\mathfrak{H}_\Delta(n)^+$ is generated by $(u_i^+)^{(m)}$ and $u_{\mathbf{a}}^+$, for $i \in I$, $m \geqslant 1$, and sincere $\mathbf{a} \in \mathbb{N}^n$. Since $n > r$, it follows from Corollary 3.5.8 that $\xi_{r,\mathcal{Z}}^+(u_{\mathbf{a}}^+) = 0$, for all sincere $\mathbf{a} \in \mathbb{N}^n$. Thus, $\xi_{r,\mathcal{Z}}^+$ gives rise to a surjective $\mathcal{Z}$-algebra homomorphism

$$\theta_r^+ : \mathfrak{C}_\Delta(n)^+ \longrightarrow \mathcal{S}_\Delta(n,r)^+,$$

where $\mathfrak{C}_\Delta(n)^+$ is the $\mathcal{Z}$-subalgebra of $\mathfrak{D}_\Delta(n)$ generated by the $(u_i^+)^{(m)}$. Similarly, we obtain a surjective $\mathcal{Z}$-algebra homomorphism

$$\theta_r^- : \mathfrak{C}_\Delta(n)^- \longrightarrow \mathcal{S}_\Delta(n,r)^-,$$

where $\mathfrak{C}_\Delta(n)^-$ is the $\mathcal{Z}$-subalgebra of $\mathfrak{D}_\Delta(n)$ generated by the $(u_i^-)^{(m)}$. By (2.4.4.2),

$$U_\Delta(n) = \mathfrak{C}_\Delta(n)^+ \, \mathfrak{D}_\Delta(n)^0 \, \mathfrak{C}_\Delta(n)^-.$$

The assertion then follows from the triangular decomposition of $\mathcal{S}_\Delta(n,r)$ given in Theorem 3.7.7. □

If $z \in \mathbb{C}^*$ is not a root of unity, and $\mathfrak{D}_{\Delta\mathbb{C}}(n)$ is the double Ringel–Hall algebra over $\mathbb{C}$ with parameter z considered in Remark 2.1.4, then we have algebra homomorphisms $\xi_{r,\mathbb{C}}$ and $\xi_{r,\mathbb{C}}^\vee$ as given in (3.5.10.1). The proof of the theorem above gives the following.

[5] Lusztig constructed a canonical basis in [**56**, §4.1] for $\mathcal{S}_\Delta(n,r)$ and proved that those canonical basis elements labeled by aperiodic matrices form a basis for $U_\Delta(n,r) := \xi_r(U_\Delta(n))$.

Corollary 3.8.3. *The* $\mathbb{C}$*-algebra homomorphism*

$$\xi_{r,\mathbb{C}} : \mathfrak{D}_{\Delta,\mathbb{C}}(n) \longrightarrow \mathcal{S}_\Delta(n,r)_\mathbb{C}$$

is surjective. If $n \geqslant r$, $\xi^\vee_{r,\mathbb{C}} : \mathcal{H}_\Delta(r)_\mathbb{C} \longrightarrow \mathrm{End}_{\mathfrak{D}_{\Delta,\mathbb{C}}(n)}(\Omega_\mathbb{C}^{\otimes r})^{\mathrm{op}}$ *is an isomorphism.*

The above corollary together with [**61**, Th. 8.1] shows the following Schur–Weyl reciprocity in the affine quantum case.

Corollary 3.8.4. *Let* q *be a prime power. By specializing* v *to* $\sqrt{q}$, *the* $\mathfrak{D}_{\Delta,\mathbb{C}}(n)$*-*$\mathcal{H}_\Delta(r)_\mathbb{C}$*-bimodule* $\Omega_\mathbb{C}^{\otimes r}$ *induces algebra homomorphisms*

$$\xi_{r,\mathbb{C}} : \mathfrak{D}_{\Delta,\mathbb{C}}(n) \longrightarrow \mathrm{End}_\mathbb{C}(\Omega_\mathbb{C}^{\otimes r}) \ \text{and}\ \xi^\vee_{r,\mathbb{C}} : \mathcal{H}_\Delta(r)_\mathbb{C} \longrightarrow \mathrm{End}_\mathbb{C}(\Omega_\mathbb{C}^{\otimes r})^{\mathrm{op}}$$

such that

$$\mathrm{Im}\,(\xi_{r,\mathbb{C}}) = \mathrm{End}_{\mathcal{H}_\Delta(r)_\mathbb{C}}(\Omega_\mathbb{C}^{\otimes r}) = \mathcal{S}_\Delta(n,r)_\mathbb{C} \ \text{and}\ \mathrm{Im}\,(\xi^\vee_{r,\mathbb{C}}) = \mathrm{End}_{\mathcal{S}_\Delta(n,r)_\mathbb{C}}(\Omega_\mathbb{C}^{\otimes r})^{\mathrm{op}}.$$

Remarks 3.8.5. (1) As established in §2.5, $\boldsymbol{\mathfrak{D}}_\Delta(n) \cong \mathbf{U}(\widehat{\mathfrak{gl}}_n)$, the quantum loop algebra. Hence, ξ_r induces a surjective algebra homomorphism $\mathbf{U}(\widehat{\mathfrak{gl}}_n) \to \boldsymbol{\mathcal{S}}_\Delta(n,r)$. Similarly, $\xi_{r,\mathbb{C}}$ induces an algebra epimorphism $\mathrm{U}_\mathbb{C}(\widehat{\mathfrak{gl}}_n) \to \mathcal{S}_\Delta(n,r)_\mathbb{C}$. Here $\mathrm{U}_\mathbb{C}(\widehat{\mathfrak{gl}}_n)$ is the quantum loop algebra over $\mathbb{C}$ defined in Definition 2.5.1, which is isomorphic to $\mathfrak{D}_{\Delta,\mathbb{C}}(n)$ by Remark 2.5.5(3); cf. Theorem 2.5.3. It would be interesting to find explicit formulas for the action of generators of $\mathbf{U}(\widehat{\mathfrak{gl}}_n)$ on the tensor space $\boldsymbol{\Omega}^{\otimes r}$.

(2) In [**75**, Th. 2], Vasserot has also constructed a surjective map Ψ_z from $\mathrm{U}_\mathbb{C}(\widehat{\mathfrak{gl}}_n)$ to the K-theoretic construction $K^G(Z)_z$ of $\mathcal{S}_\Delta(n,r)_\mathbb{C}$. It would also be interesting to know if Ψ_z is equivalent to the epimorphism $\xi_{r,\mathbb{C}}$, namely, if $g_r \circ \Psi_z = f \circ \xi_{r,\mathbb{C}}$ under the isomorphisms $f : \mathrm{U}_\mathbb{C}(\widehat{\mathfrak{gl}}_n) \xrightarrow{\sim} \mathfrak{D}_{\Delta,\mathbb{C}}(n)$ and $g_r : K^G(Z)_z \xrightarrow{\sim} \mathcal{S}_\Delta(n,r)_\mathbb{C}$ (see [**32**, (9.4)]).

(3) By the epimorphisms ξ_r and $\xi_{r,\mathbb{C}}$, different types of generators for double Ringel–Hall algebras (see Remark 2.3.6(1)) give rise to corresponding generators for affine quantum Schur algebras. Thus, we may speak of semisimple generators, homogeneous indecomposable generators, etc., for $\boldsymbol{\mathcal{S}}_\Delta(n,r)$. See §§5.4, 6.2.

We end this section with a few conjectures.

In the proof of the surjectivity of ξ_r in Theorem 3.8.1(1), we proved that the restriction of ξ_r to the free $\mathcal{Z}$-module $\widetilde{\mathfrak{D}}_\Delta(n)$ defined in (3.8.1.1) maps onto the (integral) algebra $\mathcal{S}_\Delta(n,r)$. Since $\mathfrak{H}_\Delta(n)^\pm$ are generated by semisimple generators (see [**13**, Th. 5.2(ii)]), it is natural to expect that the commutator formulas given in Corollary 2.6.7 hold in $\widetilde{\mathfrak{D}}_\Delta(n)$. Naturally, if the following conjecture was true, we would call $\widetilde{\mathfrak{D}}_\Delta(n)$ an *integral form of Lusztig type* for $\boldsymbol{\mathfrak{D}}_\Delta(n)$.

Conjecture 3.8.6. The $\mathcal{Z}$-module $\widetilde{\mathfrak{D}}_\triangle(n)$ is a subalgebra of $\boldsymbol{\mathfrak{D}}_\triangle(n)$.

Since both $\mathfrak{H}_\triangle(n)^+\mathfrak{D}_\triangle(n)^0$ and $\mathfrak{D}_\triangle(n)^0\mathfrak{H}_\triangle(n)^-$ are subalgebras, this conjecture is equivalent to proving that all coefficients $c(\lambda, \mu, \alpha) := \sum_{0\leqslant\gamma\leqslant\alpha} x_{\alpha,\gamma}\widetilde{K}_{2\gamma-\alpha}$ appearing in Corollary 2.6.7, are in $\mathfrak{D}_\triangle(n)^0$.

We make some comparisons with the integral form $\mathfrak{D}_\triangle(n)$ for $\boldsymbol{\mathfrak{D}}_\triangle(n)$ introduced in Definition 2.4.4 and the restricted integral form discussed in [**28**, §7.2].

Remarks 3.8.7. (1) Since the integral composition algebra $\mathfrak{C}_\triangle(n)^\pm$ is a $\mathcal{Z}$-subalgebra of the integral Ringel–Hall algebra $\mathfrak{H}_\triangle(n)^\pm$ which also contains the central generators $\mathsf{z}_m^\pm$ by Corollary 3.7.5, it follows that

$$\mathfrak{D}_\triangle(n) \subset \widetilde{\mathfrak{D}}_\triangle(n) \subset \boldsymbol{\mathfrak{D}}_\triangle(n). \tag{3.8.7.1}$$

However, we will see in Remark 5.3.8 that the restriction to $\mathfrak{D}_\triangle(n)$ of the homomorphism ξ_r in general does not map onto the integral affine quantum Schur algebra $\mathcal{S}_\triangle(n,r)$. Thus, $\mathfrak{D}_\triangle(n) \neq \widetilde{\mathfrak{D}}_\triangle(n)$, and we cannot use this integral form to get the Schur–Weyl theory at the roots of unity.

(2) The restricted integral form $U_v^{\text{res}}(\widehat{\mathfrak{gl}}_n)$ is the $\mathbb{C}[v, v^{-1}]$-subalgebra of $\mathbf{U}(\widehat{\mathfrak{gl}}_n)$ generated by divided powers $(\mathsf{x}_{i,s}^\pm)^{(m)}$, $\mathsf{k}_i^\pm$, $\left[{\mathsf{k}_i;0 \atop t}\right]$, and $\frac{\mathsf{g}_{i,m}}{[m]}$ (see [**28**, §7.2]). If we identify $\boldsymbol{\mathfrak{D}}_\triangle(n)$ with $\mathbf{U}(\widehat{\mathfrak{gl}}_n)$ under the isomorphism $\mathcal{E}_\mathrm{H}$ given in (2.5.2.1) (see Theorem 2.5.3), (2.5.1.1) implies that $\mathsf{z}_m^\pm = \theta_{\pm m} \in U_v^{\text{res}}(\widehat{\mathfrak{gl}}_n)$ for all $m \geqslant 1$. Thus, $\mathfrak{D}_\triangle(n) \otimes_\mathbb{Z} \mathbb{C}$ is a subalgebra of $U_v^{\text{res}}(\widehat{\mathfrak{gl}}_n)$. However, it is not clear if $U_v^{\text{res}}(\widehat{\mathfrak{gl}}_n)$ is a subalgebra of $\widetilde{\mathfrak{D}}_\triangle(n) \otimes_\mathbb{Z} \mathbb{C}$. Also, if we assume the conjecture, then both $\mathfrak{D}_\triangle(n)$ and $\widetilde{\mathfrak{D}}_\triangle(n)$ are Hopf subalgebras. But, as pointed out in [**28**], it is not known if $U_v^{\text{res}}(\widehat{\mathfrak{gl}}_n)$ is a Hopf subalgebra.

The surjective homomorphism $\xi_{r,\mathbb{C}}^\vee : \mathcal{H}_\triangle(r)_\mathbb{C} \to \mathrm{End}_{\mathcal{S}_\triangle(n,r)_\mathbb{C}}(\Omega_\mathbb{C}^{\otimes r})^{\mathrm{op}}$ for $v = \sqrt{q}$ was established by a geometric method. We do not know in general if the surjectivity holds over $\mathbb{Q}(v)$. Since both $\xi_r : \boldsymbol{\mathfrak{D}}_\triangle(n) \to \boldsymbol{\mathcal{S}}_\triangle(n,r)$ and $\xi_{r,\mathbb{C}} : \mathfrak{D}_{\triangle,\mathbb{C}}(n) \to \mathcal{S}_\triangle(n,r)_\mathbb{C}$ are surjective, the following conjecture gives the affine Schur–Weyl reciprocity over $\mathbb{Q}(v)$ and $\mathbb{C}$ for a non-root-of-unity specialization.

Conjecture 3.8.8. For $n < r$, the algebra homomorphisms

$$\xi_r^\vee : \boldsymbol{\mathcal{H}}_\triangle(r) \longrightarrow \mathrm{End}_{\boldsymbol{\mathcal{S}}_\triangle(n,r)}(\boldsymbol{\Omega}^{\otimes r})^{\mathrm{op}} \text{ and } \xi_{r,\mathbb{C}}^\vee : \mathcal{H}_\triangle(r)_\mathbb{C} \longrightarrow \mathrm{End}_{\mathcal{S}_\triangle(n,r)_\mathbb{C}}(\Omega_\mathbb{C}^{\otimes r})^{\mathrm{op}}$$

are surjective, where a base change to $\mathbb{C}$ is obtained by specializing v to a non-root-of-unity $z \in \mathbb{C}$.

With the truth of Conjecture 3.8.6, specializing v to any element in a field $\mathbb{F}$ (of any characteristic) results in a surjective homomorphism $\widetilde{\mathfrak{D}}_\Delta(n) \otimes \mathbb{F} \to \mathcal{S}_\Delta(n,r) \otimes \mathbb{F}$. It is natural to further expect that the affine Schur–Weyl reciprocity holds at roots of unity.

Conjecture 3.8.9. The affine quantum Schur–Weyl reciprocity over any field $\mathbb{F}$ holds.

3.9. Polynomial identity arising from semisimple generators

In this last section we will give an application of our theory. We use the commutator formulas in Theorem 2.6.3(5) to derive a certain polynomial identity which seems to be interesting in its own right.

For $\lambda \in \mathbb{N}^n_\Delta$, set as in (1.2.0.1)

$$S_\lambda = \bigoplus_{i=1}^{n} \lambda_i S_i \quad \text{and} \quad A_\lambda = \sum_{i=1}^{n} \lambda_i E^\Delta_{i,i+1} \in \Theta^+_\Delta(n).$$

Then, $\mathfrak{d}(A_\lambda) = \dim S_\lambda = \sigma(\lambda)$ and $M(A_\lambda) = S_\lambda$. Furthermore, for $\lambda, \alpha, \beta \in \mathbb{N}^n_\Delta$, set $\varphi^\lambda_{\alpha,\beta}$ to be the Hall polynomial $\varphi^{A_\lambda}_{A_\alpha,A_\beta}$ and

$$\left[\!\!\left[\begin{matrix} \lambda \\ \alpha \end{matrix} \right]\!\!\right] = \left[\!\!\left[\begin{matrix} \lambda_1 \\ \alpha_1 \end{matrix} \right]\!\!\right] \left[\!\!\left[\begin{matrix} \lambda_2 \\ \alpha_2 \end{matrix} \right]\!\!\right] \cdots \left[\!\!\left[\begin{matrix} \lambda_n \\ \alpha_n \end{matrix} \right]\!\!\right] \qquad \text{(cf. (2.6.6.1))}.$$

Recall from (1.1.0.3) that, for $\lambda = (\lambda_i)$, $\mu = (\mu_i) \in \mathbb{N}^n_\Delta$, $\lambda \leqslant \mu$ means $\lambda_i \leqslant \mu_i$ for all i. Recall also from (1.2.0.6) and (1.2.0.3) the number d'_A and the polynomial $\mathfrak{a}_A$. For semisimple modules, we have the following easy formulas.

Lemma 3.9.1. *Let $\lambda, \alpha \in \mathbb{N}^n_\Delta$ satisfy $\alpha \leqslant \lambda$. Then*

$$\varphi^\lambda_{\alpha,\lambda-\alpha} = \left[\!\!\left[\begin{matrix} \lambda \\ \alpha \end{matrix} \right]\!\!\right], \quad d'_{A_\lambda} = \sum_{1\leqslant i\leqslant n} (\lambda_i^2 - \lambda_i), \quad \text{and} \quad \mathfrak{a}_\lambda := \mathfrak{a}_{A_\lambda} = \prod_{\substack{1\leqslant i\leqslant n \\ 0\leqslant s\leqslant \lambda_i-1}} (v^{2\lambda_i} - v^{2s}).$$

In particular, $\varphi^\lambda_{\lambda-\alpha,\alpha} = \varphi^\lambda_{\alpha,\lambda-\alpha} = v^{2\sum_{i=1}^n \alpha_i(\alpha_i-\lambda_i)} \frac{\mathfrak{a}_\lambda}{\mathfrak{a}_\alpha \mathfrak{a}_{\lambda-\alpha}}$.

We will also use the abbreviation for the elements

$$\varphi^{A_1,B_1}_{A,B} = \frac{\mathfrak{a}_{A_1}\mathfrak{a}_{B_1}}{\mathfrak{a}_A \mathfrak{a}_B} \sum_{A_2 \in \Theta^+_\Delta(n)} v^{2\mathfrak{d}(A_2)} \mathfrak{a}_{A_2} \varphi^A_{A_1,A_2} \varphi^B_{B_1,A_2} \quad \text{and}$$

$$\widetilde{\varphi^{A_1,B_1}_{A,B}} = \frac{\mathfrak{a}_{A_1}\mathfrak{a}_{B_1}}{\mathfrak{a}_A \mathfrak{a}_B} \sum_{A_2 \in \Theta^+_\Delta(n)} v^{2\mathfrak{d}(A_2)} \mathfrak{a}_{A_2} \varphi^A_{A_2,A_1} \varphi^B_{A_2,B_1},$$

defined in (2.6.3.1) by setting, for $\lambda, \mu, \alpha, \beta \in \mathbb{N}^n_\Delta$,

$$\varphi^{\alpha,\beta}_{\lambda,\mu} = \varphi^{A_\alpha,A_\beta}_{A_\lambda,A_\mu} \text{ and } \widetilde{\varphi^{\alpha,\beta}_{\lambda,\mu}} = \widetilde{\varphi^{A_\alpha,A_\beta}_{A_\lambda,A_\mu}}.$$

Proposition 3.9.2. *For $\lambda, \mu \in \mathbb{N}^n_\Delta$ and $A, B \in \Theta^+_\Delta(n)$, if $\varphi^{A,B}_{A_\lambda,A_\mu}$ or $\widetilde{\varphi^{A,B}_{A_\lambda,A_\mu}}$ is non-zero, then there exist $\alpha, \beta \in \mathbb{N}^n_\Delta$ such that $A = A_\alpha$, $B = A_\beta$, $\lambda - \alpha = \mu - \beta \geqslant \mathbf{0}$, and*

$$\varphi^{\alpha,\beta}_{\lambda,\mu} = \widetilde{\varphi^{\alpha,\beta}_{\lambda,\mu}} = \frac{1}{\mathfrak{a}_{\lambda-\alpha}} v^{2\sigma(\lambda-\alpha)+\sum_{1\leqslant i\leqslant n} 2(\alpha_i(\alpha_i-\lambda_i)+\beta_i(\beta_i-\mu_i))}.$$

Proof. If either $\varphi^{A,B}_{A_\lambda,A_\mu} \neq 0$ or $\widetilde{\varphi^{A,B}_{A_\lambda,A_\mu}} \neq 0$, then $\varphi^{A_\lambda}_{A,C}\varphi^{A_\mu}_{B,C} \neq 0$ or $\varphi^{A_\lambda}_{C,A}\varphi^{A_\mu}_{C,B} \neq 0$ for some C. In either case, both $M(A)$ and $M(B)$, as submodules or quotient modules of a semisimple module, are semisimple, and $\lambda - \mathbf{d}(A) = \mathbf{d}(C) = \mu - \mathbf{d}(B)$. Write $A = A_\alpha$, $B = A_\beta$ for some $\alpha, \beta \in \mathbb{N}^n_\Delta$. Then $\lambda - \alpha = \mu - \beta$ and so $\mathfrak{a}_{\lambda-\alpha} = \mathfrak{a}_{\mu-\beta}$. By the last assertion of Lemma 3.9.1, one sees immediately that

$$\begin{aligned}\varphi^{A,B}_{A_\lambda,A_\mu} = \widetilde{\varphi^{A,B}_{A_\lambda,A_\mu}} &= \frac{\mathfrak{a}_\alpha\mathfrak{a}_\beta}{\mathfrak{a}_\lambda\mathfrak{a}_\mu}\left[\!\left[\begin{matrix}\lambda\\ \alpha\end{matrix}\right]\!\right]\left[\!\left[\begin{matrix}\mu\\ \beta\end{matrix}\right]\!\right]\mathfrak{a}_{\lambda-\alpha}v^{2\sigma(\lambda-\alpha)}\\ &= \frac{1}{\mathfrak{a}_{\lambda-\alpha}}v^{2\sigma(\lambda-\alpha)+\sum_{1\leqslant i\leqslant n} 2(\alpha_i(\alpha_i-\lambda_i)+\beta_i(\beta_i-\mu_i))},\end{aligned}$$

as required. □

Recall the surjective homomorphism $\xi_r : \mathfrak{D}_\Delta(n) \to \boldsymbol{\mathcal{S}}_\Delta(n,r)$ as explicitly described in Theorem 3.6.3. For $A, B \in \Theta^+_\Delta(n)$, let $X_{A,B} := \xi_r(L_{A,B})$ and $Y_{A,B} := \xi_r(R_{A,B})$, where $L_{A,B}$ (resp., $R_{A,B}$) are the LHS (resp., RHS) of the commutator relation given in Theorem 2.6.3(5). Then $X_{A,B} = Y_{A,B}$. In fact, since these commutator formulas continue to hold in $\mathfrak{D}_{\Delta\mathbb{C}}(n)$, where v is specialized to a non-root-of-unity in $\mathbb{C}$ (see Remark 2.1.4), $X_{A,B} = Y_{A,B}$ holds in $\mathcal{S}_\Delta(n,r)_\mathbb{C}$. In particular, for each prime power q, by specializing v to $\sqrt{q}$, we will view both $X_{A,B}$ and $Y_{A,B}$ as elements in the convolution algebra $\mathbb{C}_G(\mathscr{Y} \times \mathscr{Y}) \cong \mathcal{S}_\Delta(n,r)_\mathbb{C}$. In this case, denote $X_{A,B}$ and $Y_{A,B}$ by $X^q_{A,B}$ and $Y^q_{A,B}$, respectively. Thus, we have $X^q_{A,B} = Y^q_{A,B}$, where

$$\begin{aligned}X^q_{A,B} &= \sqrt{q}^{\langle\mathbf{d}(B),\mathbf{d}(B)\rangle}\sum_{A_1,B_1}\varphi^{A_1,B_1}_{A,B}(q)\sqrt{q}^{\langle\mathbf{d}(B_1),\mathbf{d}(A)+\mathbf{d}(B)-\mathbf{d}(B_1)\rangle-d'_{A_1}-d'_{B_1}}\\ &\qquad\times\widetilde{\mathfrak{k}}_{\mathbf{d}(B)-\mathbf{d}(B_1)} * ({}^tB_1)(\mathbf{0},r) * A_1(\mathbf{0},r) \quad\text{and}\\ Y^q_{A,B} &= \sqrt{q}^{\langle\mathbf{d}(B),\mathbf{d}(A)\rangle}\sum_{A_1,B_1}\widetilde{\varphi^{A_1,B_1}_{A,B}}(q)\sqrt{q}^{\langle\mathbf{d}(B)-\mathbf{d}(B_1),\mathbf{d}(A_1)\rangle+\langle\mathbf{d}(B),\mathbf{d}(B_1)\rangle-d'_{A_1}-d'_{B_1}}\\ &\qquad\times\widetilde{\mathfrak{k}}_{\mathbf{d}(B_1)-\mathbf{d}(B)} * A_1(\mathbf{0},r) * ({}^tB_1)(\mathbf{0},r).\end{aligned} \tag{3.9.2.1}$$

Here $\widetilde{\mathfrak{k}}_{\mathbf{a}} = \prod_{i=1}^n \mathfrak{k}_i^{a_i}\mathfrak{k}_{i+1}^{-a_i}$, for any $\mathbf{a} = (a_i) \in \mathbb{N}_\Delta^n$, with $\mathfrak{k}_i = \xi_r(K_i) = 0(\boldsymbol{e}_i^\Delta, r)$, for any $i \in I$.

In the following we are going to use the equality $X_{A,B}^q = Y_{A,B}^q$ to derive some interesting polynomial identities. In the rest of this section, we fix the finite field $\mathbb{F}$ with q elements. As in §3.1, let $\mathscr{Y} = \mathscr{F}_\Delta(q)$ be the set of all cyclic flags $\mathbf{L} = (L_i)_{i\in\mathbb{Z}}$ of lattices in a fixed $\mathbb{F}[\varepsilon, \varepsilon^{-1}]$-free module V of rank $r \geqslant 1$.

Lemma 3.9.3. *For* $\mathbf{a} \in \mathbb{N}_\Delta^n$ *and* $(\mathbf{L}, \mathbf{L}') \in \mathscr{Y} \times \mathscr{Y}$,

$$\widetilde{\mathfrak{k}}_{\mathbf{a}}(\mathbf{L}, \mathbf{L}') = \begin{cases} q^{\frac{1}{2}\langle \mathbf{a}, \lambda\rangle}, & \text{if } \mathbf{L} = \mathbf{L}' \text{ and } (\mathbf{L}, \mathbf{L}) \in \mathcal{O}_{\mathrm{diag}(\lambda)} \text{ for } \lambda \in \Lambda_\Delta(n, r); \\ 0, & \text{otherwise.} \end{cases}$$

Proof. Let $\lambda \in \Lambda_\Delta(n, r)$. By Lemma 3.7.1(1)&(2), $(\mathbf{L}, \mathbf{L}') \in \mathcal{O}_{\mathrm{diag}(\lambda)}$ if and only if $\mathbf{L} = \mathbf{L}'$ and $\lambda_i = \dim(L_i/L_{i-1})$ for all i. Thus, if $\widetilde{\mathfrak{k}}_{\mathbf{a}}(\mathbf{L}, \mathbf{L}') \neq 0$, then $\mathbf{L} = \mathbf{L}'$. Now we assume $\mathbf{L} = \mathbf{L}'$ and $(\mathbf{L}, \mathbf{L}) \in \mathcal{O}_{\mathrm{diag}(\lambda)}$. Since $\widetilde{\mathfrak{k}}_{\mathbf{a}} = \sum_{\mu\in\Lambda_\Delta(n,r)} q^{\frac{1}{2}\sum_{1\leqslant i\leqslant n}(a_i - a_{i-1})\mu_i}[\mathrm{diag}(\mu)]$ and $[\mathrm{diag}(\mu)] = \chi_{\mathcal{O}_{\mathrm{diag}(\mu)}}$, it follows that $\widetilde{\mathfrak{k}}_{\mathbf{a}}(\mathbf{L}, \mathbf{L}) = q^{\frac{1}{2}\sum_{1\leqslant i\leqslant n}(a_i - a_{i-1})\lambda_i} = q^{\frac{1}{2}\langle \mathbf{a}, \lambda\rangle}$. □

For integers N, t with $t \geqslant 0$, let $\left[\!\left[{N \atop t} \right]\!\right]_q = \prod_{1\leqslant i\leqslant t} \frac{q^{N-i+1}-1}{q^i - 1}$. For $\alpha, \beta \in \mathbb{N}_\Delta^n$ and $(\mathbf{L}, \mathbf{L}'') \in \mathscr{Y} \times \mathscr{Y}$, consider the subsets of $\mathscr{Y}$:

$$X(\alpha, \beta, \mathbf{L}, \mathbf{L}'') := \{\mathbf{L}' \in \mathscr{Y} \mid \mathbf{L}, \mathbf{L}'' \subseteq \mathbf{L}', \mathbf{L}'/\mathbf{L} \cong S_\beta, \mathbf{L}'/\mathbf{L}'' \cong S_\alpha\} \text{ and}$$
$$Y(\alpha, \beta, \mathbf{L}, \mathbf{L}'') := \{\mathbf{L}' \in \mathscr{Y} \mid \mathbf{L}' \subseteq \mathbf{L}, \mathbf{L}'', \mathbf{L}/\mathbf{L}' \cong S_\alpha, \mathbf{L}''/\mathbf{L}' \cong S_\beta\}.$$

Here, $\mathbf{L}, \mathbf{L}'' \subseteq \mathbf{L}'$ and $\mathbf{L}' \subseteq \mathbf{L}, \mathbf{L}''$ are the short form of $\mathbf{L} \subseteq \mathbf{L}'$, $\mathbf{L}'' \subseteq \mathbf{L}'$ and $\mathbf{L}' \subseteq \mathbf{L}$, $\mathbf{L}' \subseteq \mathbf{L}''$, respectively.

Lemma 3.9.4. *For* $\alpha, \beta \in \mathscr{Y}$ *and* $(\mathbf{L}, \mathbf{L}'') \in \mathscr{Y} \times \mathscr{Y}$, *if* $X(\alpha, \beta, \mathbf{L}, \mathbf{L}'') \neq \emptyset$ *or* $Y(\alpha, \beta, \mathbf{L}, \mathbf{L}'') \neq \emptyset$, *then* $L_i + L_i'' \subseteq L_{i+1} \cap L_{i+1}''$ *and*

$$\beta_i - \dim((L_i + L_i'')/L_i) = \alpha_i - \dim((L_i + L_i'')/L_i'') \geqslant 0, \ \text{for all } i \in \mathbb{Z}.$$

Moreover, in this case,

$$|X(\alpha, \beta, \mathbf{L}, \mathbf{L}'')| = \prod_{1\leqslant i\leqslant n} \left[\!\!\left[{\dim(L_{i+1} \cap L_{i+1}''/(L_i + L_i'')) \atop \alpha_i - \dim((L_i + L_i'')/L_i'')} \right]\!\!\right]_q \quad \text{and}$$
$$|Y(\alpha, \beta, \mathbf{L}, \mathbf{L}'')| = \prod_{1\leqslant i\leqslant n} \left[\!\!\left[{\dim(L_i \cap L_i''/(L_{i-1} + L_{i-1}'')) \atop \alpha_i - \dim((L_i + L_i'')/L_i'')} \right]\!\!\right]_q.$$

Proof. If, as representations of $\Delta(n)$, both $\mathbf{L}'/\mathbf{L} = (L_i'/L_i, f_i)$ and $\mathbf{L}'/\mathbf{L}'' = (L_i'/L_i'', f_i'')$ are semisimple, then the linear maps f_i, f_i'' induced from the inclusion $L'_i \subseteq L'_{i+1}$ are zero maps. This forces $L'_i \subseteq L_{i+1}$ and $L'_i \subseteq L''_{i+1}$ for all i. Hence,

$$X(\alpha,\beta,\mathbf{L},\mathbf{L}'') \neq \emptyset \iff \exists \mathbf{L}' \in X(\alpha,\beta,\mathbf{L},\mathbf{L}'')$$
$$\iff \begin{cases} L_t + L''_t \subseteq L'_t \subseteq L_{t+1} \cap L''_{t+1}, & \text{for } t \in \mathbb{Z}; \\ \dim(L'_t/L_t) = \beta_t,\ \dim(L'_t/L''_t) = \alpha_t, & \text{for } t \in \mathbb{Z} \end{cases}$$
$$\iff \begin{cases} L_t + L''_t \subseteq L'_t \subseteq L_{t+1} \cap L''_{t+1}, & \text{for } t \in \mathbb{Z}; \\ \dim(L'_t/(L_t + L''_t)) = \beta_t - \dim((L_t + L''_t)/L_t), & \text{for } t \in \mathbb{Z}; \\ \dim(L'_t/(L_t + L''_t)) = \alpha_t - \dim((L_t + L''_t)/L''_t), & \text{for } t \in \mathbb{Z}, \end{cases}$$

and

$$Y(\alpha,\beta,\mathbf{L},\mathbf{L}'') \neq \emptyset \iff \exists \mathbf{L}' \in Y(\alpha,\beta,\mathbf{L},\mathbf{L}'')$$
$$\iff \begin{cases} L'_t \subseteq L_t \subseteq L'_{t+1},\ \dim(L_t/L'_t) = \alpha_t, & \text{for } t \in \mathbb{Z}; \\ L'_t \subseteq L''_t \subseteq L'_{t+1},\ \dim(L''_t/L'_t) = \beta_t, & \text{for } t \in \mathbb{Z} \end{cases}$$
$$\iff \begin{cases} L_{t-1} \subseteq L'_t \subseteq L_t,\ \dim(L_t/L'_t) = \alpha_t, & \text{for } t \in \mathbb{Z}; \\ L''_{t-1} \subseteq L'_t \subseteq L''_t,\ \dim(L''_t/L'_t) = \beta_t, & \text{for } t \in \mathbb{Z} \end{cases}$$
$$\iff \begin{cases} L_{t-1} + L''_{t-1} \subseteq L'_t \subseteq L_t \cap L''_t, & \text{for } t \in \mathbb{Z}; \\ \dim(L_t \cap L''_t/L'_t) = \alpha_t - \dim((L_t + L''_t)/L''_t), & \text{for } t \in \mathbb{Z}; \\ \dim(L_t \cap L''_t/L'_t) = \beta_t - \dim((L_t + L''_t)/L_t), & \text{for } t \in \mathbb{Z}. \end{cases}$$

The rest of the proof is clear. □

For any $\lambda = (\lambda_i)_{i\in\mathbb{Z}}, z = (z_i)_{i\in\mathbb{Z}} \in \mathbb{N}^n_\Delta$, define the polynomials in v^2 over $\mathbb{Z}$:

$$P_{\lambda,z}(v^2) = \sum_{\substack{0\leqslant \nu\leqslant\lambda \\ \nu\in\mathbb{N}^n_\Delta}} v^{2\sum_{1\leqslant i\leqslant n}\left(\frac{\nu_i^2-\nu_i}{2}+(\lambda_i-\nu_i)(z_i-\nu_{i-1})\right)} \times \prod_{i=1}^n (v^2-1)^{\nu_i} [\![\nu_i]\!]^! \left[\!\left[\begin{matrix}\lambda_i\\ \nu_i\end{matrix}\right]\!\right] \left[\!\left[\begin{matrix}z_{i+1}\\ \nu_i\end{matrix}\right]\!\right]$$

and

$$P'_{\lambda,z}(v^2) = \sum_{\substack{0\leqslant \nu\leqslant\lambda \\ \nu\in\mathbb{N}^n_\Delta}} v^{2\sum_{1\leqslant i\leqslant n}\left(\frac{\nu_i^2-\nu_i}{2}+(\lambda_i-\nu_i)(z_{i+1}-\nu_{i+1})\right)} \times \prod_{i=1}^n (v^2-1)^{\nu_i} [\![\nu_i]\!]^! \left[\!\left[\begin{matrix}\lambda_i\\ \nu_i\end{matrix}\right]\!\right] \left[\!\left[\begin{matrix}z_i\\ \nu_i\end{matrix}\right]\!\right].$$

We now prove that these polynomials occur naturally in the coefficients of $X_{A,B}, Y_{A,B}$ for $A, B \in \Theta_\triangle^+(n)^{ss}$ when they are written as a linear combination of e_C, $C \in \Theta_\triangle(n,r)$.

Theorem 3.9.5. *For $x, y \in \Lambda_\triangle(n,r)$, $(\mathbf{L}, \mathbf{L}'') \in \mathscr{F}_{\triangle,x}(q) \times \mathscr{F}_{\triangle,y}(q)$, and $\lambda, \mu \in \mathbb{N}_\triangle^n$, let $\widehat{\lambda} := \lambda - \gamma$ and $\widehat{\mu} := \mu - \delta$, where*

$$\gamma = (\dim(L_i + L_i'')/L_i'')_{i\in\mathbb{Z}} \text{ and } \delta = (\dim(L_i + L_i'')/L_i)_{i\in\mathbb{Z}}.$$

If $X_{\lambda,\mu}^q(\mathbf{L}, \mathbf{L}'') := X_{A_\lambda,A_\mu}^q(\mathbf{L}, \mathbf{L}'') \neq 0$ or $Y_{\lambda,\mu}^q(\mathbf{L}, \mathbf{L}'') := Y_{A_\lambda,A_\mu}^q(\mathbf{L}, \mathbf{L}'') \neq 0$, then $L_i + L_i'' \subseteq L_{i+1} \cap L_{i+1}''$ for all $i \in \mathbb{Z}$ and $\widehat{\lambda} = \widehat{\mu} \geqslant 0$. Moreover, putting $z = (z_i)$ with $z_i = \dim(L_i \cap L_i''/(L_{i-1} + L_{i-1}''))$, for all $i \in \mathbb{Z}$, we have in this case:

$$X_{\lambda,\mu}^q(\mathbf{L}, \mathbf{L}'') = q^{\frac{1}{2}(f_{\widehat{\lambda}} + \sum_{1\leqslant i\leqslant n}(\lambda_i + \mu_i + \widehat{\lambda}_i - \widehat{\lambda}_i^2))} \prod_{\substack{1\leqslant i\leqslant n\\ 1\leqslant s\leqslant \widehat{\lambda}_i}} \frac{1}{q^s - 1} \cdot P_{\widehat{\lambda},z}(q) \text{ and}$$

$$Y_{\lambda,\mu}^q(\mathbf{L}, \mathbf{L}'') = q^{\frac{1}{2}(f_{\widehat{\lambda}} + \sum_{1\leqslant i\leqslant n}(\lambda_i + \mu_i + \widehat{\lambda}_i - \widehat{\lambda}_i^2))} \prod_{\substack{1\leqslant i\leqslant n\\ 1\leqslant s\leqslant \widehat{\lambda}_i}} \frac{1}{q^s - 1} \cdot P'_{\widehat{\lambda},z}(q),$$

where $f_{\widehat{\lambda}} = \sum_{1\leqslant i\leqslant n}\big(\lambda_i(\lambda_{i-1} - \lambda_i - y_i) - x_{i+1}(\delta_i + \widehat{\lambda}_i)\big) + \langle\mu,\mu\rangle + \langle\delta + \widehat{\lambda}, \lambda\rangle$.

Proof. We need to compute the value of (3.9.2.1) at $(\mathbf{L}, \mathbf{L}'')$ for $A = A_\lambda$, $B = A_\mu$ (hence, $A_1 = A_\alpha$ and $B_1 = A_\beta$). By Corollary 3.6.2 and Lemma 3.9.3, and noting $\dim \mathfrak{O}_{A_\nu} = 0$ for $\nu \in \mathbb{N}_\triangle^n$,

$$\begin{aligned}(\widetilde{\mathfrak{k}}_{\mu-\beta} * {}^tA_\beta(\mathbf{0}, r) * A_\alpha(\mathbf{0}, r))(\mathbf{L}, \mathbf{L}'')\\ &= \sum_{\mathbf{L}'\in\mathscr{Y}} (\widetilde{\mathfrak{k}}_{\mu-\beta} * {}^tA_\beta(\mathbf{0}, r))(\mathbf{L}, \mathbf{L}')A_\alpha(\mathbf{0}, r)(\mathbf{L}', \mathbf{L}'')\\ &= \sum_{\substack{\mathbf{L}'\in\mathscr{Y},\mathbf{L},\mathbf{L}''\subseteq\mathbf{L}'\\ \mathbf{L}'/\mathbf{L}\cong S_\beta,\mathbf{L}'/\mathbf{L}''\cong S_\alpha}} q^{\frac{1}{2}(\langle\mu-\beta,x\rangle - a(\mathbf{L},\mathbf{L}') - c(\mathbf{L}',\mathbf{L}''))},\end{aligned}$$

where $a(\mathbf{L}, \mathbf{L}')$, $c(\mathbf{L}', \mathbf{L}'')$ are defined in (3.6.0.1). Thus, if $X_{\lambda,\mu}^q(\mathbf{L}, \mathbf{L}'') \neq 0$ or $Y_{\lambda,\mu}^q(\mathbf{L}, \mathbf{L}'') \neq 0$, then some $\varphi_{A_\lambda,A_\mu}^{A,B} = \widetilde{\varphi_{A_\lambda,A_\mu}^{A,B}} \neq 0$. So applying the first assertion of Proposition 3.9.2 yields

$$\begin{aligned}X_{\lambda,\mu}^q(\mathbf{L}, \mathbf{L}'') = &\sum_{\substack{\alpha,\beta\in\mathbb{N}_\triangle^n\\ \lambda-\alpha=\mu-\beta\geqslant 0}} \varphi_{\lambda,\mu}^{\alpha,\beta}(q) q^{\frac{1}{2}(\langle\mu,\mu\rangle + \langle\beta,\mu-\beta\rangle + \langle\beta,\lambda\rangle - d'_{A_\beta} - d'_{A_\alpha})}\\ &\times \sum_{\substack{\mathbf{L}'\in\mathscr{Y},\mathbf{L},\mathbf{L}''\subseteq\mathbf{L}'\\ \mathbf{L}'/\mathbf{L}\cong S_\beta,\mathbf{L}'/\mathbf{L}''\cong S_\alpha}} q^{\frac{1}{2}(\langle\mu-\beta,x\rangle - a(\mathbf{L},\mathbf{L}') - c(\mathbf{L}',\mathbf{L}''))}.\end{aligned}$$

Similarly,

$$Y^q_{\lambda,\mu}(\mathbf{L},\mathbf{L}'') = \sum_{\substack{\alpha,\beta\in\mathbb{N}^n_\Delta \\ \lambda-\alpha=\mu-\beta\geqslant 0}} \varphi^{\alpha,\beta}_{\lambda,\mu}(q) q^{\frac{1}{2}(\langle\mu,\lambda\rangle+\langle\mu-\beta,\alpha\rangle+\langle\mu,\beta\rangle-d'_{A_\beta}-d'_{A_\alpha})}$$
$$\times \sum_{\substack{\mathbf{L}'\in\mathscr{Y},\mathbf{L}'\subseteq\mathbf{L},\mathbf{L}'' \\ \mathbf{L}/\mathbf{L}'\cong S_\alpha,\mathbf{L}''/\mathbf{L}'\cong S_\beta}} q^{\frac{1}{2}(\langle\beta-\mu,x\rangle-a(\mathbf{L}',\mathbf{L}'')-c(\mathbf{L},\mathbf{L}'))}.$$

If $\mathbf{L}'\in\mathscr{Y}$ satisfies $\mathbf{L},\mathbf{L}''\subseteq\mathbf{L}'$, $\mathbf{L}'/\mathbf{L}\cong S_\beta$, and $\mathbf{L}'/\mathbf{L}''\cong S_\alpha$, then

$$a(\mathbf{L},\mathbf{L}')+c(\mathbf{L}',\mathbf{L}'') = \sum_{1\leqslant i\leqslant n}\beta_i(x_{i+1}-\beta_i)+\sum_{1\leqslant i\leqslant n}\alpha_{i+1}(y_{i+1}-\alpha_i) =: \heartsuit.$$

Likewise, if $\mathbf{L}'\in\mathscr{Y}$ satisfies $\mathbf{L}'\subseteq\mathbf{L},\mathbf{L}''$, $\mathbf{L}/\mathbf{L}'\cong S_\alpha$, and $\mathbf{L}''/\mathbf{L}'\cong S_\beta$, then $L_i\subseteq L'_{i+1}$ and $L''_i\subseteq L'_{i+1}$ for all i. Thus,

$$(L'_{i+1}/L'_i)/(L_i/L'_i)\cong L'_{i+1}/L_i \quad\text{and}\quad (L_{i+1}/L_i)/(L'_{i+1}/L_i)\cong L_{i+1}/L'_{i+1},$$

and so

$$\dim(L'_{i+1}/L'_i) = x_{i+1}-\alpha_{i+1}+\alpha_i.$$

Similarly, $\dim(L'_{i+1}/L'_i) = y_{i+1}-\beta_{i+1}+\beta_i$. Hence,

$$\begin{aligned}
&a(\mathbf{L}',\mathbf{L}'')+c(\mathbf{L},\mathbf{L}')\\
&= \sum_{1\leqslant i\leqslant n}\beta_i(\dim(L'_{i+1}/L'_i)-\beta_i)+\sum_{1\leqslant i\leqslant n}\alpha_{i+1}(\dim(L'_{i+1}/L'_i)-\alpha_i)\\
&= \sum_{1\leqslant i\leqslant n}\beta_i(x_{i+1}-\alpha_{i+1}+\alpha_i-\beta_i)+\sum_{1\leqslant i\leqslant n}\alpha_{i+1}(y_{i+1}-\beta_{i+1}+\beta_i-\alpha_i)\\
&= \sum_{1\leqslant i\leqslant n}\beta_i(x_{i+1}-\beta_i)+\sum_{1\leqslant i\leqslant n}\alpha_{i+1}(y_{i+1}-\alpha_i) = \heartsuit,
\end{aligned}$$

and consequently,

$$X^q_{\lambda,\mu}(\mathbf{L},\mathbf{L}'') = \sum_{\substack{\alpha,\beta\in\mathbb{N}^n_\Delta \\ \lambda-\alpha=\mu-\beta\geqslant 0}} \varphi^{\alpha,\beta}_{\lambda,\mu}(q) q^{\frac{1}{2}(-d'_{A_\alpha}-d'_{A_\beta}-\heartsuit)}$$
$$\times\, q^{\frac{1}{2}(\langle\mu,\mu\rangle+\langle\beta,\mu-\beta\rangle+\langle\beta,\lambda\rangle+\langle\mu-\beta,x\rangle)}|X(\alpha,\beta,\mathbf{L},\mathbf{L}'')|$$

and

$$Y^q_{\lambda,\mu}(\mathbf{L},\mathbf{L}'') = \sum_{\substack{\alpha,\beta\in\mathbb{N}^n_\Delta \\ \lambda-\alpha=\mu-\beta\geqslant 0}} \varphi^{\alpha,\beta}_{\lambda,\mu}(q) q^{\frac{1}{2}(-d'_{A_\alpha}-d'_{A_\beta}-\heartsuit)}$$
$$\times\, q^{\frac{1}{2}(\langle\mu,\lambda\rangle+\langle\mu-\beta,\alpha\rangle+\langle\mu,\beta\rangle+\langle\beta-\mu,x\rangle)}|Y(\alpha,\beta,\mathbf{L},\mathbf{L}'')|.$$

Thus, $X^q_{\lambda,\mu}(\mathbf{L},\mathbf{L}'')\neq 0$ or $Y^q_{\lambda,\mu}(\mathbf{L},\mathbf{L}'')\neq 0$ implies some $X(\alpha,\beta,\mathbf{L},\mathbf{L}'')\neq\emptyset$ or $Y(\alpha,\beta,\mathbf{L},\mathbf{L}'')\neq\emptyset$. Hence, by Lemma 3.9.4, $L_i+L''_i\subseteq L_{i+1}\cap L''_{i+1}$ for all i and $\beta-\delta=\alpha-\gamma\geqslant 0$. The latter together with $\lambda-\alpha=\mu-\beta\geqslant 0$ implies

$$\widehat{\lambda}-\widehat{\mu}=(\lambda-\mu)-(\gamma-\delta)=(\alpha-\beta)-(\gamma-\delta)=0 \text{ and}$$
$$\widehat{\lambda}=\lambda-\gamma\geqslant\alpha-\gamma\geqslant 0.$$

So we have proved the first assertion.

It remains to simplify $X^q_{\lambda,\mu}(\mathbf{L},\mathbf{L}'')$ and $Y^q_{\lambda,\mu}(\mathbf{L},\mathbf{L}'')$ under the assumption that $L_i+L''_i\subseteq L_{i+1}\cap L''_{i+1}$ for all i and $\widehat{\lambda}=\widehat{\mu}\geqslant 0$. First, for $z_i=\dim(L_i\cap L''_i/(L_{i-1}+L''_{i-1}))$, Lemma 3.9.4 gives

$$|X(\alpha,\beta,\mathbf{L},\mathbf{L}'')|=\prod_{1\leqslant i\leqslant n}\begin{bmatrix} z_i+1\\ \alpha_i-\gamma_i\end{bmatrix}_q.$$

Second, for $\alpha,\beta\in\mathbb{N}^n_\Delta$ satisfying $\lambda-\alpha=\mu-\beta\geqslant 0$, Lemma 3.9.1 and Proposition 3.9.2 imply

$$\begin{aligned}
&\varphi^{\alpha,\beta}_{\lambda,\mu}(q)q^{\frac12(-d'_{A_\alpha}-d'_{A_\beta}-\heartsuit)}\\
&=\frac{q^{\sum_{1\leqslant i\leqslant n}(\lambda_i-\alpha_i+\alpha_i(\alpha_i-\lambda_i)+\beta_i(\beta_i-\mu_i))}}{|\operatorname{Aut}(S_{\lambda-\alpha})|}\\
&\quad\times q^{\frac12\sum_{1\leqslant i\leqslant n}\left((\beta_i-\beta_i^2+\alpha_i-\alpha_i^2)-\beta_i(x_{i+1}-\beta_i)-\alpha_i(y_i-\alpha_{i-1})\right)}\\
&=q^{\frac12\sum_{1\leqslant i\leqslant n}(\lambda_i+\mu_i)}\frac{q^{\frac12\sum_{1\leqslant i\leqslant n}\left(\alpha_i(\alpha_i+\alpha_{i-1}-2\lambda_i-y_i)+\beta_i(2\beta_i-2\mu_i-x_{i+1})\right)}}{|\operatorname{Aut}(S_{\lambda-\alpha})|},
\end{aligned}$$

since $2\lambda_i-\alpha_i+\beta_i=\lambda_i+\mu_i$ for all i. Hence,

$$\begin{aligned}
&X^q_{\lambda,\mu}(\mathbf{L},\mathbf{L}'')\\
&=q^{\frac12\sum_{1\leqslant i\leqslant n}(\lambda_i+\mu_i)}\sum_{\substack{\alpha,\beta\in\mathbb{N}^n_\Delta\\ \lambda-\alpha=\mu-\beta\geqslant 0}}\frac{q^{\frac12\sum_{1\leqslant i\leqslant n}\left(\alpha_i(\alpha_i+\alpha_{i-1}-2\lambda_i-y_i)+\beta_i(2\beta_i-2\mu_i-x_{i+1})\right)}}{|\operatorname{Aut}(S_{\lambda-\alpha})|}\\
&\qquad\times q^{\frac12(\langle\mu,\mu\rangle+\langle\beta,\mu-\beta\rangle+\langle\beta,\lambda\rangle+\langle\mu-\beta,x\rangle)}\prod_{1\leqslant i\leqslant n}\begin{bmatrix} z_i+1\\ \alpha_i-\gamma_i\end{bmatrix}_q.
\end{aligned}$$

Here we have implicitly assumed $\alpha\geqslant\gamma$ (or equivalently, $\beta\geqslant\delta$). Setting $\nu=\alpha-\gamma=\beta-\delta$ gives

$$X^q_{\lambda,\mu}(\mathbf{L},\mathbf{L}'')=q^{\frac12\sum_{1\leqslant i\leqslant n}(\lambda_i+\mu_i)}\sum_{\substack{\nu\in\mathbb{N}^n_\Delta\\ 0\leqslant\nu\leqslant\lambda-\gamma}}\frac{q^{\frac12 f_\nu}}{|\operatorname{Aut}(S_{\lambda-\gamma-\nu})|}\prod_{1\leqslant i\leqslant n}\begin{bmatrix} z_i+1\\ \nu_i\end{bmatrix}_q,$$

where

$$f_\nu = \sum_{1\leqslant i\leqslant n} \big((\gamma_i+\nu_i)(\gamma_i+\nu_i+\gamma_{i-1}+\nu_{i-1}-2\lambda_i-y_i) + (\delta_i+\nu_i)(2(\gamma_i+\nu_i-\lambda_i)-x_{i+1})\big) + \langle\mu,\mu\rangle + \langle\delta+\nu,\lambda-(\gamma+\nu)\rangle + \langle\delta+\nu,\lambda\rangle + \langle\lambda-(\gamma+\nu),x\rangle, \tag{3.9.5.1}$$

for $\nu\in\mathbb{N}^n_\Delta$ with $0\leqslant\nu\leqslant\widehat{\lambda}$. Similarly,

$$\begin{aligned} &Y^q_{\lambda,\mu}(\mathbf{L},\mathbf{L}'') \\ &= q^{\frac12\sum_{1\leqslant i\leqslant n}(\lambda_i+\mu_i)} \sum_{\substack{\alpha,\beta\in\mathbb{N}^n_\Delta\\ \lambda-\alpha=\mu-\beta\geqslant 0}} \frac{q^{\frac12\sum_{1\leqslant i\leqslant n}\left(\alpha_i(\alpha_i+\alpha_{i-1}-2\lambda_i-y_i)+\beta_i(2\beta_i-2\mu_i-x_{i+1})\right)}}{|\operatorname{Aut}(S_{\lambda-\alpha})|} \\ &\qquad \times q^{\frac12(\langle\mu,\lambda\rangle+\langle\mu-\beta,\alpha\rangle+\langle\mu,\beta\rangle+\langle\beta-\mu,x\rangle)} \prod_{1\leqslant i\leqslant n}\begin{bmatrix} z_i\\ \alpha_i-\gamma_i\end{bmatrix}_q \\ &= q^{\frac12\sum_{1\leqslant i\leqslant n}(\lambda_i+\mu_i)} \sum_{\substack{\nu\in\mathbb{N}^n_\Delta\\ 0\leqslant\nu\leqslant\lambda-\gamma}} \frac{q^{\frac12 g_\nu}}{|\operatorname{Aut}(S_{\lambda-\gamma-\nu})|}\prod_{1\leqslant i\leqslant n}\begin{bmatrix} z_i\\ \nu_i\end{bmatrix}_q, \end{aligned}$$

where

$$g_\nu = \sum_{1\leqslant i\leqslant n} \big((\gamma_i+\nu_i)(\gamma_i+\nu_i+\gamma_{i-1}+\nu_{i-1}-2\lambda_i-y_i) + (\delta_i+\nu_i)(2(\gamma_i+\nu_i-\lambda_i)-x_{i+1})\big) + \langle\mu,\lambda\rangle + \langle\lambda-\gamma-\nu,\gamma+\nu\rangle + \langle\mu,\delta+\nu\rangle + \langle\gamma+\nu-\lambda,x\rangle. \tag{3.9.5.2}$$

Simplifying $f_{\widehat{\lambda}}-f_\nu$ and $g_{\widehat{\lambda}}-g_\nu$ and substituting, we obtain, by Lemma 3.10.3 in the appendix,

$$\begin{aligned} &\frac{X^q_{\lambda,\mu}(\mathbf{L},\mathbf{L}'')}{q^{\frac12\sum_{1\leqslant i\leqslant n}(\lambda_i+\mu_i)+\frac12 f_{\widehat{\lambda}}}} \\ &= \sum_{\substack{0\leqslant\nu\leqslant\widehat{\lambda}\\ \nu\in\mathbb{N}^n_\Delta}} \frac{q^{\sum_{1\leqslant i\leqslant n}(\widehat{\lambda}_i z_i-\nu_i(z_i+\widehat{\lambda}_i+\widehat{\lambda}_{i+1}-\nu_i-\nu_{i+1}))}}{|\operatorname{Aut}(S_{\widehat{\lambda}-\nu})|}\prod_{1\leqslant i\leqslant n}\begin{bmatrix} z_{i+1}\\ \nu_i\end{bmatrix}_q \\ &= \sum_{\substack{0\leqslant\nu\leqslant\widehat{\lambda}\\ \nu\in\mathbb{N}^n_\Delta}} h_\nu q^{\sum_{1\leqslant i\leqslant n}(\widehat{\lambda}_i z_i-\nu_i z_i-\nu_i\widehat{\lambda}_{i+1})} \prod_{\substack{1\leqslant i\leqslant n\\ 1\leqslant s\leqslant\nu_i}}(q^{z_{i+1}-s+1}-1) \end{aligned}$$

and

$$\frac{Y^q_{\lambda,\mu}(\mathbf{L}, , \mathbf{L}'')}{q^{\frac{1}{2}\sum_{1\leqslant i\leqslant n}(\lambda_i+\mu_i)+\frac{1}{2}f_{\widehat{\lambda}}}}$$
$$=\sum_{\substack{0\leqslant\nu\leqslant\widehat{\lambda}\\ \nu\in\mathbb{N}^n_\Delta}}\frac{q^{\sum_{1\leqslant i\leqslant n}(\widehat{\lambda}_i z_{i+1}-\nu_i(z_{i+1}+\widehat{\lambda}_i+\widehat{\lambda}_{i-1}-\nu_i-\nu_{i+1}))}}{|\operatorname{Aut}(S_{\widehat{\lambda}-\nu})|}\prod_{1\leqslant i\leqslant n}\begin{bmatrix} z_i\\ \nu_i\end{bmatrix}_q$$
$$=\sum_{\substack{0\leqslant\nu\leqslant\widehat{\lambda}\\ \nu\in\mathbb{N}^n_\Delta}}h_\nu q^{\sum_{1\leqslant i\leqslant n}(\widehat{\lambda}_i z_{i+1}-\nu_i z_{i+1}-\nu_i\widehat{\lambda}_{i-1})}\prod_{\substack{1\leqslant i\leqslant n\\ 1\leqslant s\leqslant\nu_i}}(q^{z_i-s+1}-1),$$

where

$$h_\nu=\frac{q^{-\sum_{1\leqslant i\leqslant n}\nu_i(\widehat{\lambda}_i-\nu_i-\nu_{i+1})}}{|\operatorname{Aut}(S_{\widehat{\lambda}-\nu})|}\prod_{\substack{1\leqslant i\leqslant n\\ 1\leqslant s\leqslant\nu_i}}\frac{1}{q^s-1}.$$

Further simplification by Lemma 3.9.1 gives

$$h_\nu=\frac{q^{-\sum_{1\leqslant i\leqslant n}\nu_i(\widehat{\lambda}_i-\nu_i-\nu_{i+1})}}{q^{\sum_{1\leqslant i\leqslant n}\frac{1}{2}(\widehat{\lambda}_i-\nu_i-1)(\widehat{\lambda}_i-\nu_i)}}\prod_{\substack{1\leqslant i\leqslant n\\ 1\leqslant s\leqslant\widehat{\lambda}_i-\nu_i}}\frac{1}{q^s-1}\prod_{\substack{1\leqslant i\leqslant n\\ 1\leqslant s\leqslant\nu_i}}\frac{1}{q^s-1}$$
$$=q^{\sum_{1\leqslant i\leqslant n}\frac{1}{2}(\widehat{\lambda}_i-\widehat{\lambda}_i^2)}\prod_{\substack{1\leqslant i\leqslant n\\ 1\leqslant s\leqslant\widehat{\lambda}_i}}\frac{1}{q^s-1}q^{\sum_{1\leqslant i\leqslant n}(\frac{\nu_i^2-\nu_i}{2}+\nu_i\nu_{i+1})}\prod_{1\leqslant i\leqslant n}\begin{bmatrix}\widehat{\lambda}_i\\ \nu_i\end{bmatrix}_q.$$

Thus,

$$X^q_{\lambda,\mu}(\mathbf{L},\mathbf{L}'')=q^{\frac{1}{2}(f_{\widehat{\lambda}}+\sum_{1\leqslant i\leqslant n}(\lambda_i+\mu_i+\widehat{\lambda}_i-\widehat{\lambda}_i^2))}\prod_{\substack{1\leqslant i\leqslant n\\ 1\leqslant s\leqslant\widehat{\lambda}_i}}\frac{1}{q^s-1}\cdot Q_{\widehat{\lambda},z}\ \text{ and}$$
$$Y^q_{\lambda,\mu}(\mathbf{L},\mathbf{L}'')=q^{\frac{1}{2}(f_{\widehat{\lambda}}+\sum_{1\leqslant i\leqslant n}(\lambda_i+\mu_i+\widehat{\lambda}_i-\widehat{\lambda}_i^2))}\prod_{\substack{1\leqslant i\leqslant n\\ 1\leqslant s\leqslant\widehat{\lambda}_i}}\frac{1}{q^s-1}\cdot Q'_{\widehat{\lambda},z},$$

where, for $p_{\nu,\lambda}(q)=q^{\sum_{1\leqslant i\leqslant n}(\frac{\nu_i^2-\nu_i}{2}+\nu_i\nu_{i+1})}\prod_{1\leqslant i\leqslant n}\begin{bmatrix}\lambda_i\\ \nu_i\end{bmatrix}_q$,

$$Q_{\widehat{\lambda},z}=\sum_{\substack{0\leqslant\nu\leqslant\widehat{\lambda}\\ \nu\in\mathbb{N}^n_\Delta}}p_{\nu,\widehat{\lambda}}(q)\cdot q^{\sum_{1\leqslant i\leqslant n}(\widehat{\lambda}_i z_i-\nu_i z_i-\nu_i\widehat{\lambda}_{i+1})}\prod_{\substack{1\leqslant i\leqslant n\\ 1\leqslant s\leqslant\nu_i}}(q^{z_{i+1}-s+1}-1)\ \text{ and}$$
$$Q'_{\widehat{\lambda},z}=\sum_{\substack{0\leqslant\nu\leqslant\widehat{\lambda}\\ \nu\in\mathbb{N}^n_\Delta}}p_{\nu,\widehat{\lambda}}(q)\cdot q^{\sum_{1\leqslant i\leqslant n}(\widehat{\lambda}_i z_{i+1}-\nu_i z_{i+1}-\nu_{i+1}\widehat{\lambda}_i)}\prod_{\substack{1\leqslant i\leqslant n\\ 1\leqslant s\leqslant\nu_i}}(q^{z_i-s+1}-1).$$

Finally, it is clear to see that $P_{\widehat{\lambda},z}=Q_{\widehat{\lambda},z}$ and $P'_{\widehat{\lambda},z}=Q'_{\widehat{\lambda},z}$. □

The fact that ξ_r is an algebra homomorphism immediately gives the following polynomial identity.

Corollary 3.9.6. *For any* $\lambda = (\lambda_i)_{i\in\mathbb{Z}}, z = (z_i)_{i\in\mathbb{Z}} \in \mathbb{N}^n_\Delta$, $P_{\lambda,z}(v^2) = P'_{\lambda,z}(v^2)$.

Proof. If $z = \mathbf{0}$, then the equality holds trivially. Now suppose $z \neq 0$ and set $r = \sum_{i=1}^n z_i$. Let $\mathbb{F}$ be a finite field with q elements. By Theorems 3.6.3 and 2.6.3(5), we have $X^q_{\lambda,\mu}(\mathbf{L}, \mathbf{L}'') = Y^q_{\lambda,\mu}(\mathbf{L}, \mathbf{L}'')$ for all λ, μ and $(\mathbf{L}, \mathbf{L}'')$. Take $\mathbf{L} = (L_i)_{i\in\mathbb{Z}} = \mathbf{L}'' \in \mathscr{Y}$ (thus, $\widehat{\lambda} = \lambda = \mu = \widehat{\mu}$) such that $\dim L_i/L_{i-1} = z_i$, for all $i \in \mathbb{Z}$. Applying Theorem 3.9.5 to the equality $X^q_{\lambda,\lambda}(\mathbf{L}, \mathbf{L}) = Y^q_{\lambda,\lambda}(\mathbf{L}, \mathbf{L})$ gives the equality $P_{\lambda,z}(q) = P'_{\lambda,z}(q)$. □

Remark 3.9.7. We point out that the polynomial identity $P_{\lambda,z}(v^2) = P'_{\lambda,z}(v^2)$ is equivalent to the fact that the algebra homomorphisms $\zeta_r^\pm : \mathfrak{D}_\Delta(n)^\pm \to \boldsymbol{\mathcal{S}}_\Delta(n, r)$ constructed by Varagnolo and Vasserot **[73]** (see (3.6.2.1)), which have easy extensions $\zeta_r^{\geqslant 0} : \mathfrak{D}_\Delta(n)^{\geqslant 0} \to \boldsymbol{\mathcal{S}}_\Delta(n, r)$ and $\zeta_r^{\leqslant 0} : \mathfrak{D}_\Delta(n)^{\leqslant 0} \to \boldsymbol{\mathcal{S}}_\Delta(n, r)$, can be extended to an algebra homomorphism $\mathfrak{D}_\Delta(n) \to \boldsymbol{\mathcal{S}}_\Delta(n, r)$.

In fact, there is an obvious linear extension $\widetilde{\xi}_r$ of $\zeta_r^{\geqslant 0}$ and $\zeta_r^{\leqslant 0}$ which is an algebra homomorphism if and only if $\widetilde{\xi}_r$ preserves the commutator relations in Theorem 2.6.3(5) on semisimple generators (see Lemma 2.6.1). This is the case by Theorem 3.9.5 if $P_{\lambda,z}(q) = P'_{\lambda,z}(q)$ for every prime power q and λ, z.

For every $\lambda \in \mathbb{N}^n_\Delta$ and $r \geqslant 0$, define functions $f_{\lambda,r}, f'_{\lambda,r} : \Lambda_\Delta(n, r) \to \mathbb{Z}[v^2]$ such that $f_{\lambda,r}(z) = P_{\lambda,z}(v^2)$ and $f'_{\lambda,r}(z) = P'_{\lambda,z}(v^2)$. The polynomial identity shows that the two functions are identical.

Problem 3.9.8. Give a direct (or combinatorial) proof of the identity $P_{\lambda,z}(v^2) = P'_{\lambda,z}(v^2)$, or equivalently, $f_{\lambda,r} = f'_{\lambda,r}$.

3.10. Appendix

In this appendix, we prove a few lemmas which have been used in the previous sections. The first one reflects some affine phenomenon for the length of the longest element $w_{0,\lambda}$ of $\mathfrak{S}_\lambda$. This is used in the proof of Corollary 3.2.4.

Lemma 3.10.1. *For* $\lambda \in \Lambda_\Delta(n, r)$ *and* $d \in \mathscr{D}^\Delta_\lambda$, *let*

$$Y = \{(s, t) \in \mathbb{Z}^2 \mid 1 \leqslant d^{-1}(s) \leqslant r, s < t, s, t \in R^\lambda_k \text{ for some } k \in \mathbb{Z}\} \text{ and}$$

$$Z = \{(s, t) \in \mathbb{Z}^2 \mid 1 \leqslant s \leqslant r, s < t, s, t \in R^\lambda_k \text{ for some } 1 \leqslant k \leqslant n\}.$$

Then $|Y| = |Z| = \ell(w_{0,\lambda})$.

Proof. For $a \in \mathbb{Z}$ let $Y_a = \{b \in \mathbb{Z} \mid a < b,\ a, b \in R_k^\lambda \text{ for some } k \in \mathbb{Z}\}$. We first claim $|Y_{a_1}| = |Y_{a_2}|$ whenever $a_1 \equiv a_2 \mod r$. Indeed, write $a_2 = a_1 + cr$ and assume $a_1 \in R_k^\lambda$ (see (3.2.1.4)). Then $a_2 \in R_{k+cn}^\lambda$. By definition,

$$Y_{a_2} = \{b \in \mathbb{Z} \mid a_2 < b,\ b \in R_{k+cn}^\lambda\} = \{b \in \mathbb{Z} \mid a_1 < b - cr,\ b - cr \in R_k^\lambda\}.$$

Hence, there is a bijection from Y_{a_2} to Y_{a_1} defined by sending b to $b - cr$, proving the claim.

Since the remainders of $d(i)$ when divided by r are all distinct and

$$Y = \bigcup_{1 \leqslant s \leqslant r} \{(d(s), j) \mid j \in Y_{d(s)}\},$$

it follows from the claim that

$$\begin{aligned} |Y| &= \sum_{1 \leqslant s \leqslant r} |Y_{d(s)}| = \sum_{1 \leqslant s \leqslant r} |Y_s| \\ &= \sum_{1 \leqslant s \leqslant r} |\{t \in \mathbb{Z} \mid s < t,\ s, t \in R_k^\lambda \text{ for some } k \in \mathbb{Z}\}| \\ &= |Z| = \ell(w_{0,\lambda}), \end{aligned}$$

as desired. □

The next lemma is used in the proof of Proposition 3.7.3.

Lemma 3.10.2. *Keep the notation* $A, A^+, A^-, \lambda, \mu, \nu$ *used in* (3.7.3.1). *We have*

$$d_{A+\mathrm{diag}(\lambda)} = d_{A^+ + \mathrm{diag}(\mu)} + d_{A^- + \mathrm{diag}(\nu)}.$$

Proof. Recall from Lemma 3.7.2(2) that μ, ν are uniquely determined by the conditions

$$(\mathbf{L}, \mathbf{L}') \in \mathcal{O}_{A^+ + \mathrm{diag}(\mu)} \quad \text{and} \quad (\mathbf{L}', \mathbf{L}'') \in \mathcal{O}_{A^- + \mathrm{diag}(\nu)},$$

whenever $(\mathbf{L}, \mathbf{L}'') \in \mathcal{O}_{A+\mathrm{diag}(\lambda)}$ and $\mathbf{L}' = \mathbf{L} \cap \mathbf{L}''$. By Lemma 3.7.1(3), for $i \in \mathbb{Z}$,

$$\begin{aligned} \lambda_i + \sum_{k, k \neq i} a_{i,k} &= \dim(L_i / L_{i-1}) = \mu_i + \sum_{k, i<k} a_{i,k} \quad \text{and} \\ \lambda_i + \sum_{k, k \neq i} a_{k,i} &= \dim(L''_i / L''_{i-1}) = \nu_i + \sum_{k, k>i} a_{k,i}. \end{aligned}$$

Thus, $\mu_i = \lambda_i + \sum_{k<i} a_{i,k}$ and $\nu_i = \lambda_i + \sum_{k<i} a_{k,i}$ for all i. Moreover, since for such μ, ν,

$$d_{A+\mathrm{diag}(\lambda)} = \sum_{\substack{1 \leqslant i \leqslant n \\ i \geqslant k;\, j<l}} a_{i,j} a_{k,l} + \sum_{\substack{1 \leqslant i \leqslant n \\ k \leqslant i < l}} \lambda_i a_{k,l} + \sum_{j<k \leqslant i} \lambda_k a_{i,j},$$

$$d_{A^+ + \mathrm{diag}(\mu)} = \sum_{\substack{1 \leqslant i \leqslant n \\ i \geqslant k;\, j < l \\ i < j;\, k < l}} a_{i,j}a_{k,l} + \sum_{\substack{1\leqslant i\leqslant n \\ k\leqslant i<l}} \mu_i a_{k,l}$$

$$= \sum_{\substack{1 \leqslant i \leqslant n \\ i \geqslant k;\, j < l \\ i < j;\, k < l}} a_{i,j}a_{k,l} + \sum_{\substack{1\leqslant i\leqslant n \\ k\leqslant i<l}} \lambda_i a_{k,l} + \sum_{\substack{1 \leqslant i \leqslant n \\ k \leqslant i < l \\ i > j}} a_{i,j}a_{k,l}, \quad \text{and}$$

$$d_{A^- + \mathrm{diag}(\nu)} = \sum_{\substack{1 \leqslant i \leqslant n \\ i \geqslant k;\, j < l \\ i > j;\, k > l}} a_{i,j}a_{k,l} + \sum_{\substack{1\leqslant i\leqslant n \\ j<k\leqslant i}} \nu_k a_{i,j}$$

$$= \sum_{\substack{1 \leqslant i \leqslant n \\ i \geqslant k;\, j < l \\ i > j;\, k > l}} a_{i,j}a_{k,l} + \sum_{\substack{1\leqslant i\leqslant n \\ j<k\leqslant i}} \lambda_k a_{i,j} + \sum_{\substack{1 \leqslant i \leqslant n \\ j < l \leqslant i \\ k < l}} a_{i,j}a_{k,l},$$

it follows that

$$\begin{aligned}
& d_{A+\mathrm{diag}(\lambda)} - d_{A^+ + \mathrm{diag}(\mu)} - d_{A^- + \mathrm{diag}(\nu)} \\
&= \sum_{\substack{1\leqslant i\leqslant n \\ i\geqslant k; j<l}} a_{i,j}a_{k,l} - \sum_{\substack{1 \leqslant i \leqslant n \\ i \geqslant k,\, j < l \\ i < j,\, k < l}} a_{i,j}a_{k,l} - \sum_{\substack{1 \leqslant i \leqslant n \\ i \geqslant k;\, j < l \\ i > j;\, k > l}} a_{i,j}a_{k,l} \\
&\quad - \sum_{\substack{1 \leqslant i \leqslant n \\ k \leqslant i < l \\ i > j}} a_{i,j}a_{k,l} - \sum_{\substack{1 \leqslant i \leqslant n \\ j < l \leqslant i \\ k < l}} a_{i,j}a_{k,l} \\
&= \sum_{\substack{1 \leqslant i \leqslant n \\ i \geqslant k,\, j < l \\ i > j,\, k < l}} a_{i,j}a_{k,l} - \sum_{\substack{1 \leqslant i \leqslant n \\ k \leqslant i < l \\ i > j}} a_{i,j}a_{k,l} - \sum_{\substack{1 \leqslant i \leqslant n \\ j < l \leqslant i \\ k < l}} a_{i,j}a_{k,l} \\
&= 0,
\end{aligned}$$

as required. □

Let f_ν, g_ν be the numbers defined as in (3.9.5.1) and (3.9.5.2). The following lemma establishes a certain relationship between these numbers and is used in the proof of Theorem 3.9.5.

Lemma 3.10.3. *Maintain the notation in the proof of Theorem 3.9.5. For any* $\mathbf{0} \leqslant \nu \leqslant \widehat{\lambda}$, *we have*

$$f_{\widehat{\lambda}} = g_{\widehat{\lambda}},$$

$$f_{\widehat{\lambda}} - f_\nu = -2\sum_{1\leqslant i\leqslant n} \widehat{\lambda}_i z_i + \sum_{1\leqslant i\leqslant n} 2\nu_i(z_i + \widehat{\lambda}_i + \widehat{\lambda}_{i+1} - \nu_i - \nu_{i+1}), \quad \textit{and}$$

$$g_{\widehat{\lambda}} - g_\nu = -2\sum_{1\leqslant i\leqslant n} \widehat{\lambda}_i z_{i+1} + \sum_{1\leqslant i\leqslant n} 2\nu_i(z_{i+1} + \widehat{\lambda}_i + \widehat{\lambda}_{i-1} - \nu_i - \nu_{i+1}).$$

Proof. Since $\widehat{\lambda} = \lambda - \gamma = \mu - \delta = \widehat{\mu}$, it follows that

$$\begin{aligned} g_{\widehat{\lambda}} &= \sum_{1\leqslant i\leqslant n} \big(\lambda_i(\lambda_{i-1} - \lambda_i - y_i) - x_{i+1}(\delta_i + \widehat{\lambda}_i)\big) + \langle \mu, \lambda\rangle + \langle \mu, \delta + \widehat{\lambda}\rangle \\ &= \sum_{1\leqslant i\leqslant n} \big(\lambda_i(\lambda_{i-1} - \lambda_i - y_i) - x_{i+1}(\delta_i + \widehat{\lambda}_i)\big) + \langle \widehat{\lambda} + \delta, \lambda\rangle + \langle \mu, \mu\rangle \\ &= f_{\widehat{\lambda}}. \end{aligned}$$

By definition, for $0 \leqslant \nu \leqslant \widehat{\lambda} = \lambda - \gamma$,

$$\begin{aligned} f_{\widehat{\lambda}} - f_\nu = &\sum_{1\leqslant i\leqslant n} (\widehat{\lambda}_i + \gamma_i)(\widehat{\lambda}_{i-1} + \gamma_{i-1} - (\gamma_i + \widehat{\lambda}_i) - y_i) \\ &- \sum_{1\leqslant i\leqslant n} (\delta_i + \widehat{\mu}_i)x_{i+1} + \langle \delta + \widehat{\mu}, \gamma + \widehat{\lambda}\rangle \\ &- \sum_{1\leqslant i\leqslant n} (\gamma_i + \nu_i)(-\gamma_i + \nu_i + \gamma_{i-1} + \nu_{i-1} - 2\widehat{\lambda}_i - y_i) \\ &- \sum_{1\leqslant i\leqslant n} (\delta_i + \nu_i)(2(\nu_i - \widehat{\lambda}_i) - x_{i+1}) \\ &- \langle \delta + \nu, \widehat{\lambda} - \nu\rangle - \langle \delta + \nu, \gamma + \widehat{\lambda}\rangle - \langle \widehat{\lambda} - \nu, x\rangle \\ = &\sum_{1\leqslant i\leqslant n} \big(\widehat{\lambda}_i\widehat{\lambda}_{i-1} + \widehat{\lambda}_i\gamma_{i-1} - \widehat{\lambda}_i^2 - \widehat{\lambda}_i y_i + \gamma_i\widehat{\lambda}_{i-1} - x_{i+1}\widehat{\mu}_i \\ &\qquad - \gamma_i\nu_{i-1} - 3\nu_i^2 - \nu_i\gamma_{i-1} - \nu_i\nu_{i-1}\big) \\ &+ \sum_{1\leqslant i\leqslant n} (4\nu_i\widehat{\lambda}_i + \nu_i y_i - 2\delta_i\nu_i + 2\delta_i\widehat{\lambda}_i + \nu_i x_{i+1}) \\ &+ \langle \widehat{\mu} - \nu, \gamma + \widehat{\lambda}\rangle - \langle \delta + \nu, \widehat{\lambda} - \nu\rangle - \langle \widehat{\lambda} - \nu, x\rangle = f'_\nu + f''_\nu, \end{aligned}$$

where

$$\begin{aligned} f'_\nu = &\sum_{1\leqslant i\leqslant n} \big(\widehat{\lambda}_i\widehat{\lambda}_{i-1} + \widehat{\lambda}_i\gamma_{i-1} - \widehat{\lambda}_i^2 - \widehat{\lambda}_i y_i + \gamma_i\widehat{\lambda}_{i-1} - x_{i+1}\widehat{\mu}_i + \delta_i\widehat{\lambda}_i \\ &+ \widehat{\mu}_i\widehat{\lambda}_i + \widehat{\mu}_i\gamma_i - \widehat{\mu}_i\gamma_{i+1} - \widehat{\mu}_i\widehat{\lambda}_{i+1}\big) + \sum_{1\leqslant i\leqslant n} (\delta_i\widehat{\lambda}_{i+1} - x_i\widehat{\lambda}_i + x_{i+1}\widehat{\lambda}_i) \end{aligned}$$

and

$$\begin{aligned} f''_\nu = &\sum_{1\leqslant i\leqslant n} \big(- \gamma_i\nu_i - \gamma_i\nu_{i-1} - 2\nu_i^2 - \nu_i\gamma_{i-1} - \nu_i\nu_{i-1} + 2\nu_i\widehat{\lambda}_i + \nu_i y_i \\ &- \delta_i\nu_i + \nu_i\gamma_{i+1} + 2\nu_i\widehat{\lambda}_{i+1}\big) + \sum_{1\leqslant i\leqslant n} (-\delta_i\nu_{i+1} - \nu_i\nu_{i+1} + x_i\nu_i). \end{aligned}$$

Since $\widehat{\mu} = \widehat{\lambda}$ and $\gamma_{i-1} + \gamma_i - y_i - x_i + \delta_{i-1} + \delta_i = -2z_i$ for all $i \in \mathbb{Z}$, we have

$$\begin{aligned} f'_\nu &= \sum_{1\leqslant i\leqslant n} \widehat{\lambda}_i(\gamma_{i-1} - y_i + \delta_i + \gamma_i - \gamma_{i+1} - x_i) + \sum_{1\leqslant i\leqslant n} \gamma_i\widehat{\lambda}_{i-1} + \sum_{1\leqslant i\leqslant n} \delta_i\widehat{\lambda}_{i+1} \\ &= \sum_{1\leqslant i\leqslant n} \widehat{\lambda}_i(\gamma_{i-1} + \gamma_i - y_i - x_i + \delta_{i-1} + \delta_i) \\ &= -2\sum_{1\leqslant i\leqslant n} \widehat{\lambda}_i z_i \end{aligned}$$

and

$$\begin{aligned} f''_\nu &= \sum_{1\leqslant i\leqslant n} \nu_i\big(-(\gamma_{i-1} + \gamma_i - y_i - x_i + \delta_{i-1} + \delta_i) \\ &\qquad\qquad - 2\nu_i - \nu_{i-1} + 2\widehat{\lambda}_i + 2\widehat{\lambda}_{i+1} - \nu_{i+1}\big) \\ &= \sum_{1\leqslant i\leqslant n} 2\nu_i(z_i + \widehat{\lambda}_i + \widehat{\lambda}_{i+1} - \nu_i - \nu_{i+1}). \end{aligned}$$

Consequently, for $0 \leqslant \nu \leqslant \widehat{\lambda}$, we obtain that

$$f_{\widehat{\lambda}} - f_\nu = -2\sum_{1\leqslant i\leqslant n} \widehat{\lambda}_i z_i + \sum_{1\leqslant i\leqslant n} 2\nu_i(z_i + \widehat{\lambda}_i + \widehat{\lambda}_{i+1} - \nu_i - \nu_{i+1}).$$

Since $\widehat{\mu} = \widehat{\lambda}$, for $0 \leqslant \nu \leqslant \widehat{\lambda}$, we have

$$\begin{aligned} g_{\widehat{\lambda}} - g_\nu &= \sum_{1\leqslant i\leqslant n} (\widehat{\lambda}_i + \gamma_i)(\gamma_{i-1} + \widehat{\lambda}_{i-1} - \gamma_i - \widehat{\lambda}_i - y_i) \\ &\quad - \sum_{1\leqslant i\leqslant n} (\delta_i + \widehat{\lambda}_i)x_{i+1} + \langle \delta + \widehat{\lambda}, \delta + \widehat{\lambda}\rangle \\ &\quad - \sum_{1\leqslant i\leqslant n} (\gamma_i + \nu_i)(-\gamma_i + \gamma_{i-1} + \nu_i + \nu_{i-1} - 2\widehat{\lambda}_i - y_i) \\ &\quad - \sum_{1\leqslant i\leqslant n} (\delta_i + \nu_i)(2(\nu - \widehat{\lambda}_i) - x_{i+1}) \\ &\quad - \langle \widehat{\lambda} - \nu, \gamma + \nu\rangle - \langle \delta + \widehat{\lambda}, \delta + \nu\rangle - \langle \nu - \widehat{\lambda}, x\rangle \\ &= g'_\nu + g''_\nu, \end{aligned}$$

where

$$\begin{aligned} g'_\nu = \sum_{1\leqslant i\leqslant n} \big(&\widehat{\lambda}_i\gamma_{i-1} - \widehat{\lambda}_i\gamma_i - \widehat{\lambda}_i y_i + \gamma_i\widehat{\lambda}_{i-1} - 2x_{i+1}\widehat{\lambda}_i \\ &+ 3\delta_i\widehat{\lambda}_i - \delta_i\widehat{\lambda}_{i+1} + \widehat{\lambda}_i\gamma_{i+1} + x_i\widehat{\lambda}_i\big) \end{aligned}$$

and

$$g''_\nu = \sum_{1\leqslant i\leqslant n} \big(\gamma_i\nu_i - \gamma_i\nu_{i-1} - 2\nu_i^2 - \nu_i\gamma_{i-1} - \nu_i\nu_{i-1} + 2\nu_i\widehat{\lambda}_i + \nu_i y_i - 3\delta_i\nu_i$$
$$+ 2\nu_i x_{i+1} + \delta_i\nu_{i+1} + 2\widehat{\lambda}_i\nu_{i+1} - \nu_i\gamma_{i+1} - \nu_i\nu_{i+1} - x_i\nu_i\big).$$

It is easy to check that for all $i \in \mathbb{Z}$,

$$\gamma_{i-1} - \gamma_i + 2\gamma_{i+1} + x_i - y_i - 2x_{i+1} + 3\delta_i - \delta_{i-1} = -2z_{i+1}.$$

Therefore,

$$\begin{aligned} g'_\nu &= \sum_{1\leqslant i\leqslant n} \widehat{\lambda}_i(\gamma_{i-1} - \gamma_i + 2\gamma_{i+1} + x_i - y_i - 2x_{i+1} + 3\delta_i - \delta_{i-1}) \\ &= -2\sum_{1\leqslant i\leqslant n} \widehat{\lambda}_i z_{i+1} \end{aligned}$$

and

$$\begin{aligned} g''_\nu &= \sum_{1\leqslant i\leqslant n} \big(-(\gamma_{i-1} - \gamma_i + 2\gamma_{i+1} + x_i - y_i - 2x_{i+1} + 3\delta_i - \delta_{i-1}) \\ &\qquad - 2\nu_i - \nu_{i-1} - \nu_{i+1} + 2\widehat{\lambda}_i + 2\widehat{\lambda}_{i-1}\big) \\ &= \sum_{1\leqslant i\leqslant n} 2\nu_i(z_{i+1} + \widehat{\lambda}_i + \widehat{\lambda}_{i-1} - \nu_i - \nu_{i+1}). \end{aligned}$$

Thus, for $0 \leqslant \nu \leqslant \widehat{\lambda}$,

$$g_{\widehat{\lambda}} - g_\nu = -2\sum_{1\leqslant i\leqslant n} \widehat{\lambda}_i z_{i+1} + \sum_{1\leqslant i\leqslant n} 2\nu_i(z_{i+1} + \widehat{\lambda}_i + \widehat{\lambda}_{i-1} - \nu_i - \nu_{i+1}).$$

□

4

Representations of affine quantum Schur algebras

The quantum Schur–Weyl duality links representations of quantum $\mathfrak{gl}_n$ with those of Hecke algebras of symmetric groups. More precisely, there are category equivalences between level r representations of quantum $\mathfrak{gl}_n$ and representations of the Hecke algebra of $\mathfrak{S}_r$. This relationship provides in turn two approaches to representations of quantum Schur algebras. First, one uses the polynomial representation theory of quantum $\mathfrak{gl}_n$ to determine representations of quantum Schur algebras. This is a *downward* approach. Second, representations of quantum Schur algebras can also be determined by those of Hecke algebras. We call this an *upward* approach. In this chapter, we will investigate the affine versions of the two approaches.

When the parameter v is specialized to a non-root-of-unity $z \in \mathbb{C}$, classification of finite dimensional simple (or irreducible) polynomial representations of the quantum loop algebra $\mathrm{U}_{\mathbb{C}}(\widehat{\mathfrak{gl}}_n)$ was completed by Frenkel–Mukhin [**28**], based on Chari–Pressley's classification of finite dimensional simple representations of the quantum loop algebra $\mathrm{U}_{\mathbb{C}}(\widehat{\mathfrak{sl}}_n)$ [**6, 7**]. On the other hand, Zelevinsky [**81**] and Rogawski [**66**] classified simple representations of the Hecke algebra $\mathcal{H}_\Delta(r)_{\mathbb{C}}$. Moreover, Chari–Pressley have also established a category equivalence between the module category $\mathcal{H}_\Delta(r)_{\mathbb{C}}$-mod and a certain full subcategory of $\mathrm{U}_{\mathbb{C}}(\widehat{\mathfrak{sl}}_n)$-mod when $n > r$ without using affine quantum Schur algebras.

We will first establish in §4.1 a category equivalence between the categories $\mathcal{S}_\Delta(n,r)_{\mathbb{F}}$-Mod and $\mathcal{H}_\Delta(r)_{\mathbb{F}}$-Mod when $n \geqslant r$ (Theorem 4.1.3). As an immediate application of this equivalence, we prove that every simple representation in $\mathcal{S}_\Delta(n,r)_{\mathbb{F}}$-Mod is finite dimensional (Theorem 4.1.6). We will briefly review the classification theorems of Chari–Pressley and Frenkel–Mukhin in §4.2 and of Rogawski in §4.3. We will then use the category equivalence to classify simple representations of affine quantum Schur algebras via Rogawski's classification of simple representations of $\mathcal{H}_\Delta(r)_{\mathbb{C}}$ (Theorem 4.3.4).

From §4.4 onwards, we will investigate the downward approach. Using the surjective homomorphism $\xi'_{r,\mathbb{C}} : \mathrm{U}_{\mathbb{C}}(\widehat{\mathfrak{gl}}_n) \to \mathcal{S}_{\vartriangle}(n,r)_{\mathbb{C}}$, every simple $\mathcal{S}_{\vartriangle}(n,r)_{\mathbb{C}}$-module is inflated to a simple $\mathrm{U}_{\mathbb{C}}(\widehat{\mathfrak{gl}}_n)$-module. Our question is how to identify those inflated simple modules. We use a two-step approach. First, motivated by a result of Chari–Pressley, we identify in the $n > r$ case the simple $\mathcal{S}_{\vartriangle}(n,r)_{\mathbb{C}}$-modules arising from simple $\mathcal{H}_{\vartriangle}(r)_{\mathbb{C}}$-modules in terms of simple polynomial representations of $\mathrm{U}_{\mathbb{C}}(\widehat{\mathfrak{gl}}_n)$ (Theorem 4.4.2). Then, using this identification, we will classify all simple $\mathcal{S}_{\vartriangle}(n,r)_{\mathbb{C}}$-modules through simple polynomial representations of $\mathrm{U}_{\mathbb{C}}(\widehat{\mathfrak{gl}}_n)$ (Theorem 4.6.8). Thus, all finite dimensional simple polynomial representations of $\mathrm{U}_{\mathbb{C}}(\widehat{\mathfrak{gl}}_n)$ are inflations of simple representations of affine quantum Schur algebras. In this way, like the classical theory, affine quantum Schur algebras play a bridging role between polynomial representations of quantum affine $\mathfrak{gl}_n$ and those of affine Hecke algebras of $\mathfrak{S}_r$. We end the chapter by presenting a classification of finite dimensional simple $U_{\vartriangle}(n,r)_{\mathbb{C}}$-modules (Theorem 4.7.5).

Throughout the chapter, all representations are defined over a ground field $\mathbb{F}$ with a fixed specialization $\mathcal{Z} \to \mathbb{F}$ by sending v to a non-root-of-unity $z \in \mathbb{F}$. From §4.2 onwards, $\mathbb{F}$ will be the complex number field $\mathbb{C}$. The algebras $\mathrm{U}_{\mathbb{C}}(\widehat{\mathfrak{sl}}_n)$ and $\mathrm{U}_{\mathbb{C}}(\widehat{\mathfrak{gl}}_n)$ are defined in §2.5 with Drinfeld's new presentation. For any algebra $\mathscr{A}$ considered in this chapter, the notation $\mathscr{A}$-mod represents the category of all finite dimensional left $\mathscr{A}$-modules. We also use $\mathscr{A}$-Mod to denote the category of all left $\mathscr{A}$-modules.

4.1. Affine quantum Schur–Weyl duality, II

In this section we establish the affine version of the category equivalences mentioned in the introduction above. As a by-product, we prove that every irreducible representation over the affine quantum Schur algebra is finite dimensional.

Recall from §2.3 (see footnote 2 there) the extended affine $\mathfrak{sl}_n$, $\mathrm{U}_{\vartriangle}(n)$, generated by E_i, F_i, $K_i^{\pm}$ subject to relations (QGL1)–(QGL5) given in Definition 2.3.1. By dropping the subscripts $\vartriangle$ from $\mathbf{U}_{\vartriangle}(n)$, $\boldsymbol{\mathcal{S}}_{\vartriangle}(n,r)$, and $\boldsymbol{\mathcal{H}}_{\vartriangle}(r)$, we obtain the notation $\mathbf{U}(n)$, $\boldsymbol{\mathcal{S}}(n,r)$, and $\boldsymbol{\mathcal{H}}(r)$ for quantum $\mathfrak{gl}_n$, quantum Schur algebras, and Hecke algebras of type A. The algebras $\mathbf{U}(n)$, $\boldsymbol{\mathcal{S}}(n,r)$, and $\boldsymbol{\mathcal{H}}(r)$ will be naturally viewed as subalgebras of $\mathbf{U}_{\vartriangle}(n)$, $\boldsymbol{\mathcal{S}}_{\vartriangle}(n,r)$, and $\boldsymbol{\mathcal{H}}_{\vartriangle}(r)$, respectively. Moreover, we may also regard $\mathbf{U}(n)$ as a subalgebra of $\mathfrak{D}_{\vartriangle}(n) = \mathbf{U}(\widehat{\mathfrak{gl}}_n)$. Let $U_{\vartriangle}(n)$ be the $\mathcal{Z}$-form of $\mathbf{U}_{\vartriangle}(n)$ defined in (2.4.4.2) and let $U(n)$, $\mathcal{S}(n,r)$, and $\mathcal{H}(r)$ denote the corresponding $\mathcal{Z}$-free subalgebras of $U_{\vartriangle}(n)$, $\mathcal{S}_{\vartriangle}(n,r)$, and $\mathcal{H}_{\vartriangle}(r)$.

The tensor space $\Omega^{\otimes r}$ over $\mathcal{Z}$ is a $U_\Delta(n)$-$\mathcal{H}_\Delta(r)$-bimodule via the actions (3.5.0.1) and (3.3.0.4), where the affine $\mathcal{H}_\Delta(r)$-action is a natural extension from the $\mathcal{H}(r)$-action on $\Omega_n^{\otimes r}$, where Ω_n is the free $\mathcal{Z}$-submodule of Ω spanned by the elements ω_i $(1 \leqslant i \leqslant n)$; see (3.3.0.4). Thus, $\Omega_n^{\otimes r}$ is a $U(n)$-$\mathcal{H}(r)$-subbimodule of $\Omega^{\otimes r}$ by restriction.

Lemma 4.1.1. *There is a $U(n)$-$\mathcal{H}_\Delta(r)$-bimodule isomorphism*

$$\Omega_n^{\otimes r} \otimes_{\mathcal{H}(r)} \mathcal{H}_\Delta(r) \xrightarrow{\sim} \Omega^{\otimes r}, \ x \otimes h \longmapsto xh.$$

Proof. Clearly, there is a $U(n)$-$\mathcal{H}_\Delta(r)$-bimodule homomorphism

$$\varphi : \Omega_n^{\otimes r} \otimes_{\mathcal{H}(r)} \mathcal{H}_\Delta(r) \longrightarrow \Omega^{\otimes r}, \ x \otimes h \longmapsto xh.$$

Since the set $\{T_w X_1^{a_1} \cdots X_r^{a_r} \mid w \in \mathfrak{S}_r,\ a_i \in \mathbb{Z},\ 1 \leqslant i \leqslant r\}$ forms a $\mathcal{Z}$-basis for $\mathcal{H}_\Delta(r)$, it follows that the set

$$\mathcal{X} := \{\omega_{\mathbf{i}} \otimes X_1^{a_1} \cdots X_r^{a_r} \mid \mathbf{i} \in I(n,r),\ a_i \in \mathbb{Z},\ 1 \leqslant i \leqslant r\}$$

forms a $\mathcal{Z}$-basis for $\Omega_n^{\otimes r} \otimes_{\mathcal{H}(r)} \mathcal{H}_\Delta(r)$. Furthermore, by (3.3.0.4),

$$\varphi(\omega_{\mathbf{i}} \otimes X_1^{a_1} \cdots X_r^{a_r}) = \omega_{\mathbf{i}} X_1^{a_1} \cdots X_r^{a_r} = \omega_{\mathbf{i}-n(a_1,a_2,\ldots,a_r)},$$

for all $(a_1, a_2, \ldots, a_r) \in \mathbb{Z}^r$ and $\mathbf{i} \in I(n,r)$. Thus, the bijection (3.3.0.3) from $I_\Delta(n,r)$ to $\mathbb{Z}^r = I(n,r) + n\mathbb{Z}^r$ implies that $\varphi(\mathcal{X}) = \{\omega_{\mathbf{i}} \mid \mathbf{i} \in I_\Delta(n,r)\}$ forms a $\mathcal{Z}$-basis for $\Omega^{\otimes r}$. Hence, φ is a $U(n)$-$\mathcal{H}_\Delta(r)$-bimodule isomorphism. □

For $N = \max\{n, r\}$, let $\omega \in \Lambda_\Delta(N,r)$ be defined as in (3.1.3.1). Thus, if $\omega = (\omega_i)$, then $\omega_i = 1$ for $1 \leqslant i \leqslant r$, and $\omega_i = 0$ if $n > r$ and $r < i \leqslant n$. Let $e_\omega = e_{\mathrm{diag}(\omega)}$.

Corollary 4.1.2. *If $n \geqslant r$, then there is an $\mathcal{S}(n,r)$-$\mathcal{H}_\Delta(r)$-bimodule isomorphism*

$$\mathcal{S}_\Delta(n,r)e_\omega \cong \mathcal{S}(n,r)e_\omega \otimes_{\mathcal{H}(r)} \mathcal{H}_\Delta(r).$$

Proof. By the proof of Lemma 3.1.3, there is an $\mathcal{S}_\Delta(n,r)$-$\mathcal{H}_\Delta(r)$-bimodule isomorphism $\mathcal{S}_\Delta(n,r)e_\omega \cong \mathcal{T}_\Delta(n,r)$. By Proposition 3.3.1, this isomorphism extends to a bimodule isomorphism $\mathcal{S}_\Delta(n,r)e_\omega \cong \Omega^{\otimes r}$ (cf. the proof of Theorem 3.8.1(3)). Now, the required result follows from the lemma above. □

We are now ready to establish a category equivalence. *For the rest of this section, we assume $\mathbb{F}$ is a (large enough) field which is a $\mathcal{Z}$-algebra such that the image $z \in \mathbb{F}$ of v is not a root of unity.* Thus, both $\mathcal{H}(r)_\mathbb{F}$ and $\mathcal{S}(n,r)_\mathbb{F}$ are semisimple $\mathbb{F}$-algebras, and every finite dimensional $U(n)_\mathbb{F}$-module is completely reducible. The following result generalizes the category equivalence between $\mathcal{S}(n,r)_\mathbb{F}$-mod and $\mathcal{H}(r)_\mathbb{F}$-mod $(n \geqslant r)$ to the affine case.

Theorem 4.1.3. *Assume that $n \geqslant r$. The categories $\mathcal{S}_\Delta(n,r)_\mathbb{F}$-Mod and $\mathcal{H}_\Delta(r)_\mathbb{F}$-Mod are equivalent. It also induces a category equivalence between $\mathcal{S}_\Delta(n,r)_\mathbb{F}$-mod and $\mathcal{H}_\Delta(r)_\mathbb{F}$-mod. Hence, in this case, the algebras $\mathcal{S}_\Delta(n,r)_\mathbb{F}$ and $\mathcal{H}_\Delta(r)_\mathbb{F}$ are Morita equivalent.*

Proof. The $\mathcal{S}_\Delta(n,r)_\mathbb{F}$-$\mathcal{H}_\Delta(r)_\mathbb{F}$-bimodule $\Omega_\mathbb{F}^{\otimes r}$ induces a functor

$$\mathsf{F} : \mathcal{H}_\Delta(r)_\mathbb{F}\text{-Mod} \longrightarrow \mathcal{S}_\Delta(n,r)_\mathbb{F}\text{-Mod},\ L \longmapsto \Omega_\mathbb{F}^{\otimes r} \otimes_{\mathcal{H}_\Delta(r)_\mathbb{F}} L. \qquad (4.1.3.1)$$

Since $n \geqslant r$, there is a functor, the Schur functor,

$$\mathsf{G} : \mathcal{S}_\Delta(n,r)_\mathbb{F}\text{-Mod} \longrightarrow \mathcal{H}_\Delta(r)_\mathbb{F}\text{-Mod},\ M \longmapsto e_\omega M.$$

Here we have identified $e_\omega \mathcal{S}_\Delta(n,r)_\mathbb{F} e_\omega$ with $\mathcal{H}_\Delta(r)_\mathbb{F}$. We need to prove that there are natural isomorphisms $\mathsf{G} \circ \mathsf{F} \cong \mathrm{id}_{\mathcal{H}_\Delta(r)_\mathbb{F}\text{-Mod}}$, which is clear (see, e.g., [**33**, (6.2d)]), and $\mathsf{F} \circ \mathsf{G} \cong \mathrm{id}_{\mathcal{S}_\Delta(n,r)_\mathbb{F}\text{-Mod}}$.

By the semisimplicity, for any $\mathcal{S}(n,r)_\mathbb{F}$-module M, there is a left $\mathcal{S}(n,r)_\mathbb{F}$-module isomorphism

$$f : \mathcal{S}(n,r)_\mathbb{F} e_\omega \otimes_{\mathcal{H}(r)_\mathbb{F}} e_\omega M \cong M \qquad (4.1.3.2)$$

defined by $f(x \otimes m) = xm$ for any $x \in \mathcal{S}(n,r)_\mathbb{F} e_\omega$ and $m \in e_\omega M$. By Corollary 4.1.2, (4.1.3.2) induces a left $\mathcal{S}(n,r)_\mathbb{F}$-module isomorphism

$$g : \mathcal{S}_\Delta(n,r)_\mathbb{F} e_\omega \otimes_{\mathcal{H}_\Delta(r)_\mathbb{F}} e_\omega M \cong M$$

satisfying $g(x \otimes m) = xm$, for any $x \in \mathcal{S}(n,r)_\mathbb{F} e_\omega$ and $m \in e_\omega M$.

We now claim that g is an $\mathcal{S}_\Delta(n,r)_\mathbb{F}$-module isomorphism. Indeed, by identifying $\mathcal{S}_\Delta(n,r)_\mathbb{F}$ with $\mathcal{S}_\Delta^\mathcal{H}(n,r)$ under the isomorphism given in Proposition 3.2.8 and considering the basis $\{\phi_{\lambda,\mu}^d\}$, we have $e_\omega = \phi_{\omega,\omega}^1$ and

$$\mathcal{S}_\Delta(n,r)_\mathbb{F} e_\omega = \bigoplus_{\lambda \in \Lambda_\Delta(n,r)} \mathrm{Hom}_{\mathcal{H}_\Delta(r)_\mathbb{F}}(\mathcal{H}_\Delta(r)_\mathbb{F}, x_\lambda \mathcal{H}_\Delta(r)_\mathbb{F}) = \bigoplus_{\lambda \in \Lambda_\Delta(n,r)} \phi_{\lambda,\omega}^1 \mathcal{H}_\Delta(r)_\mathbb{F},$$

where $\phi_{\lambda,\omega}^1 \in \mathcal{S}(n,r)_\mathbb{F}$. Since $g(\phi_{\lambda,\omega}^1 \otimes m) = \phi_{\lambda,\omega}^1 m$, for all $\lambda \in \Lambda_\Delta(n,r)$ and $m \in e_\omega M$, it follows that, for any $\lambda \in \Lambda_\Delta(n,r)$, $h \in \mathcal{H}_\Delta(r)_\mathbb{F}$, and $m \in e_\omega M$,

$$g(\phi_{\lambda,\omega}^1 h \otimes m) = g(\phi_{\lambda,\omega}^1 \otimes hm) = \phi_{\lambda,\omega}^1 (hm) = (\phi_{\lambda,\omega}^1 h)m.$$

Hence, $g(x \otimes m) = xm$, for all $x \in \mathcal{S}_\Delta(n,r)_\mathbb{F} e_\omega$ and $m \in e_\omega M$. Thus, g is an $\mathcal{S}_\Delta(n,r)_\mathbb{F}$-module isomorphism. Therefore, for any $M \in \mathcal{S}_\Delta(n,r)_\mathbb{F}$-Mod, there is a natural isomorphism

$$\mathsf{F} \circ \mathsf{G}(M) \cong \mathcal{S}_\Delta(n,r)_\mathbb{F} e_\omega \otimes_{\mathcal{H}_\Delta(r)_\mathbb{F}} e_\omega M \cong M,$$

proving $\mathsf{F} \circ \mathsf{G} \cong \mathrm{id}_{\mathcal{S}_\Delta(n,r)_\mathbb{F}\text{-Mod}}$.

The last assertion follows from the fact that, if N is a finite dimensional $\mathcal{H}_\Delta(r)_{\mathbb{F}}$-module, then Lemma 4.1.1 implies that $\mathsf{F}(N) = \Omega_{\mathbb{F}}^{\otimes r} \otimes_{\mathcal{H}_\Delta(r)_{\mathbb{F}}} N$ is also finite dimensional. □

Remarks 4.1.4. (1) For the case $\mathbb{F} = \mathbb{C}$, a direct category equivalence from $\mathcal{H}_\Delta(r)_{\mathbb{C}}$-mod to a full subcategory of $\mathrm{U}_{\mathbb{C}}(\widehat{\mathfrak{gl}}_n)$-mod has been established in [**31**, Th. 6.8] when $n \geqslant r$. The construction there is geometric, using intersection cohomology complexes.

(2) See [**76**] for a similar equivalence in the context of representations of p-adic groups. See also [**74**] for a connection with representations of double affine Hecke algebras of type A.

Following [**9**, 2.5], a finite dimensional $U(n)_{\mathbb{F}}$-module M is said to be of *level r* if every irreducible component of M is isomorphic to an irreducible component of $\Omega_{n,\mathbb{F}}^{\otimes r}$. In other words, a $U(n)_{\mathbb{F}}$-module M has level r if and only if it is an $\mathcal{S}(n,r)_{\mathbb{F}}$-module. We will generalize this definition to the affine case; see Corollary 4.6.9 and Remark 4.6.10 below. Note that, since levels are defined for modules of quantum $\mathfrak{sl}_n$ in [**9**, 2.5], the condition $n > r$ there is necessary.

Recall that a $U_\Delta(n)_{\mathbb{F}}$-module M is called of *type 1*, if it is a direct sum $M = \oplus_{\lambda\in\mathbb{Z}^n} M_\lambda$ of its weight spaces M_λ which have the form

$$M_\lambda = \{x \in M \mid K_i x = z^{\lambda_i} x\}.$$

In other words, a $U_\Delta(n)_{\mathbb{F}}$-module M is of type 1 if it is of type 1 as a $U(n)_{\mathbb{F}}$-module.

Theorem 4.1.3 immediately implies the following category equivalence due to Chari–Pressley [**9**]. However, a different functor is used in [**9**, Th. 4.2]. We will make a comparison of the two functors in the next section.

Corollary 4.1.5. *If $n > r$, then the functor* F *induces a category equivalence between the category of finite dimensional $U_\Delta(n)_{\mathbb{F}}$-modules of type 1 which are of level r when restricted to $U(n)_{\mathbb{F}}$-modules and the category of finite dimensional $\mathcal{H}_\Delta(r)_{\mathbb{F}}$-modules.*

Proof. By Corollary 3.5.9 (see also Lemma 5.2.1 below), the condition $n > r$ implies $\mathcal{S}_\Delta(n,r)_{\mathbb{F}} = U_\Delta(n,r)_{\mathbb{F}}$ is a homomorphic image of $U_\Delta(n)_{\mathbb{F}}$. Thus, level r representations considered are precisely $\mathcal{S}_\Delta(n,r)_{\mathbb{F}}$-modules. Now the result follows immediately from the theorem above. □

If $N \geqslant n$, then there is a natural injective map

$$\widetilde{\ }: \Theta_\Delta(n) \longrightarrow \Theta_\Delta(N), \quad A = (a_{i,j}) \longmapsto \widetilde{A} = (\widetilde{a}_{i,j}),$$

where $\widetilde{A}$ is defined in (3.1.2.1). Similarly, there is an injective map

$$\widetilde{\ }: \mathbb{Z}^n_\Delta \longrightarrow \mathbb{Z}^N_\Delta, \quad \lambda \longmapsto \widetilde{\lambda}, \tag{4.1.5.1}$$

where $\widetilde{\lambda}_i = \lambda_i$ for $1 \leqslant i \leqslant n$ and $\widetilde{\lambda}_i = 0$ for $n+1 \leqslant i \leqslant N$. See the proof of Lemma 3.1.3.

These two maps induce naturally by Lemma 3.1.3 an injective algebra homomorphism (not sending 1 to 1)

$$\iota_r = \iota_{n,N,r} : \mathcal{S}_\Delta(n,r) \longrightarrow \mathcal{S}_\Delta(N,r), \quad [A] \longmapsto [\widetilde{A}] \text{ for } A \in \Theta_\Delta(n,r).$$

In other words, we may identify $\mathcal{S}_\Delta(n,r)$ as the centralizer subalgebra $e\mathcal{S}_\Delta(N,r)e$ of $\mathcal{S}_\Delta(N,r)$, where $e = \sum_{\lambda\in\Lambda_\Delta(n,r)} \phi^1_{\widetilde{\lambda},\widetilde{\lambda}} = \sum_{\lambda\in\Lambda_\Delta(n,r)}[\operatorname{diag}(\widetilde{\lambda})] \in \mathcal{S}_\Delta(N,r)$.

Now, the fact that every simple module of an affine Hecke algebra is finite dimensional implies immediately that the same is true for affine quantum Schur algebras.

Theorem 4.1.6. *Every simple $\mathcal{S}_\Delta(n,r)_\mathbb{F}$-module is finite dimensional.*

Proof. If $n \geqslant r$, then the algebras $\mathcal{S}_\Delta(n,r)_\mathbb{F}$ and $\mathcal{H}_\Delta(r)_\mathbb{F}$ are Morita equivalent. The assertion follows from the fact that every simple $\mathcal{H}_\Delta(r)_\mathbb{F}$-module is finite dimensional.

If $n < r$, then, by taking $N = r$, $\mathcal{S}_\Delta(n,r)_\mathbb{F} \cong e\mathcal{S}_\Delta(r,r)_\mathbb{F}e$, where $e = \sum_{\lambda\in\Lambda_\Delta(n,r)} \phi^1_{\widetilde{\lambda},\widetilde{\lambda}}$. It follows that each simple $\mathcal{S}_\Delta(n,r)_\mathbb{F}$-module is isomorphic to eL for a simple $\mathcal{S}_\Delta(r,r)_\mathbb{F}$-module L; see, for example, [**33**, 6.2(g)]. As shown above, all simple $\mathcal{S}_\Delta(r,r)_\mathbb{F}$-modules are finite dimensional. Hence, so are all simple $\mathcal{S}_\Delta(n,r)_\mathbb{F}$-modules. □

4.2. Chari–Pressley category equivalence and classification

We first prove that the functor F defined in (4.1.3.1) coincides with the functor $\mathcal{F}$ defined in [**9**, Th. 4.2]. Then, we describe the Chari–Pressley classification of finite dimensional simple $\mathrm{U}_\mathbb{C}(\widehat{\mathfrak{sl}}_n)$-modules and its generalization to $\mathrm{U}_\mathbb{C}(\widehat{\mathfrak{gl}}_n)$ for later use.

From now on, we assume the ground field $\mathbb{F} = \mathbb{C}$, the complex number field, which is a $\mathcal{Z}$-module where v is mapped to z with $z^m \neq 1$ for all $m \geqslant 1$.

Following [**9**], let E_θ and F_θ be the operators on $\Omega_{n,\mathbb{C}}$ defined respectively by

$$E_\theta \cdot \omega_i = \delta_{i,n}\omega_1 \text{ and } F_\theta \cdot \omega_i = \delta_{i,1}\omega_n. \tag{4.2.0.1}$$

Let $\widetilde{K}_\theta = \widetilde{K}_1 \cdots \widetilde{K}_{n-1} = \widetilde{K}_0^{-1} = K_0^{-1}K_1$.[1] For a left $\mathcal{H}_\Delta(r)_F$-module M, define

$$\mathcal{F}(M) = \Omega_{n,\mathbb{C}}^{\otimes r} \otimes_{\mathcal{H}(r)_\mathbb{C}} M.$$

Then $\mathcal{F}(M)$ is equipped with the $U(n)_\mathbb{C}$-module structure induced by that on $\Omega_{n,\mathbb{C}}^{\otimes r}$. By [**9**, Th. 4.2], $\mathcal{F}(M)$ becomes a $U_\Delta(n)_\mathbb{C}$-module via the action:

$$E_0(w \otimes m) = \sum_{j=1}^{r} (Y_j^+ w) \otimes (X_j m) \text{ and}$$

$$F_0(w \otimes m) = \sum_{j=1}^{r} (Y_j^- w) \otimes (X_j^{-1} m),$$

where $w \in \Omega_{n,\mathbb{C}}^{\otimes r}$, $m \in M$ and the operators $Y_j^\pm \in \mathrm{End}_\mathbb{C}(\Omega_{n,\mathbb{C}}^{\otimes r})$ $(1 \leqslant j \leqslant r)$ are defined by

$$Y_j^+ = 1^{\otimes(j-1)} \otimes F_\theta \otimes (\widetilde{K}_\theta^{-1})^{\otimes(r-j)} = 1^{\otimes(j-1)} \otimes F_\theta \otimes (\widetilde{K}_0)^{\otimes(r-j)} \text{ and}$$
$$Y_j^- = (\widetilde{K}_\theta)^{\otimes(j-1)} \otimes E_\theta \otimes 1^{\otimes(r-j)} = (\widetilde{K}_0^{-1})^{\otimes(j-1)} \otimes E_\theta \otimes 1^{\otimes(r-j)}.$$

Hence, we obtain a functor

$$\mathcal{F} : \mathcal{H}_\Delta(r)_\mathbb{C}\text{-mod} \longrightarrow U_\Delta(n)_\mathbb{C}\text{-mod}.$$

When $n > r$, Chari–Pressley used this functor to establish a category equivalence between $\mathcal{H}_\Delta(r)_\mathbb{C}$-mod and the full subcategory of $U_\Delta(n)_\mathbb{C}$-mod consisting of finite dimensional $U_\Delta(n)_\mathbb{C}$-modules which are of level r when restricted to $U(n)_\mathbb{C}$-modules; see Corollary 4.1.5.

On the other hand, the algebra homomorphism $\xi_{r,\mathbb{C}} : U_\Delta(n)_\mathbb{C} \to \mathcal{S}_\Delta(n,r)_\mathbb{C}$ (cf. Remark 3.5.10) gives an inflation functor

$$\Upsilon : \mathcal{S}_\Delta(n,r)_\mathbb{C}\text{-mod} \longrightarrow U_\Delta(n)_\mathbb{C}\text{-mod}.$$

Thus, we obtain the functor

$$\Upsilon\mathsf{F} = \Upsilon \circ \mathsf{F} : \mathcal{H}_\Delta(r)_\mathbb{C}\text{-mod} \longrightarrow U_\Delta(n)_\mathbb{C}\text{-mod}.$$

Proposition 4.2.1. *There is a natural isomorphism of functors*

$$\varphi : \mathcal{F} \rightsquigarrow \Upsilon\mathsf{F}.$$

In other words, for any $\mathcal{H}_\Delta(r)_F$-modules M, M' and homomorphism $f : M \to M'$, there is a $U_\Delta(n)_\mathbb{C}$-module isomorphism

$$\varphi_M : \mathcal{F}(M) \xrightarrow{\sim} \Upsilon\mathsf{F}(M), \quad w \otimes m \longmapsto w \otimes_\Delta m,$$

[1] The K_0 and E_0, F_0 below should be regarded as K_n, E_n, F_n, respectively, if the index set $I = \mathbb{Z}/n\mathbb{Z}$ is identified as $\{1, 2, \ldots, n\}$.

for all $w \in \Omega_{n,\mathbb{C}}^{\otimes r}$ *and* $m \in M$ *such that* $\Upsilon\mathsf{F}(f)\varphi_M = \varphi_{M'}\mathcal{F}(f)$. *Here* $\otimes = \otimes_{\mathcal{H}(r)_\mathbb{C}}$ *and* $\otimes_\Delta = \otimes_{\mathcal{H}_\Delta(r)_\mathbb{C}}$.

Proof. We first recall from Remark 3.5.10 that the representation $U_\Delta(n)_\mathbb{C} \to \mathrm{End}_\mathbb{C}(\Omega_\mathbb{C}^{\otimes r})$ factors through the algebra homomorphism $\xi_{r,\mathbb{C}} : U_\Delta(n)_\mathbb{C} \to \mathcal{S}_\Delta(n,r)_\mathbb{C}$. Thus, by Lemma 4.1.1, there is a $U(n)_\mathbb{C}$-module isomorphism

$$\begin{aligned}\varphi = \varphi_M : \mathcal{F}(M) &\longrightarrow \Upsilon\mathsf{F}(M)\\ w \otimes m &\longmapsto w \otimes_\Delta m \quad (\text{for all } w \in \Omega_{n,\mathbb{C}}^{\otimes r},\ m \in M).\end{aligned}$$

It remains to prove that $\varphi(E_0(\omega_\mathbf{i} \otimes m)) = E_0(\omega_\mathbf{i} \otimes_\Delta m)$ and $\varphi(F_0(\omega_\mathbf{i} \otimes m)) = F_0(\omega_\mathbf{i} \otimes_\Delta m)$, for all $\mathbf{i} \in I(n,r)$ and $m \in M$.

By the definition above, for $\mathbf{i} \in I(n,r)$ and $m \in M$,

$$\begin{aligned}\varphi(E_0(\omega_\mathbf{i} \otimes m)) &= \varphi\Big(\sum_{j=1}^{r}(Y_j^+\omega_\mathbf{i}) \otimes (X_j m)\Big)\\ &= \sum_{j=1}^{r} Y_j^+\omega_\mathbf{i} \otimes_\Delta X_j m = \sum_{j=1}^{r}(Y_j^+\omega_\mathbf{i})X_j \otimes_\Delta m,\end{aligned} \tag{4.2.1.1}$$

where, by (4.2.0.1) and (3.3.0.4),

$$\begin{aligned}(Y_j^+\omega_\mathbf{i})X_j &= v^{\sum_{j+1\leqslant k\leqslant r}(\delta_{i_k,n}-\delta_{i_k,1})}\delta_{i_j,1}(\omega_{i_1}\cdots\omega_{i_{j-1}}\omega_n\omega_{i_{j+1}}\cdots\omega_{i_r})X_j\\ &= v^{\sum_{j+1\leqslant k\leqslant r}(\delta_{i_k,n}-\delta_{i_k,1})}\delta_{i_j,1}\omega_{i_1}\cdots\omega_{i_{j-1}}\omega_0\omega_{i_{j+1}}\cdots\omega_{i_r}.\end{aligned} \tag{4.2.1.2}$$

On the other hand, E_0 acts on $\Omega_\mathbb{C}^{\otimes r}$ via the comultiplication Δ as described in Corollary 2.3.5. Since

$$\Delta^{(r-1)}(E_0) = \sum_{j=1}^{r} 1^{\otimes(j-1)} \otimes E_0 \otimes \widetilde{K}_0^{\otimes(r-j)},$$

this together with (3.5.0.1) gives

$$\begin{aligned}E_0\omega_\mathbf{i} &= \sum_{j=1}^{r} v^{\sum_{j+1\leqslant k\leqslant r}(\delta_{i_k,n}-\delta_{i_k,1})}\delta_{i_j,1}\omega_{i_1}\cdots\omega_{i_{j-1}}\omega_0\omega_{i_{j+1}}\cdots\omega_{i_r}\\ &= \sum_{j=1}^{r}(Y_j^+\omega_\mathbf{i})X_j.\end{aligned}$$

(Note that, for $1 \leqslant i, j \leqslant n$, the values $\delta_{i,j}$ are unchanged regardless of viewing i, j as elements in $\mathbb{Z}$ or in $I = \mathbb{Z}/n\mathbb{Z}$.) Hence, by (4.2.1.1),

$$E_0\varphi(\omega_\mathbf{i} \otimes m) = E_0\omega_\mathbf{i} \otimes_\Delta m = \varphi(E_0(\omega_\mathbf{i} \otimes m)).$$

Similarly, we can prove that for $\mathbf{i} \in I(n,r)$ and $m \in M$,

$$\begin{aligned} &F_0\varphi(\omega_{\mathbf{i}} \otimes m) \\ &= \sum_{j=1}^{r} v^{\sum_{1\leqslant k\leqslant j-1}(\delta_{i_k,1}-\delta_{i_k,n})}\delta_{i_j,n}\omega_{i_1}\cdots\omega_{i_{j-1}}\omega_{n+1}\omega_{i_{j+1}}\cdots\omega_{i_r}\otimes_\Delta m \\ &= \varphi(F_0(\omega_{\mathbf{i}} \otimes m)), \end{aligned}$$

as required. The commutativity relation $\Upsilon\mathsf{F}(f)\circ\varphi_M = \varphi_{M'}\circ\mathcal{F}(f)$ is clear. □

We now recall another theorem of Chari–Pressley which classifies finite dimensional simple $\mathrm{U}_{\mathbb{C}}(\widehat{\mathfrak{sl}}_n)$-modules and its generalization by Frenkel–Mukhin to the classification of finite dimensional irreducible polynomial representations of $\mathrm{U}_{\mathbb{C}}(\widehat{\mathfrak{gl}}_n)$.

For $1 \leqslant j \leqslant n-1$ and $s \in \mathbb{Z}$, define the elements $\mathscr{P}_{j,s} \in \mathrm{U}_{\mathbb{C}}(\widehat{\mathfrak{sl}}_n)$ through the generating functions

$$\begin{aligned} \mathscr{P}_j^{\pm}(u) &:= \exp\left(-\sum_{t\geqslant 1}\frac{1}{[t]_z}\mathrm{h}_{j,\pm t}(zu)^{\pm t}\right) \\ &= \sum_{s\geqslant 0}\mathscr{P}_{j,\pm s}u^{\pm s} \in \mathrm{U}_{\mathbb{C}}(\widehat{\mathfrak{sl}}_n)[[u,u^{-1}]]. \end{aligned}$$

Here we may view $\mathscr{P}_j^{\pm}(u)$ as formal power series (at 0 or ∞) with coefficients in $\mathrm{U}_{\mathbb{C}}(\widehat{\mathfrak{sl}}_n)$. Note that these elements are related to the elements $\phi_{j,s}^{\pm}$ used in the definition of $\mathrm{U}_{\mathbb{C}}(\widehat{\mathfrak{sl}}_n)$ (see Definition 2.5.1(2)) through the formula

$$\Phi_j^{\pm}(u) = \widetilde{\mathrm{k}}_j^{\pm 1}\frac{\mathscr{P}_j^{\pm}(z^{-2}u)}{\mathscr{P}_j^{\pm}(u)}. \tag{4.2.1.3}$$

(For a proof, see, e.g., [**10**, p. 291].)[2]

For any polynomial $f(u) = \prod_{1\leqslant i\leqslant m}(1-a_iu) \in \mathbb{C}[u]$ with constant term 1 and $a_i \in \mathbb{C}^*$, define $f^{\pm}(u)$ (so that $f^+(u) = f(u)$) as follows:

$$f^{\pm}(u) = \prod_{1\leqslant i\leqslant m}(1-a_i^{\pm 1}u^{\pm 1}). \tag{4.2.1.4}$$

Note that $f^-(u) = (\prod_{i=1}^m(-a_iu)^{-1})f^+(u)$.

Let V be a finite dimensional representation of $\mathrm{U}_{\mathbb{C}}(\widehat{\mathfrak{sl}}_n)$ of type 1. Then $V = \oplus_{\lambda\in\mathbb{Z}^{n-1}}V_\lambda$, where

$$V_\lambda = \{x \in V \mid \widetilde{\mathrm{k}}_ix = z^{\lambda_i}x, 1 \leqslant i \leqslant n-1\},$$

[2] If $\mathcal{P}_j^{\pm}(u)$ denote the elements defined in line 2 above [**28**, (4.1)], then $\mathscr{P}_j^{\pm}(u) = \mathcal{P}_j^{\pm}(uz)$. We have corrected a typo by removing the + sign in $\exp(\mp\sum\ldots)$. If the + sign is there, then the − case in $\Phi_i^{\pm}(u) = k_i^{\pm 1}\mathcal{P}_j^{\pm}(uz^{-1})/\mathcal{P}_j^{\pm}(uz)$ as given in [**28**, (4.1)] is no longer true. Also, (4.2.1.3) shows that + sign is unnecessary.

and, since all $\mathscr{P}_{i,s}$ commute with the $\widetilde{\mathrm{k}}_j$, each V_λ is a direct sum of generalized eigenspaces of the form

$$V_{\lambda,\gamma} = \{x \in V_\lambda \mid (\mathscr{P}_{i,s} - \gamma_{i,s})^p x = 0 \text{ for some } p\,(1 \leqslant i \leqslant n-1, s \in \mathbb{Z})\}, \tag{4.2.1.5}$$

where $\gamma = (\gamma_{i,s})$ with $\gamma_{i,s} \in \mathbb{C}$. If we put $\Gamma_j^\pm(u) = \sum_{s\geqslant 0} \gamma_{j,\pm s} u^{\pm s}$, then by [**29**, Prop. 1], there exist polynomials $f_i(u) = \prod_{1\leqslant j\leqslant m_i}(1 - a_{j,i}u)$ and $g_i(u) = \prod_{1\leqslant j\leqslant n_i}(1 - b_{j,i}u)$ in $\mathbb{C}[u]$ such that

$$\Gamma_i^\pm(u) = \frac{f_i^\pm(u)}{g_i^\pm(u)} \quad \text{and} \quad \lambda_i = m_i - n_i. \tag{4.2.1.6}$$

Following [**7**, 12.2.4], a non-zero (μ-weight) vector $w \in V$ is called a *pseudo-highest weight vector* if there exist some $P_{j,s} \in \mathbb{C}$ such that

$$\mathrm{x}_{j,s}^+ w = 0, \quad \mathscr{P}_{j,s} w = P_{j,s} w, \quad \text{and} \quad \widetilde{\mathrm{k}}_j w = z^{\mu_j} w,$$

for all $1 \leqslant j \leqslant n-1$ and $s \in \mathbb{Z}$. The module V is called a *pseudo-highest weight module*[3] if $V = \mathrm{U}_\mathbb{C}(\widehat{\mathfrak{sl}}_n)w$ for some pseudo-highest weight vector w.

For notational simplicity, the expressions $\mathscr{P}_{j,s} w = P_{j,s} w$, for all $s \in \mathbb{Z}$, will be written as a single expression

$$\mathscr{P}_j^\pm(u) w = P_j^\pm(u) w \in V[[u, u^{-1}]],$$

where

$$P_j^\pm(u) = \sum_{s\geqslant 0} P_{j,\pm s} u^{\pm s}. \tag{4.2.1.7}$$

Let $\mathcal{P}(n)$ be the set of $(n-1)$-tuples of polynomials with constant terms 1. For $\mathbf{P} = (P_1(u), \ldots, P_{n-1}(u)) \in \mathcal{P}(n)$, define $P_{j,s} \in \mathbb{C}$, for $1 \leqslant j \leqslant n-1$ and $s \in \mathbb{Z}$, as in $P_j^\pm(u) = \sum_{s\geqslant 0} P_{j,\pm s} u^{\pm s}$, where $P_j^\pm(u)$ is defined by (4.2.1.4).

Let $\bar{I}(\mathbf{P})$ be the left ideal of $\mathrm{U}_\mathbb{C}(\widehat{\mathfrak{sl}}_n)$ generated by $\mathrm{x}_{j,s}^+$, $\mathscr{P}_{j,s} - P_{j,s}$, and $\widetilde{\mathrm{k}}_j - z^{\mu_j}$, for $1 \leqslant j \leqslant n-1$ and $s \in \mathbb{Z}$, where $\mu_j = \deg P_j(u)$, and define the "Verma module"

$$\bar{M}(\mathbf{P}) = \mathrm{U}_\mathbb{C}(\widehat{\mathfrak{sl}}_n)/\bar{I}(\mathbf{P}).$$

Then $\bar{M}(\mathbf{P})$ has a unique simple quotient, denoted by $\bar{L}(\mathbf{P})$. The polynomials $P_i(u)$ are called *Drinfeld polynomials* associated with $\bar{L}(\mathbf{P})$.

The following result is due to Chari–Pressley (see [**6, 7**] or [**8**, pp.7–8]).

Theorem 4.2.2. *The modules $\bar{L}(\mathbf{P})$ with $\mathbf{P} \in \mathcal{P}(n)$ are all non-isomorphic finite dimensional simple $\mathrm{U}_\mathbb{C}(\widehat{\mathfrak{sl}}_n)$-modules of type* 1.

[3] There are simple highest weight integrable modules Λ_λ considered in [**54**, 6.2.3] (defined in [**54**, 3.5.6]) which, in general, are infinite dimensional.

Let $U_{\mathbb{C}}(\widehat{\mathfrak{gl}}_n)$ be the algebra over $\mathbb{C}$ generated by $x^{\pm}_{i,s}$ $(1 \leqslant i < n, s \in \mathbb{Z})$, $k_i^{\pm 1}$, and $\mathrm{g}_{i,t}$ $(1 \leqslant i \leqslant n, t \in \mathbb{Z}\backslash\{0\})$ with relations similar to (QLA1)–(QLA7) as defined in §2.5. Chari–Pressley's classification is easily generalized to that of simple $U_{\mathbb{C}}(\widehat{\mathfrak{gl}}_n)$-modules as follows; see [**28**, §4].

For $1 \leqslant i \leqslant n$ and $s \in \mathbb{Z}$, define the elements $\mathcal{Q}_{i,s} \in U_{\mathbb{C}}(\widehat{\mathfrak{gl}}_n)$ through the generating functions

$$\mathcal{Q}_i^{\pm}(u) := \exp\left(-\sum_{t \geqslant 1} \frac{1}{[t]_z} \mathrm{g}_{i,\pm t}(zu)^{\pm t}\right) = \sum_{s \geqslant 0} \mathcal{Q}_{i,\pm s} u^{\pm s} \in U_{\mathbb{C}}(\widehat{\mathfrak{gl}}_n)[[u, u^{-1}]].$$

Since $\mathrm{h}_{i,m} = z^{(i-1)m}\mathrm{g}_{i,m} - z^{(i+1)m}\mathrm{g}_{i+1,m}$, it follows from the definitions that

$$\mathcal{P}_j^{\pm}(u) = \frac{\mathcal{Q}_j^{\pm}(uz^{j-1})}{\mathcal{Q}_{j+1}^{\pm}(uz^{j+1})},$$

for $1 \leqslant j \leqslant n-1$.

As in the case of $U_{\mathbb{C}}(\widehat{\mathfrak{sl}}_n)$, for a representation V of $U_{\mathbb{C}}(\widehat{\mathfrak{gl}}_n)$, a non-zero ($\lambda$-weight) vector $w \in V$ is called a *pseudo-highest weight vector* if there exist some $Q_{i,s} \in \mathbb{C}$ such that

$$x^{+}_{j,s} w = 0, \quad \mathcal{Q}_{i,s} w = Q_{i,s} w, \quad \text{and} \quad k_i w = z^{\lambda_i} w, \tag{4.2.2.1}$$

for all $1 \leqslant i \leqslant n$, $1 \leqslant j \leqslant n-1$, and $s \in \mathbb{Z}$. The module V is called a *pseudo-highest weight module* if $V = U_{\mathbb{C}}(\widehat{\mathfrak{gl}}_n)w$ for some pseudo-highest weight vector w. Associated to the sequence $(Q_{i,s})_{s\in\mathbb{Z}}$, define two formal power series by

$$Q_i^{\pm}(u) = \sum_{s \geqslant 0} Q_{i,\pm s} u^{\pm s}. \tag{4.2.2.2}$$

We also write the short form $\mathcal{Q}_i^{\pm}(u)w = Q_i^{\pm}(u)w$ for the relations $\mathcal{Q}_{i,s}w = Q_{i,s}w$ $(s \in \mathbb{Z})$.

A finite dimensional $U_{\mathbb{C}}(\widehat{\mathfrak{gl}}_n)$-module V is called a *polynomial representation* if the restriction of V to $U_{\mathbb{C}}(\mathfrak{gl}_n)$ is a polynomial representation of type 1 and, for every weight $\lambda = (\lambda_1, \ldots, \lambda_n) \in \mathbb{N}^n$ of V, the formal power series $\Gamma_i^{\pm}(u)$ associated to the eigenvalues $(\gamma_{i,s})_{s\in\mathbb{Z}}$ defining the generalized eigenspaces $V_{\lambda,\gamma}$ as given in (4.2.1.5), where $\mathcal{P}_{i,s}$ is replaced by $\mathcal{Q}_{i,s}$ and $n-1$ by n, are polynomials in $u^{\pm}$ of degree λ_i so that the zeros of the functions $\Gamma_i^{+}(u)$ and $\Gamma_i^{-}(u)$ are the same.

Following [**28**], an n-tuple of polynomials $\mathbf{Q} = (Q_1(u), \ldots, Q_n(u))$ with constant terms 1 is called *dominant* if, for each $1 \leqslant i \leqslant n-1$, the ratio $Q_i(uz^{i-1})/Q_{i+1}(uz^{i+1})$ is a polynomial. Let $\mathcal{Q}(n)$ be the set of dominant n-tuples of polynomials.

For $\mathbf{Q} = (Q_1(u), \ldots, Q_n(u)) \in \mathcal{Q}(n)$, define $Q_{i,s} \in \mathbb{C}$, for $1 \leqslant i \leqslant n$ and $s \in \mathbb{Z}$, by the following formula

$$Q_i^{\pm}(u) = \sum_{s \geqslant 0} Q_{i,\pm s} u^{\pm s},$$

where $Q_i^{\pm}(u)$ is defined using (4.2.1.4). Let $I(\mathbf{Q})$ be the left ideal of $\mathrm{U}_{\mathbb{C}}(\widehat{\mathfrak{gl}}_n)$ generated by $\mathsf{x}_{j,s}^{+}$, $\mathscr{Q}_{i,s} - Q_{i,s}$, and $\mathsf{k}_i - z^{\lambda_i}$, for $1 \leqslant j \leqslant n-1$, $1 \leqslant i \leqslant n$, and $s \in \mathbb{Z}$, where $\lambda_i = \deg Q_i(u)$, and define

$$M(\mathbf{Q}) = \mathrm{U}_{\mathbb{C}}(\widehat{\mathfrak{gl}}_n)/I(\mathbf{Q}).$$

Then $M(\mathbf{Q})$ has a unique simple quotient, denoted by $L(\mathbf{Q})$. The polynomials $Q_i(u)$ are called *Drinfeld polynomials* associated with $L(\mathbf{Q})$.

Theorem 4.2.3. ([**28**]) *The* $\mathrm{U}_{\mathbb{C}}(\widehat{\mathfrak{gl}}_n)$*-modules* $L(\mathbf{Q})$ *with* $\mathbf{Q} \in \mathcal{Q}(n)$ *are all non-isomorphic finite dimensional simple polynomial representations of* $\mathrm{U}_{\mathbb{C}}(\widehat{\mathfrak{gl}}_n)$. *Moreover,*

$$L(\mathbf{Q})|_{\mathrm{U}_{\mathbb{C}}(\widehat{\mathfrak{sl}}_n)} \cong \bar{L}(\mathbf{P}),$$

where $\mathbf{P} = (P_1(u), \ldots, P_{n-1}(u))$ *with* $P_i(u) = Q_i(uz^{i-1})/Q_{i+1}(uz^{i+1})$.

4.3. Classification of simple $\mathcal{S}_{\vartriangle}(n,r)_{\mathbb{C}}$-modules: the upward approach

We first recall the classification of irreducible representations of $\mathcal{H}_{\vartriangle}(r)_{\mathbb{C}}$ or equivalently, simple $\mathcal{H}_{\vartriangle}(r)_{\mathbb{C}}$-modules.

For $\mathbf{a} = (a_1, \ldots, a_r) \in (\mathbb{C}^*)^r$, let $M_{\mathbf{a}} = \mathcal{H}_{\vartriangle}(r)_{\mathbb{C}}/J_{\mathbf{a}}$, where $J_{\mathbf{a}}$ is the left ideal of $\mathcal{H}_{\vartriangle}(r)_{\mathbb{C}}$ generated by $X_j - a_j$ for $1 \leqslant j \leqslant r$. Then $M_{\mathbf{a}}$ is an $\mathcal{H}_{\vartriangle}(r)_{\mathbb{C}}$-module of dimension $r!$ and, regarded as an $\mathcal{H}(r)_{\mathbb{C}}$-module by restriction, $M_{\mathbf{a}}$ is isomorphic to the regular representation of $\mathcal{H}(r)_{\mathbb{C}}$.

Applying the functor F to $M_{\mathbf{a}}$ yields an $\mathcal{S}_{\vartriangle}(n,r)_{\mathbb{C}}$-module $\Omega_{\mathbb{C}}^{\otimes r} \otimes_{\mathcal{H}_{\vartriangle}(r)_{\mathbb{C}}} M_{\mathbf{a}}$ and hence, a $\mathfrak{D}_{\vartriangle\mathbb{C}}(n)$-module inflated by the homomorphism $\xi_{r,\mathbb{C}}$ given in (3.5.10.1). The following result, which will be used in the next section, tells how the central generators $\mathsf{z}_t^{\pm}$ of $\mathfrak{D}_{\vartriangle\mathbb{C}}(n)$ defined in (2.2.1.2) act on this module.

Lemma 4.3.1. *Let* $\mathbf{a} = (a_1, \ldots, a_r) \in (\mathbb{C}^*)^r$ *and* $w \in \Omega_{\mathbb{C}}^{\otimes r} \otimes_{\mathcal{H}_{\vartriangle}(r)_{\mathbb{C}}} M_{\mathbf{a}}$. *Then we have* $\mathsf{z}_t^{\pm} w = \sum_{1 \leqslant s \leqslant r} a_s^{\pm t} w$ *for* $t \geqslant 1$.

Proof. We may assume $w = \omega_{\mathbf{i}} \otimes \bar{h}$ for some $\mathbf{i} \in I_{\vartriangle}(n,r)$ and $\bar{h} \in M_{\mathbf{a}}$. Since for each $t \geqslant 1$, $\sum_{1 \leqslant s \leqslant r} X_s^{\pm t}$ is a central element in $\mathcal{H}_{\vartriangle}(r)_{\mathbb{C}}$, it follows from

(3.5.5.2) that

$$\begin{aligned} z_t^{\pm} w &= \omega_{\mathbf{i}} \sum_{1 \leqslant s \leqslant r} X_s^{\pm t} \otimes \bar{h} = \omega_{\mathbf{i}} \otimes h \sum_{1 \leqslant s \leqslant r} X_s^{\pm t} \bar{1} \\ &= \sum_{1 \leqslant s \leqslant r} a_s^{\pm t} \omega_{\mathbf{i}} \otimes h\bar{1} = \sum_{1 \leqslant s \leqslant r} a_s^{\pm t} w. \end{aligned}$$

□

A *segment* s with center $a \in \mathbb{C}^*$ is by definition an ordered sequence

$$\mathsf{s} = (az^{-k+1}, az^{-k+3}, \ldots, az^{k-1}) \in (\mathbb{C}^*)^k.$$

Here k is called the length of the segment, denoted by $|\mathsf{s}|$. If $\mathbf{s} = \{\mathsf{s}_1, \ldots, \mathsf{s}_p\}$ is an unordered collection of segments, define $\wp(\mathbf{s})$ to be the partition associated with the sequence $(|\mathsf{s}_1|, \ldots, |\mathsf{s}_p|)$. That is, $\wp(\mathbf{s}) = (|\mathsf{s}_{i_1}|, \ldots, |\mathsf{s}_{i_p}|)$ with $|\mathsf{s}_{i_1}| \geqslant \cdots \geqslant |\mathsf{s}_{i_p}|$, where $|\mathsf{s}_{i_1}|, \ldots, |\mathsf{s}_{i_p}|$ is a permutation of $|\mathsf{s}_1|, \ldots, |\mathsf{s}_p|$. We also call $|\mathbf{s}| := \sigma(\wp(\mathbf{s}))$ the length of $\mathbf{s}$.

Let $\mathscr{S}_r$ be the set of unordered collections of segments $\mathbf{s}$ with $|\mathbf{s}| = r$. Then $\mathscr{S}_r = \cup_{\lambda \in \Lambda^+(r)} \mathscr{S}_{r,\lambda}$, where $\mathscr{S}_{r,\lambda} = \{\mathbf{s} \in \mathscr{S}_r \mid \wp(\mathbf{s}) = \lambda\}$ and $\Lambda^+(r)$ is the set of partitions of r.

For $\mathbf{s} = \{\mathsf{s}_1, \ldots, \mathsf{s}_p\} \in \mathscr{S}_{r,\lambda}$, let

$$\mathbf{a}(\mathbf{s}) = (\mathsf{s}_1, \ldots, \mathsf{s}_p) \in (\mathbb{C}^*)^r$$

be the r-tuple obtained by juxtaposing the segments in $\mathbf{s}$. Then the element

$$y_\lambda = \sum_{w \in \mathfrak{S}_\lambda} (-z^2)^{-\ell(w)} T_w \in \mathcal{H}_\Delta(r)_\mathbb{C}$$

generates the submodule $\mathcal{H}_\Delta(r)_\mathbb{C}\bar{y}_\lambda$ of $M_{\mathbf{a}(\mathbf{s})}$ which, as an $\mathcal{H}(r)_\mathbb{C}$-module, is isomorphic to $\mathcal{H}(r)_\mathbb{C} y_\lambda$. However,

$$\mathcal{H}(r)_\mathbb{C} y_\lambda \cong E_\lambda \oplus \Big(\bigoplus_{\mu \vdash r, \mu \rhd \lambda} m_{\mu,\lambda} E_\mu \Big) = S_{\lambda'} \oplus \Big(\bigoplus_{\mu \vdash r, \mu \rhd \lambda} m_{\mu,\lambda} S_{\mu'} \Big), \quad (4.3.1.1)$$

where $S_{\nu'} := E_\nu$ is the left cell module defined by the Kazhdan–Lusztig C-basis [**50**] associated with the left cell containing $w_{0,\nu}$. The notation $S_{\nu'}$ indicates that E_ν is the Specht module associated with the dual partition ν' of ν (i.e., isomorphic to $\mathcal{H}(r)_\mathbb{C} y_\nu T_{w_\nu} x_{\nu'}$ in the notation of [**15**, p.25]).

Let $V_\mathbf{s}$ be the unique composition factor of the $\mathcal{H}_\Delta(r)_\mathbb{C}$-module $\mathcal{H}_\Delta(r)_\mathbb{C}\bar{y}_\lambda$ such that the multiplicity of $S_{\lambda'}$ in $V_\mathbf{s}$ as an $\mathcal{H}(r)_\mathbb{C}$-module is non-zero.

We can now state the following classification theorem due to Zelevinsky [**81**] and Rogawski [**66**]. The construction above follows [**66**].

Theorem 4.3.2. *Let* $\mathrm{Irr}(\mathcal{H}_\Delta(r)_\mathbb{C})$ *be the set of isoclasses of all simple* $\mathcal{H}_\Delta(r)_\mathbb{C}$*-modules. Then the correspondence* $\mathbf{s} \mapsto V_\mathbf{s}$ *defines a bijection from* $\mathscr{S}_r$ *to* $\mathrm{Irr}(\mathcal{H}_\Delta(r)_\mathbb{C})$.

We record the following general result.

Lemma 4.3.3. *Let S and H be two algebras over a field $\mathbb{F}$. If V is an S-H-bimodule, M is a left H-module, and $e \in S$ is an idempotent element, then eV is an eSe-H-bimodule and there is an eSe-module isomorphism*

$$(eV) \otimes_H M \cong e(V \otimes_H M) \quad (ew \otimes m \longmapsto e(w \otimes m)).$$

Proof. There is a natural map $\alpha : (eV) \otimes_H M \to V \otimes_H M$ defined by sending $ew \otimes m$ to $ew \otimes m$. The map α induces a surjective map $\bar{\alpha} : (eV) \otimes_H M \to e(V \otimes_H M)$. On the other hand, there is a surjective right H-module homomorphism from V to eV defined by sending w to ew, for $w \in V$. This map induces a natural surjective map $\beta : V \otimes_H M \to (eV) \otimes_H M$ defined by sending $w \otimes m$ to $ew \otimes m$, for $w \in V$ and $m \in M$. By restriction, we get a map $\bar{\beta} : e(V \otimes_H M) \to (eV) \otimes_H M$. Since $\bar{\alpha}\bar{\beta} = \mathrm{id}$ and $\bar{\beta}\bar{\alpha} = \mathrm{id}$, the assertion follows. □

Let

$$\mathscr{S}_r^{(n)} = \{\mathbf{s} \in \mathscr{S}_r \mid \Omega_\mathbb{C}^{\otimes r} \otimes_{\mathcal{H}_\Delta(r)_\mathbb{C}} V_\mathbf{s} \neq 0\}.$$

Then $\mathscr{S}_r^{(n)} = \mathscr{S}_r$ for all $r \leqslant n$. We have the following classification theorem.

Theorem 4.3.4. *The set*

$$\{\Omega_\mathbb{C}^{\otimes r} \otimes_{\mathcal{H}_\Delta(r)_\mathbb{C}} V_\mathbf{s} \mid \mathbf{s} \in \mathscr{S}_r^{(n)}\}$$

is a complete set of non-isomorphic simple $\mathcal{S}_\Delta(n,r)_\mathbb{C}$*-modules. In particular, if* $n \geqslant r$*, then the set* $\{\Omega_\mathbb{C}^{\otimes r} \otimes_{\mathcal{H}_\Delta(r)_\mathbb{C}} V_\mathbf{s} \mid \mathbf{s} \in \mathscr{S}_r\}$ *is a complete set of non-isomorphic simple* $\mathcal{S}_\Delta(n,r)_\mathbb{C}$*-modules.*

Proof. If $n \geqslant r$, then the assertion follows from Theorem 4.1.3. Now we assume $n < r$. As in the proof of Theorem 4.1.6, we choose $N = r$ and regard $\Lambda_\Delta(n,r)$ as a subset of $\Lambda_\Delta(r,r)$ via the map $\mu \mapsto \widetilde{\mu}$ given in (4.1.5.1). Then $\mathcal{S}_\Delta(n,r)_\mathbb{C}$ identifies a centralizer subalgebra $e\mathcal{S}_\Delta(r,r)_\mathbb{C}e$ of $\mathcal{S}_\Delta(r,r)_\mathbb{C}$, where $e = \sum_{\lambda\in\Lambda_\Delta(n,r)}[\mathrm{diag}(\widetilde{\lambda})] = \sum_{\lambda\in\Lambda_\Delta(n,r)} \phi^1_{\widetilde{\lambda},\widetilde{\lambda}}$.

Recall from Proposition 3.3.1 that the tensor space $\Omega^{\otimes r}$ identifies $\mathcal{T}_\Delta(n,r)$. Thus, $\mathcal{T}_\Delta(r,r)_\mathbb{C}$ is the tensor space on which $\mathcal{S}_\Delta(r,r)_\mathbb{C}$ acts. Hence, the set

$$\{\mathcal{T}_\Delta(r,r)_\mathbb{C} \otimes_{\mathcal{H}_\Delta(r)_\mathbb{C}} V_\mathbf{s} \mid \mathbf{s} \in \mathscr{S}_r\}$$

forms a complete set of non-isomorphic simple $\mathcal{S}_\Delta(r,r)_\mathbb{C}$-modules. By [**33**, 6.2(g)] or a direct argument, the set

$$\{e(\mathcal{T}_\Delta(r,r)_\mathbb{C} \otimes_{\mathcal{H}_\Delta(r)_\mathbb{C}} V_\mathbf{s}) \mid \mathbf{s} \in \mathscr{S}_r\}\backslash\{0\}$$

forms a complete set of non-isomorphic simple $\mathcal{S}_\Delta(n,r)_\mathbb{C}$-modules. Since

$$\begin{aligned} e\mathcal{T}_\Delta(r,r)_\mathbb{C} &\cong e\Big(\bigoplus_{\mu\in\Lambda_\Delta(r,r)} x_\mu \mathcal{H}_\Delta(r)_\mathbb{C}\Big) \cong \bigoplus_{\mu\in\Lambda_\Delta(n,r)} x_{\tilde{\mu}}\mathcal{H}_\Delta(r)_\mathbb{C} \\ &\cong \bigoplus_{\mu\in\Lambda_\Delta(n,r)} x_\mu \mathcal{H}_\Delta(r)_\mathbb{C} \cong \Omega_\mathbb{C}^{\otimes r}, \end{aligned}$$

by Lemma 4.3.3, there is an $\mathcal{S}_\Delta(n,r)_\mathbb{C}$-module isomorphism

$$e(\mathcal{T}_\Delta(r,r)_\mathbb{C} \otimes_{\mathcal{H}_\Delta(r)_\mathbb{C}} V_\mathbf{s}) \cong (e\mathcal{T}_\Delta(r,r)_\mathbb{C}) \otimes_{\mathcal{H}_\Delta(r)_\mathbb{C}} V_\mathbf{s} \cong \Omega_\mathbb{C}^{\otimes r} \otimes_{\mathcal{H}_\Delta(r)_\mathbb{C}} V_\mathbf{s},$$

proving the $n < r$ case. □

From the proof above, one sees easily that, for $\Lambda^+(n,r) = \Lambda^+(r) \cap \Lambda(n,r)$,

$$\bigcup_{\lambda\in\Lambda^+(n,r)} \mathscr{S}_{r,\lambda} \subseteq \mathscr{S}_r^{(n)}.$$

We will prove in §4.5 that this inclusion is an equality and, thus, complete the classification of simple $\mathcal{S}_\Delta(n,r)_\mathbb{C}$-modules when $n < r$ in the upward approach.

The next example shows that $\mathscr{S}_r^{(n)} \neq \mathscr{S}_r$ does occur.

Example 4.3.5. For any $a \in \mathbb{C}^*$, there is an algebra homomorphism ev_a : $\mathcal{H}_\Delta(r)_\mathbb{C} \to \mathcal{H}(r)_\mathbb{C}$, called the *evaluation map* (see [**9**, 5.1]), such that

(1) $ev_a(T_i) = T_i, \qquad 1 \leqslant i \leqslant r-1;$
(2) $ev_a(X_j) = az^{-2(j-1)}T_{j-1}\cdots T_2T_1T_1T_2\cdots T_{j-1}, 1 \leqslant j \leqslant r.$

Thus, every simple $\mathcal{H}(r)_\mathbb{C}$-module S_λ is also a simple $\mathcal{H}_\Delta(r)_\mathbb{C}$-module $(S_\lambda)_a$. If $n < r$, then those $\mathbf{s}$ such that $V_\mathbf{s}$ is isomorphic to $(S_{(1^r)})_a$, where $\wp(\mathbf{s}) = (r)$, are not in $\mathscr{S}_r^{(n)}$.

We end this section with a brief discussion of a certain branching rule.

Since the parameter z for the specialization $\mathcal{Z} \to \mathbb{C}$, $v \mapsto z$ is not a root of unity, the Hecke algebra $\mathcal{H}(r)_\mathbb{C}$ is semisimple. Thus, as an $\mathcal{H}(r)_\mathbb{C}$-module, $V_\mathbf{s}$ is semisimple. By (4.3.1.1), we can decompose the $\mathcal{H}(r)_\mathbb{C}$-module

$$V_\mathbf{s}|_{\mathcal{H}(r)_\mathbb{C}} = \bigoplus_{\mu\vdash r,\mu\trianglerighteq\wp(\mathbf{s})} [V_\mathbf{s} : S_{\mu'}]S_{\mu'}. \tag{4.3.5.1}$$

Here the multiplicity $[V_{\mathbf{s}} : S_{\mu'}] \leqslant m_{\mu,\wp(\mathbf{s})}$, the multiplicity of $S_{\mu'}$ in $\mathcal{H}(r)_{\mathbb{C}} y_{\wp(\mathbf{s})}$, and $[V_{\mathbf{s}} : S_{\wp(\mathbf{s})'}] = 1$. We will call the description of the non-zero multiplicities in (4.3.5.1) the *affine-to-finite Branching Rule* or simply the *affine Branching Rule*.

Problem 4.3.6. Describe the affine Branching Rule. In other words, find a necessary and sufficient condition for the multiplicities $[V_{\mathbf{s}} : S_{\mu'}]$ given in (4.3.5.1) to be non-zero.

There is a similar branching rule for $\mathcal{S}_\Delta(n,r)_{\mathbb{C}}$ to $\mathcal{S}(n,r)_{\mathbb{C}}$. We will make a comparison between them in §4.5.

4.4. Identification of simple $\mathcal{S}_\Delta(n,r)_{\mathbb{C}}$-modules: the $n > r$ case

We now use the downward approach to identify the simple $\mathcal{S}_\Delta(n,r)_{\mathbb{C}}$-modules when $n \geqslant r$, and we will complete another classification in §4.6. More precisely, we will prove that each simple $\mathcal{S}_\Delta(n,r)_{\mathbb{C}}$-module is an irreducible polynomial representation of $\mathrm{U}_{\mathbb{C}}(\widehat{\mathfrak{gl}}_n)$ in Proposition 4.6.4 and prove that each irreducible polynomial representation of $\mathrm{U}_{\mathbb{C}}(\widehat{\mathfrak{gl}}_n)$ is a simple $\mathcal{S}_\Delta(n,r)_{\mathbb{C}}$-module for some r in Proposition 4.6.7. Combining Propositions 4.6.4 with 4.6.7, we can classify the simple modules for $\mathcal{S}_\Delta(n,r)_{\mathbb{C}}$.

Recall the $\mathbb{C}$-algebra $\mathfrak{D}_{\Delta,\mathbb{C}}(n)$ defined in Remark 2.1.4 and the Hopf $\mathbb{C}$-algebra isomorphism $\mathcal{E}_{\mathrm{H},\mathbb{C}} : \mathfrak{D}_{\Delta,\mathbb{C}}(n) \to \mathrm{U}_{\mathbb{C}}(\widehat{\mathfrak{gl}}_n)$ discussed in Theorem 2.5.3 and Remark 2.5.5(3). In order to make use of the results in [**9**], we need to adjust this isomorphism so that its restriction to $\mathrm{U}_{\mathbb{C}}(\widehat{\mathfrak{sl}}_n)$ agrees with the one given in [**9**, p.318].

By the presentation given in Theorem 2.3.1, it is easy to see that there is a Hopf algebra automorphism g of $\mathfrak{D}_{\Delta,\mathbb{C}}(n)$ satisfying

$$\begin{aligned} &g(K_i^{\pm 1}) = K_i^{\pm 1}, && g(u_i^{\pm}) = u_i^{\pm},\ (1 \leqslant i < n),\\ &g(u_n^{\pm}) = (-1)^n z^{\pm 1} u_n^{\pm}, && g(\mathsf{z}_s^{\pm}) = \mp s z^{\pm s} \mathsf{z}_s^{\pm}\ (s \geqslant 1). \end{aligned}$$

(The scalar $(-1)^n z^{\pm 1}$ is for the adjustment, while the scalar $\mp s z^{\pm s}$ becomes apparent in the computation right above (4.4.2.3).) Combining the two gives the following.

Proposition 4.4.1. *There is a Hopf $\mathbb{C}$-algebra isomorphism*

$$f = \mathcal{E}_{\mathrm{H},\mathbb{C}} \circ g : \mathfrak{D}_{\Delta,\mathbb{C}}(n) \longrightarrow \mathrm{U}_{\mathbb{C}}(\widehat{\mathfrak{gl}}_n)$$

such that

$$K_i^{\pm 1} \longmapsto \mathrm{k}_i^{\pm 1}, \qquad u_i^{\pm} \longmapsto \mathrm{x}_{i,0}^{\pm}\ (1 \leqslant i < n),$$
$$u_n^{\pm} \longmapsto (-1)^n z^{\pm 1}\varepsilon_n^{\pm}, \quad \mathrm{z}_s^{\pm} \longmapsto \mp s z^{\pm s}\theta_{\pm s}\ (s \geqslant 1).$$

Now the $\mathbb{C}$-algebra epimorphism $\xi_{r,\mathbb{C}} : \mathfrak{D}_{\Delta}\mathbb{C}(n) \to \mathcal{S}_\Delta(n,r)_\mathbb{C}$ described in Corollary 3.8.3 (and Theorem 3.8.1) together with f gives a $\mathbb{C}$-algebra epimorphism

$$\xi'_{r,\mathbb{C}} := \xi_{r,\mathbb{C}} \circ f^{-1} : \mathrm{U}_\mathbb{C}(\widehat{\mathfrak{gl}}_n) \longrightarrow \mathcal{S}_\Delta(n,r)_\mathbb{C}. \tag{4.4.1.1}$$

Thus, every $\mathcal{S}_\Delta(n,r)_\mathbb{C}$-module will be inflated into a $\mathrm{U}_\mathbb{C}(\widehat{\mathfrak{gl}}_n)$-module via this homomorphism. In particular, every simple $\mathcal{S}_\Delta(n,r)_\mathbb{C}$-module given in Theorem 4.3.4 is a simple $\mathrm{U}_\mathbb{C}(\widehat{\mathfrak{gl}}_n)$-module. We now identify them for the $n > r$ case in terms of the irreducible polynomial representations of $\mathrm{U}_\mathbb{C}(\widehat{\mathfrak{gl}}_n)$ described in Theorem 4.2.3.

Recall from §4.2 that $\mathcal{Q}(n)$ is the set of dominant n-tuples of polynomials. For $r \geqslant 1$, let

$$\mathcal{Q}(n)_r = \big\{\mathbf{Q} = (Q_1(u), \ldots, Q_n(u)) \in \mathcal{Q}(n) \mid r = \sum_{1 \leqslant i \leqslant n} \deg Q_i(u)\big\}.$$

Assume $n > r$. For $\mathbf{s} = \{\mathsf{s}_1, \ldots, \mathsf{s}_p\} \in \mathscr{S}_r$ with

$$\mathsf{s}_i = (a_i z^{-\mu_i+1}, a_i z^{-\mu_i+3}, \ldots, a_i z^{\mu_i-1}) \in (\mathbb{C}^*)^{\mu_i},$$

define

$$\mathbf{Q}_\mathbf{s} = (Q_1(u), \ldots, Q_n(u)) \tag{4.4.1.2}$$

by setting recursively

$$Q_i(u) = \begin{cases} 1, & \text{if } i = n; \\ P_i(uz^{-i+1})P_{i+1}(uz^{-i+2}) \cdots P_{n-1}(uz^{n-2i}), & \text{if } n-1 \geqslant i \geqslant 1, \end{cases}$$

where $P_i(u) = \prod_{1 \leqslant j \leqslant p,\ \mu_j = i} (1 - a_j u)$.

Since every $\mu_j \leqslant r$, $P_{r+1}(u) = \cdots = P_{n-1}(u) = 1$. Hence, every $\mathbf{Q}_s$ has the form $(Q_1(u), \ldots, Q_r(u), 1, \ldots, 1)$. Moreover, putting

$$\nu_i := \deg P_i(u) = \#\{j \in [1,p] \mid \mu_j = i\} \text{ and}$$
$$\lambda_i := \deg Q_i(u) = \#\{j \in [1,p] \mid \mu_j \geqslant i\}$$

gives a partition $\lambda = (\lambda_1, \ldots, \lambda_{n-1}, \lambda_n)$ of r dual to $(\mu_1, \ldots, \mu_p)$, and $\lambda_i - \lambda_{i+1} = \nu_i$ for all $1 \leqslant i < n$. From the definition, $\mathbf{s}$ consists of ν_i segments of length i with centers determined by the roots of $P_i(u)$ for all i.

We now have the following identification theorem.

Theorem 4.4.2. *Maintain the notation above and let $n > r$. The map $\mathbf{s} \mapsto \mathbf{Q_s}$ defines a bijection from $\mathscr{S}_r$ to $\mathcal{Q}(n)_r$, and induces $\mathrm{U}_{\mathbb{C}}(\widehat{\mathfrak{gl}}_n)$-module isomorphisms $\Omega_{\mathbb{C}}^{\otimes r} \otimes_{\mathcal{H}_\Delta(r)_{\mathbb{C}}} V_{\mathbf{s}} \cong L(\mathbf{Q_s})$, for all $\mathbf{s} \in \mathscr{S}_r$. Hence, the set*

$$\{L(\mathbf{Q}) \mid \mathbf{Q} \in \mathcal{Q}(n)_r\}$$

forms a complete set of non-isomorphic simple $\mathcal{S}_\Delta(n,r)_{\mathbb{C}}$-modules.

Proof. By the algebra homomorphism $\xi'_{r,\mathbb{C}}$ given in (4.4.1.1), every $\mathcal{S}_\Delta(n,r)_{\mathbb{C}}$-module M is regarded as a $\mathrm{U}_{\mathbb{C}}(\widehat{\mathfrak{gl}}_n)$-module. Let $[M]$ denote the isoclass of M. By Theorem 4.3.4, it suffices to prove that

(1) $\Omega_{\mathbb{C}}^{\otimes r} \otimes_{\mathcal{H}_\Delta(r)_{\mathbb{C}}} V_{\mathbf{s}} \cong L(\mathbf{Q_s})$, and
(2) $\{[L(\mathbf{Q})] \mid \mathbf{Q} \in \mathcal{Q}(n)_r\} = \{[\Omega_{\mathbb{C}}^{\otimes r} \otimes_{\mathcal{H}_\Delta(r)_{\mathbb{C}}} V_{\mathbf{s}}] \mid \mathbf{s} \in \mathscr{S}_r\}$.

We first prove (1). Let $\mathbf{s} = \{\mathsf{s}_1, \ldots, \mathsf{s}_p\} \in \mathscr{S}_r$ be an unordered collection of segments with

$$\mathsf{s}_i = (a_i z^{-\mu_i+1}, a_i z^{-\mu_i+3}, \ldots, a_i z^{\mu_i-1}) \in (\mathbb{C}^*)^{\mu_i}.$$

Thus, $r = \sum_{1 \leqslant i \leqslant p} \mu_i$. Let $\mathbf{a} = \mathbf{a}(\mathbf{s}) = (\mathsf{s}_1, \ldots, \mathsf{s}_p) \in (\mathbb{C}^*)^r$ be the sequence obtained by juxtaposing the segments in $\mathbf{s}$. Then the simple $\mathcal{S}_\Delta(n,r)_{\mathbb{C}}$-module $\Omega_{\mathbb{C}}^{\otimes r} \otimes_{\mathcal{H}_\Delta(r)_{\mathbb{C}}} V_{\mathbf{s}}$ becomes a simple $\mathrm{U}_{\mathbb{C}}(\widehat{\mathfrak{gl}}_n)$-module via (4.4.1.1). As a simple $\mathrm{U}_{\mathbb{C}}(\widehat{\mathfrak{sl}}_n)$-module, this module is isomorphic to the Chari–Pressley module $\mathcal{F}(V_{\mathbf{s}})$ by Proposition 4.2.1. Applying [**9**, 7.6] yields a $\mathrm{U}_{\mathbb{C}}(\widehat{\mathfrak{sl}}_n)$-module isomorphism $\Omega_{\mathbb{C}}^{\otimes r} \otimes_{\mathcal{H}_\Delta(r)_{\mathbb{C}}} V_{\mathbf{s}} \cong \bar{L}(\mathbf{P})$, where $\mathbf{P} = (P_1(u), \ldots, P_{n-1}(u))$ with

$$P_i^{\pm}(u) = \prod_{\substack{1 \leqslant j \leqslant p \\ \mu_j = i}} (1 - a_j^{\pm 1} u^{\pm 1}), \quad 1 \leqslant i \leqslant n-1. \tag{4.4.2.1}$$

As a simple $\mathrm{U}_{\mathbb{C}}(\widehat{\mathfrak{gl}}_n)$-module, by [**28**, Lem. 4.2], if $w_0 \in \Omega_{\mathbb{C}}^{\otimes r} \otimes_{\mathcal{H}_\Delta(r)_{\mathbb{C}}} V_{\mathbf{s}}$ is the pseudo-highest weight vector of weight $\lambda = (\lambda_1, \ldots, \lambda_n)$, then λ is a partition of r, since λ is also the highest weight of a simple submodule of the $\mathcal{S}(n,r)_{\mathbb{C}}$-module $\Omega_{\mathbb{C}}^{\otimes r} \otimes_{\mathcal{H}_\Delta(r)_{\mathbb{C}}} V_{\mathbf{s}}$, and there exist $Q_i^+(u) \in \mathbb{C}[[u]]$ and $Q_i^-(u) \in \mathbb{C}[[u^{-1}]]$, $1 \leqslant i \leqslant n$, such that

$$\mathscr{Q}_i^{\pm}(u) w_0 = Q_i^{\pm}(u) w_0, \quad \mathscr{P}_j^{\pm}(u) w_0 = P_j^{\pm}(u) w_0, \quad K_i w_0 = z^{\lambda_i} w_0,$$
$$P_j^{\pm}(u) = \frac{Q_j^{\pm}(uz^{j-1})}{Q_{j+1}^{\pm}(uz^{j+1})}, \quad \text{and } \deg P_i(u) = \lambda_i - \lambda_{i+1}. \tag{4.4.2.2}$$

We now prove that $Q_i^{\pm}(u)$ are polynomials of degree λ_i.

Since $f(\mathsf{z}_t^\pm) = \mp t z^{\mp t} \theta_{\pm t}$, for all $t \geqslant 1$, as in Proposition 4.4.1, it follows from (2.5.1.1) and Lemma 4.3.1 that

$$\frac{tz^{\pm t}}{[t]_z} \sum_{1 \leqslant i \leqslant n} \mathfrak{g}_{i,\pm t} w_0 = \mathsf{z}_t^\pm w_0 = \sum_{\substack{1 \leqslant i \leqslant p \\ 1 \leqslant k \leqslant \mu_i}} (a_i z^{-\mu_i + 2k-1})^{\pm t} w_0.$$

Thus,

$$\begin{aligned}
\prod_{1 \leqslant i \leqslant n} Q_i^\pm(u) w_0 &= \prod_{1 \leqslant i \leqslant n} \mathscr{Q}_i^\pm(u) w_0 \\
&= \exp\Big(-\sum_{t \geqslant 1} \frac{1}{[t]_z} \Big(\sum_{1 \leqslant i \leqslant n} \mathfrak{g}_{i,\pm t} \Big) (uz)^{\pm t} \Big) w_0 \\
&= \exp\Big(-\sum_{\substack{1 \leqslant i \leqslant p \\ 1 \leqslant k \leqslant \mu_i}} \sum_{t \geqslant 1} \frac{1}{t} (a_i u z^{2k-1-\mu_i})^{\pm t} \Big) w_0 \\
&= \prod_{\substack{1 \leqslant i \leqslant p \\ 1 \leqslant k \leqslant \mu_i}} \exp\Big(-\sum_{t \geqslant 1} \frac{1}{t} (a_i u z^{2k-1-\mu_i})^{\pm t} \Big) w_0 \\
&= \prod_{\substack{1 \leqslant i \leqslant p \\ 1 \leqslant k \leqslant \mu_i}} \Big(1 - (a_i u z^{2k-1-\mu_i})^{\pm 1} \Big) w_0,
\end{aligned}$$

as $-\sum_{t \geqslant 1} \frac{1}{t} (a_i u z^{2k-1-\mu_i})^{\pm t} = \log\big(1 - (a_i u z^{2k-1-\mu_i})^{\pm 1}\big)$. Hence,

$$\prod_{1 \leqslant i \leqslant n} Q_i^\pm(u) = \prod_{\substack{1 \leqslant i \leqslant p \\ 1 \leqslant k \leqslant \mu_i}} \Big(1 - (a_i u z^{2k-1-\mu_i})^{\pm 1} \Big). \tag{4.4.2.3}$$

On the other hand, by (4.4.2.2),

$$Q_i^\pm(u) = P_i^\pm(uz^{-i+1}) P_{i+1}^\pm(uz^{-i+2}) \cdots P_{n-1}^\pm(uz^{n-2i}) Q_n^\pm(uz^{2(n-i)}) \tag{4.4.2.4}$$

for all $1 \leqslant i \leqslant n$, and by (4.4.2.1),

$$\begin{aligned}
&P_i^\pm(uz^{-i+1}) P_i^\pm(uz^{-i+3}) \cdots P_i^\pm(uz^{i-1}) \\
&= \prod_{\substack{1 \leqslant j \leqslant p \\ \mu_j = i}} (1 - (a_j u z^{-\mu_j+1})^{\pm 1})(1 - (a_j u z^{-\mu_j+3})^{\pm 1}) \cdots (1 - (a_j u z^{\mu_j - 1})^{\pm 1}) \\
&= \prod_{\substack{1 \leqslant j \leqslant p, \mu_j = i \\ 1 \leqslant k \leqslant \mu_j}} (1 - (a_j u z^{2k-1-\mu_j})^{\pm 1}).
\end{aligned}$$

Thus,

$$\prod_{1\leqslant i\leqslant n-1}\Big(P_i^{\pm}(uz^{-i+1})P_i^{\pm}(uz^{-i+3})\cdots P_i^{\pm}(uz^{i-1})\Big)$$
$$=\prod_{\substack{1\leqslant j\leqslant p\\ 1\leqslant k\leqslant \mu_j}}\Big(1-\big(a_juz^{2k-1-\mu_j}\big)^{\pm 1}\Big).$$

Hence,

$$\prod_{1\leqslant i\leqslant n} Q_i^{\pm}(u)$$
$$=\prod_{1\leqslant i\leqslant n-1}\Big(P_i^{\pm}(uz^{-i+1})P_i^{\pm}(uz^{-i+3})\cdots P_i^{\pm}(uz^{i-1})\Big)\prod_{0\leqslant l\leqslant n-1} Q_n^{\pm}(uz^{2l})$$
$$=\prod_{\substack{1\leqslant i\leqslant p\\ 1\leqslant k\leqslant \mu_i}}\Big(1-\big(a_iuz^{2k-1-\mu_i}\big)^{\pm 1}\Big)\prod_{0\leqslant l\leqslant n-1} Q_n^{\pm}(uz^{2l}).$$

Combining this with (4.4.2.3) yields

$$\prod_{0\leqslant l\leqslant n-1} Q_n^{\pm}(uz^{2l})=1.$$

So we have

$$\exp\Big(-\sum_{t\geqslant 1}\frac{1}{[t]_z}\mathtt{g}_{n,\pm t}\Big(\sum_{0\leqslant l\leqslant n-1} z^{\pm 2lt}(uz)^{\pm t}\Big)\Big)w_0$$
$$=\prod_{0\leqslant l\leqslant n-1}\mathscr{Q}_n^{\pm}(uz^{2l})w_0=w_0.$$

It follows that

$$-\sum_{t\geqslant 1}\frac{1}{[t]_z}\mathtt{g}_{n,\pm t}\Big(\sum_{0\leqslant l\leqslant n-1} z^{\pm 2lt}(uz)^{\pm t}\Big)w_0=0.$$

This forces $\mathtt{g}_{n,\pm t}w_0=0$, for all $t\geqslant 1$. Consequently,

$$Q_n^{\pm}(u)w_0=\mathscr{Q}_n^{\pm}(u)w_0=w_0,$$

and hence, $Q_n^{\pm}(u)=1$. We conclude by (4.4.2.4) that all $Q_i^{\pm}(u)$ are polynomials with constant term 1 and $\mathbf{Q_s}=(Q_1(u),\ldots,Q_n(u))\in\mathcal{Q}(n)$. Moreover,

$$\sum_{1\leqslant i\leqslant n}\deg Q_i(u)=\sum_{1\leqslant j\leqslant n-1} j\deg P_j(u)=\sum_{1\leqslant j\leqslant p}\mu_j=r=\sum_{i=1}^{n}\lambda_i.$$

This forces $\lambda_n = 0$ and consequently, $\deg Q_i(u) = \lambda_i$, for all $1 \leqslant i \leqslant n$. Therefore, $\Omega_{\mathbb{C}}^{\otimes r} \otimes_{\mathcal{H}_\Delta(r)_\mathbb{C}} V_\mathbf{s}$ is a simple polynomial representation of $\mathrm{U}_\mathbb{C}(\widehat{\mathfrak{gl}}_n)$ and $\Omega_{\mathbb{C}}^{\otimes r} \otimes_{\mathcal{H}_\Delta(r)_\mathbb{C}} V_\mathbf{s} \cong L(\mathbf{Q}_\mathbf{s})$, proving (1).

We now prove (2). For any $\mathbf{Q} = (Q_1(u), \ldots, Q_n(u)) \in \mathcal{Q}(n)$ such that $r = \sum_{1\leqslant i\leqslant n} \lambda_i$ with $\lambda_i = \deg Q_i(u)$, we now prove that $L(\mathbf{Q})$ is a simple $\mathcal{S}_\Delta(n,r)_\mathbb{C}$-module.

Since the polynomials

$$P_j(u) = \frac{Q_j(uz^{j-1})}{Q_{j+1}(uz^{j+1})} \quad (1 \leqslant j \leqslant n-1)$$

have constant term 1 and $\deg P_j(u) = \lambda_j - \lambda_{j+1} =: \nu_j$, it follows that $\lambda \in \Lambda^+(n,r)$ is a partition with at most n parts. So $n > r$ implies $\lambda_n = 0$. Moreover, we may write, for $1 \leqslant i \leqslant n-1$,

$$P_i(u) = (1 - a_{\nu_1+\cdots+\nu_{i-1}+1}u)(1 - a_{\nu_1+\cdots+\nu_{i-1}+2}u)\cdots(1 - a_{\nu_1+\cdots+\nu_{i-1}+\nu_i}u),$$

where $a_j^{-1} \in \mathbb{C}$, $1 \leqslant j \leqslant p = \sum_i \nu_i$, are the roots of $P_i(u)$. Let $\mathbf{s} = \{\mathsf{s}_1, \ldots, \mathsf{s}_p\}$, where

$$\mathsf{s}_i = (a_i z^{-\mu_i+1}, a_i z^{-\mu_i+3}, \ldots, a_i z^{\mu_i-1})$$

and $(\mu_1, \ldots, \mu_p) = (1^{\nu_1}, \ldots, (n-1)^{\nu_{n-1}})$, and let $\mathbf{a} = (\mathsf{s}_1, \ldots, \mathsf{s}_p)$. Since

$$\sum_{1\leqslant j\leqslant p} \mu_j = \sum_{1\leqslant i\leqslant n-1} i\nu_i = \sum_{1\leqslant i\leqslant n} \lambda_i = r,$$

we have $\mathbf{a} \in (\mathbb{C}^*)^r$. By the first part of the proof, we see $\mathbf{Q} = \mathbf{Q}_\mathbf{s}$ and, hence, $\Omega_{\mathbb{C}}^{\otimes r} \otimes_{\mathcal{H}_\Delta(r)_\mathbb{C}} V_\mathbf{s} \cong L(\mathbf{Q})$ as $\mathrm{U}_\mathbb{C}(\widehat{\mathfrak{gl}}_n)$-modules. In other words, $L(\mathbf{Q})$ is a simple $\mathcal{S}_\Delta(n,r)_\mathbb{C}$-module. □

4.5. Application: the set $\mathscr{S}_r^{(n)}$

We can apply Theorem 4.4.2 to determine the index set $\mathscr{S}_r^{(n)}$ used in the Classification Theorem 4.3.4. Recall from Proposition 3.7.4 that the idempotents $\mathfrak{l}_\lambda = [\mathrm{diag}(\lambda)]$, for $\lambda \in \Lambda_\Delta(n,r)$, which form a basis for $\mathcal{S}_\Delta(n,r)_\mathbb{C}^0$. Note also that $\mathcal{S}_\Delta(n,r)_\mathbb{C}^0 = \mathcal{S}(n,r)_\mathbb{C}^0$ and $\Lambda_\Delta(n,r)$ is identified with $\Lambda(n,r)$ under the map $\flat_2$ in (1.1.0.2).

Lemma 4.5.1. *If $L(\mathbf{Q})$ is a simple $\mathcal{S}_\Delta(n,r)_\mathbb{C}$-module with pseudo-highest weight λ and $\mathfrak{l}_\mu L(\mathbf{Q}) \neq 0$, then $\lambda \trianglerighteq \mu$.*

Proof. If we regard $L(\mathbf{Q})$ as a $\mathrm{U}_\mathbb{C}(\widehat{\mathfrak{gl}}_n)$-module, then $\mathfrak{l}_\mu L(\mathbf{Q}) = L(\mathbf{Q})_\mu$ is its μ-weight space. However, if w_0 is a pseudo-highest weight vector in $L(\mathbf{Q})_\lambda$, then the triangular decomposition[4] $\mathrm{U}_{\mathbb{C}}(\widehat{\mathfrak{gl}}_n) = \mathrm{U}_{\mathbb{C}}(\widehat{\mathfrak{gl}}_n)_{-}\mathrm{U}_{\mathbb{C}}(\widehat{\mathfrak{gl}}_n)_0\mathrm{U}_{\mathbb{C}}(\widehat{\mathfrak{gl}}_n)_{+}$ ([**28**, Lem. 7.4]) implies $L(\mathbf{Q}) = \mathrm{U}_{\mathbb{C}}(\widehat{\mathfrak{gl}}_n)_{-}w_0$. Here $\mathrm{U}_{\mathbb{C}}(\widehat{\mathfrak{gl}}_n)_{\pm}$ (resp., $\mathrm{U}_{\mathbb{C}}(\widehat{\mathfrak{gl}}_n)_0$) are the subalgebras generated by all $\mathrm{x}_{i,s}^{\pm}$ (resp., k_j, $\mathrm{g}_{j,t}$), for all $1 \leqslant i \leqslant n-1, s \in \mathbb{Z}$ (resp., $1 \leqslant j \leqslant n, t \in \mathbb{Z}\backslash\{0\}$). Our assertion follows from the fact that every $x_{i,s}^{-}w_0$ has weight $\lhd\lambda$; see (QLA2) in Definition 2.5.1. □

For $\mathbf{Q} \in \mathcal{Q}(n)_r$, let $\deg(\mathbf{Q}) = (\deg(Q_1(u)), \cdots, \deg(Q_n(u)))$. As usual, regard $\Lambda_\vartriangle(n, r)$ as a subset of $\Lambda_\vartriangle(N, r)$ as in (4.1.5.1).

Lemma 4.5.2. *Assume $N > r \geqslant n$. Let $e = \sum_{\mu\in\Lambda_\vartriangle(n,r)} [_\mu$. Then the set*

$$\{eL(\mathbf{Q}) \mid \mathbf{Q} \in \mathcal{Q}(N)_r,\ \deg(\mathbf{Q}) \in \Lambda_\vartriangle(n, r)\}$$

forms a complete set of non-isomorphic simple $\mathcal{S}_\vartriangle(n, r)_{\mathbb{C}}$-modules.

Proof. By Theorem 4.4.2 and [**33**, 6.2(g)], the set $\{eL(\mathbf{Q}) \neq 0 \mid \mathbf{Q} \in \mathcal{Q}(N)_r\}$ forms a complete set of non-isomorphic simple $\mathcal{S}_\vartriangle(n, r)_{\mathbb{C}}$-modules. Thus, it is enough to prove that, for $\mathbf{Q} \in \mathcal{Q}(N)_r$, $eL(\mathbf{Q}) \neq 0$ if and only if $\deg(\mathbf{Q}) \in \Lambda_\vartriangle(n, r)$, i.e., the last $N - n$ parts of $\deg(\mathbf{Q})$ are all zero.

Let $\mathbf{Q} \in \mathcal{Q}(N)_r$ and $\lambda = \deg(\mathbf{Q})$. If $\lambda \in \Lambda_\vartriangle(n, r)$, then $e([_\lambda L(\mathbf{Q})) = [_\lambda L(\mathbf{Q}) \neq 0$ and, hence, $eL(\mathbf{Q}) \neq 0$. Conversely, assume $eL(\mathbf{Q}) \neq 0$. Since $1 = \sum_{\alpha\in\Lambda_\vartriangle(N,r)} [_\alpha$,

$$eL(\mathbf{Q}) = e \bigoplus_{\alpha\in\Lambda_\vartriangle(N,r)} [_\alpha L(\mathbf{Q}) = \bigoplus_{\alpha\in\Lambda_\vartriangle(n,r)} [_\alpha L(\mathbf{Q}).$$

This together with the fact that $eL(\mathbf{Q}) \neq 0$ implies that there exists $\alpha \in \Lambda_\vartriangle(n, r)$ such that $[_\alpha L(\mathbf{Q}) \neq 0$. Since $[_\alpha L(\mathbf{Q}) \neq 0$, by Lemma 4.5.1, $\alpha \trianglelefteq \lambda$ and, hence, $r = \sum_{1\leqslant i\leqslant n} \alpha_i \leqslant \sum_{1\leqslant i\leqslant n} \lambda_i \leqslant r$, forcing $\lambda \in \Lambda_\vartriangle(n, r)$. □

Theorem 4.5.3. *We have*

$$\mathscr{S}_r^{(n)} = \{\mathbf{s} = \{\mathsf{s}_1, \dots, \mathsf{s}_p\} \in \mathscr{S}_r \mid p \geqslant 1,\ |\mathsf{s}_i| \leqslant n,\ \forall i\}.$$

Proof. We choose $N > 0$ such that $N > \max\{r, n\}$. Then, by Theorem 4.4.2, $\{L(\mathbf{Q}_\mathbf{s}) \mid \mathbf{s} \in \mathscr{S}_r\} = \{L(\mathbf{Q}) \mid \mathbf{Q} \in \mathcal{Q}(N)_r\}$ is a complete set of non-isomorphic simple $\mathcal{S}_\vartriangle(N, r)_{\mathbb{C}}$-modules. Let $e = \sum_{\mu\in\Lambda_\vartriangle(n,r)} [_\mu$. By the proof of Theorem 4.3.4 we have $eL(\mathbf{Q}_\mathbf{s}) \cong \Omega_{\mathbb{C}}^{\otimes r} \otimes_{\mathcal{H}_\vartriangle(r)_{\mathbb{C}}} V_\mathbf{s}$ for $\mathbf{s} \in \mathscr{S}_r$. Thus, by Lemma 4.5.2, the set

$$\{\Omega_{\mathbb{C}}^{\otimes r} \otimes_{\mathcal{H}_\vartriangle(r)_{\mathbb{C}}} V_\mathbf{s} \mid \mathbf{s} \in \mathscr{S}_r,\ \deg(\mathbf{Q}_\mathbf{s}) \in \Lambda_\vartriangle(n, r)\}$$

forms a complete set of non-isomorphic simple $\mathcal{S}_\vartriangle(n, r)_{\mathbb{C}}$-modules. Now, write $\mathbf{Q}_\mathbf{s} = (Q_1(u), \dots, Q_N(u))$. (Then $Q_{r+1}(u) = \cdots = Q_N(u) = 1$.) The condition $\deg(\mathbf{Q}_\mathbf{s}) \in \Lambda_\vartriangle(n, r)$, where $\mathbf{s} = \{\mathsf{s}_1, \dots, \mathsf{s}_p\} \in \mathscr{S}_r$, is equivalent to

[4] This triangular decomposition is different from the one given in Corollary 2.5.4.

$Q_{n+1}(u) = \cdots = Q_N(u) = 1$, which means $P_{n+1}(u) = \cdots = P_{N-1}(u) = 1$. By (4.4.1.2), this condition holds if and only if each segment $\mathbf{s}_i$ in $\mathbf{s}$ has length $|\mathbf{s}_i| \leqslant n$ for $1 \leqslant i \leqslant p$. □

Since $\{\Omega_{n,\mathbb{C}}^{\otimes r} \otimes_{\mathcal{H}(r)_\mathbb{C}} S_\mu\}_{\mu\in\Lambda^+(n,r)}$ forms a complete set of simple $\mathcal{S}(n,r)_\mathbb{C}$-modules, we may speak of multiplicities $[\Omega_\mathbb{C}^{\otimes r} \otimes_{\mathcal{H}_\Delta(r)_\mathbb{C}} V_\mathbf{s} : \Omega_{n,\mathbb{C}}^{\otimes r} \otimes_{\mathcal{H}(r)_\mathbb{C}} S_\mu]$ for all $\mathbf{s} \in \mathscr{S}_r^{(n)}$. An immediate consequence of the theorem above is the following multiplicity identity. Recall that λ' is the dual partition of λ.

Corollary 4.5.4. *The partition $\wp(\mathbf{s})$ associated with $\mathbf{s} \in \mathscr{S}_r^{(n)}$ gives rise to a well-defined surjective map $\wp' : \mathscr{S}_r^{(n)} \to \Lambda^+(n,r)$, $\mathbf{s} \mapsto \wp(\mathbf{s})'$. Moreover, for $\mathbf{s} \in \mathscr{S}_r^{(n)}$ and $\mu \in \Lambda^+(n,r)$, we have*

$$[\Omega_\mathbb{C}^{\otimes r} \otimes_{\mathcal{H}_\Delta(r)_\mathbb{C}} V_\mathbf{s} : \Omega_{n,\mathbb{C}}^{\otimes r} \otimes_{\mathcal{H}(r)_\mathbb{C}} S_\mu] = [V_\mathbf{s} : S_\mu].$$

Proof. By Lemma 4.1.1 and (4.3.1.1), we have $\mathcal{S}(n,r)_\mathbb{C}$-module isomorphisms

$$\Omega_\mathbb{C}^{\otimes r} \otimes_{\mathcal{H}_\Delta(r)_\mathbb{C}} V_\mathbf{s} \cong \Omega_{n,\mathbb{C}}^{\otimes r} \otimes_{\mathcal{H}(r)_\mathbb{C}} V_\mathbf{s} \cong \bigoplus_{\mu\vdash r,\mu\trianglerighteq\wp(\mathbf{s})} m_{\mu'}(\Omega_{n,\mathbb{C}}^{\otimes r} \otimes_{\mathcal{H}(r)_\mathbb{C}} S_{\mu'}),$$

where $m_{\mu'} = [V_\mathbf{s} : S_{\mu'}]$. Now, $\Omega_{n,\mathbb{C}}^{\otimes r} \otimes_{\mathcal{H}(r)_\mathbb{C}} S_{\mu'} \neq 0$ if and only if $\mu' \in \Lambda^+(n,r)$. Hence, in this case,

$$[\Omega_\mathbb{C}^{\otimes r} \otimes_{\mathcal{H}_\Delta(r)_\mathbb{C}} V_\mathbf{s} : \Omega_{n,\mathbb{C}}^{\otimes r} \otimes_{\mathcal{H}(r)_\mathbb{C}} S_{\mu'}] = m_{\mu'} = [V_\mathbf{s} : S_{\mu'}],$$

as desired. □

4.6. Classification of simple $\mathcal{S}_\Delta(n,r)_\mathbb{C}$-modules: the downward approach

We now complete the classification of simple $\mathcal{S}_\Delta(n,r)_\mathbb{C}$-modules by removing the condition $n > r$ in Theorem 4.4.2. We will continue to use the downward approach with a strategy different from that in the previous section. Throughout this section, we will identify $\mathrm{U}_\mathbb{C}(\widehat{\mathfrak{gl}}_n)$ with $\mathfrak{D}_{\Delta,\mathbb{C}}(n)$ via the isomorphism f in Proposition 4.4.1 and regard every $\mathcal{S}_\Delta(n,r)_\mathbb{C}$-module as a $\mathrm{U}_\mathbb{C}(\widehat{\mathfrak{gl}}_n)$-module via the algebra homomorphism $\xi'_{r,\mathbb{C}} : \mathrm{U}_\mathbb{C}(\widehat{\mathfrak{gl}}_n) \to \mathcal{S}_\Delta(n,r)_\mathbb{C}$; see (4.4.1.1).

Consider the $\mathrm{U}_\mathbb{C}(\widehat{\mathfrak{gl}}_n)$-module via $\xi'_{1,\mathbb{C}}$

$$\Omega_\mathbb{C}(a) := \mathsf{F}(M_a) = \Omega_\mathbb{C} \otimes_{\mathcal{H}_\Delta(1)_\mathbb{C}} M_a$$

for $a \in \mathbb{C}^*$. By Proposition 4.2.1, $\Omega_\mathbb{C}(a) \cong \Omega_{n,\mathbb{C}} \otimes_{\mathcal{H}(1)_\mathbb{C}} M_a \cong \Omega_{n,\mathbb{C}}$ as $\mathcal{S}(n,1)_\mathbb{C}$-modules. Hence, $\dim \Omega_\mathbb{C}(a) = n$. See Proposition 3.5.2 for a different construction of $\Omega_\mathbb{C}(a)$.

By Theorem 4.1.3, $\Omega_{\mathbb{C}}(a)$ is a simple $\mathrm{U}_{\mathbb{C}}(\widehat{\mathfrak{gl}}_n)$-module since $\dim_{\mathbb{C}} M_a = 1$ and $M_a = V_{\mathbf{s}}$ with $\mathbf{s} = (a) \in \mathscr{S}_1$. Since $n > 1$, by Theorem 4.4.2,

$$\Omega_{\mathbb{C}}(a) \cong L(\mathbf{Q}) \text{ with } Q_1(u) = 1 - au \text{ and } Q_i(u) = 1, \text{ for } 2 \leqslant i \leqslant n. \tag{4.6.0.1}$$

The $\mathrm{U}_{\mathbb{C}}(\widehat{\mathfrak{gl}}_n)$-modules $\Omega_{\mathbb{C}}(a)$ are very useful, and we will prove that every simple $\mathcal{S}_\vartriangle(n,r)_{\mathbb{C}}$-module is a quotient module of $\Omega_{\mathbb{C}}(a_1) \otimes_{\mathbb{C}} \cdots \otimes_{\mathbb{C}} \Omega_{\mathbb{C}}(a_r)$, for some $\mathbf{a} \in (\mathbb{C}^*)^r$, in Corollary 4.6.2.

For $i \in \mathbb{Z}$, let $\widetilde{\omega}_i = \omega_i \otimes \bar{1} \in \Omega_{\mathbb{C}}(a)$.

Lemma 4.6.1. *For any $\mathbf{a} = (a_1, \ldots, a_r) \in (\mathbb{C}^*)^r$, there is a $\mathrm{U}_{\mathbb{C}}(\widehat{\mathfrak{gl}}_n)$-module isomorphism*

$$\varphi : \Omega_{\mathbb{C}}(a_1) \otimes_{\mathbb{C}} \cdots \otimes_{\mathbb{C}} \Omega_{\mathbb{C}}(a_r) \longrightarrow \Omega_{\mathbb{C}}^{\otimes r} \otimes_{\mathcal{H}_\vartriangle(r)_{\mathbb{C}}} M_{\mathbf{a}}$$

defined by sending $\widetilde{\omega}_{\mathbf{i}}$ to $\omega_{\mathbf{i}} \otimes \bar{1}$, for $\mathbf{i} \in I_\vartriangle(n,r)$, where $\widetilde{\omega}_{\mathbf{i}} = \widetilde{\omega}_{i_1} \otimes \cdots \otimes \widetilde{\omega}_{i_r}$. Moreover, as an $\mathcal{S}(n,r)_{\mathbb{C}}$-module, $\Omega_{\mathbb{C}}(a_1) \otimes_{\mathbb{C}} \cdots \otimes_{\mathbb{C}} \Omega_{\mathbb{C}}(a_r)$ is isomorphic to the finite tensor space $\Omega_{n,\mathbb{C}}^{\otimes r}$, for all $\mathbf{a} \in (\mathbb{C}^*)^r$.

Proof. The set

$$\{\widetilde{\omega}_i \mid 1 \leqslant i \leqslant n\}$$

forms a basis of $\Omega_{\mathbb{C}}(a)$. Hence, the set

$$\{\widetilde{\omega}_{\mathbf{i}} \mid \mathbf{i} \in I(n,r)\}$$

forms a basis of $\Omega_{\mathbb{C}}(a_1) \otimes_{\mathbb{C}} \cdots \otimes_{\mathbb{C}} \Omega_{\mathbb{C}}(a_r)$.

Similarly by Proposition 4.2.1, we have

$$\Omega_{\mathbb{C}}^{\otimes r} \otimes_{\mathcal{H}_\vartriangle(r)_{\mathbb{C}}} M_{\mathbf{a}} \cong \Omega_{n,\mathbb{C}}^{\otimes r} \otimes_{\mathcal{H}(r)_{\mathbb{C}}} M_{\mathbf{a}} \cong \Omega_{n,\mathbb{C}}^{\otimes r} \tag{4.6.1.1}$$

as $\mathcal{S}(n,r)_{\mathbb{C}}$-modules. So the set

$$\{\omega_{\mathbf{i}} \otimes \bar{1} \mid \mathbf{i} \in I(n,r)\}$$

forms a basis of $\Omega_{\mathbb{C}}^{\otimes r} \otimes_{\mathcal{H}_\vartriangle(r)_{\mathbb{C}}} M_{\mathbf{a}}$. Hence, there is a linear isomorphism

$$\varphi : \Omega_{\mathbb{C}}(a_1) \otimes_{\mathbb{C}} \cdots \otimes_{\mathbb{C}} \Omega_{\mathbb{C}}(a_r) \longrightarrow \Omega_{\mathbb{C}}^{\otimes r} \otimes_{\mathcal{H}_\vartriangle(r)_{\mathbb{C}}} M_{\mathbf{a}}$$

defined by sending $\widetilde{\omega}_{\mathbf{i}}$ to $\omega_{\mathbf{i}} \otimes \bar{1}$, for all $\mathbf{i} \in I(n,r)$.

Now we assume $\mathbf{i} \in I_\vartriangle(n,r)$. We write $\mathbf{i} = \mathbf{j} + n\mathbf{t}$ with $\mathbf{j} \in I(n,r)$ and $\mathbf{t} \in \mathbb{Z}^r$. Then

$$\begin{aligned}
\varphi(\widetilde{\omega}_{\mathbf{i}}) &= \varphi((a_1^{-t_1}\widetilde{\omega}_{j_1}) \otimes \cdots \otimes (a_r^{-t_r}\widetilde{\omega}_{j_r})) \\
&= a_1^{-t_1} a_2^{-t_2} \cdots a_r^{-t_r} \omega_{\mathbf{j}} \otimes \bar{1} \\
&= \omega_{\mathbf{j}} X_1^{-t_1} X_2^{-t_2} \cdots X_r^{-t_r} \otimes \bar{1} \\
&= \omega_{\mathbf{i}} \otimes \bar{1}.
\end{aligned}$$

It follows easily that φ is a $\mathrm{U}_\mathbb{C}(\widehat{\mathfrak{gl}}_n)$-module isomorphism. The last assertion is clear from (4.6.1.1). □

Corollary 4.6.2. *Let V be a finite dimensional simple $\mathrm{U}_\mathbb{C}(\widehat{\mathfrak{gl}}_n)$-module. Then the following conditions are equivalent:*

(1) V can be regarded as an $\mathcal{S}_\Delta(n,r)_\mathbb{C}$-module via $\xi'_{r,\mathbb{C}}$;
(2) V is a quotient module of the $\mathrm{U}_\mathbb{C}(\widehat{\mathfrak{gl}}_n)$-module $\Omega_\mathbb{C}(a_1)\otimes_\mathbb{C}\cdots\otimes_\mathbb{C}\Omega_\mathbb{C}(a_r)$ for some $\mathbf{a}\in(\mathbb{C}^*)^r$;
(3) V is a quotient module of $\Omega_\mathbb{C}^{\otimes r}$;
(4) V is a subquotient module of $\Omega_\mathbb{C}^{\otimes r}$.

Proof. By the lemma above, the map

$$\Omega_\mathbb{C}^{\otimes r}\longrightarrow\Omega_\mathbb{C}(a_1)\otimes_\mathbb{C}\cdots\otimes_\mathbb{C}\Omega_\mathbb{C}(a_r),\quad \omega_\mathbf{i}\longmapsto\widetilde{\omega}_\mathbf{i}$$

is a $\mathrm{U}_\mathbb{C}(\widehat{\mathfrak{gl}}_n)$-module epimorphism (say, induced by the natural $\mathcal{H}_\Delta(r)_\mathbb{C}$-module epimorphism $\mathcal{H}_\Delta(r)_\mathbb{C}\to M_\mathbf{a}$). Hence, (2) implies (3). Certainly, (3) implies (4). Since $\Omega_\mathbb{C}^{\otimes r}$ is an $\mathcal{S}_\Delta(n,r)_\mathbb{C}$-module, (4) implies (1).

If V can be regarded as an $\mathcal{S}_\Delta(n,r)_\mathbb{C}$-module via $\xi'_{r,\mathbb{C}}$, then, by Theorem 4.3.4, $V\cong\Omega_\mathbb{C}^{\otimes r}\otimes_{\mathcal{H}_\Delta(r)_\mathbb{C}}V_\mathbf{s}$ for some $\mathbf{s}$. Since $V_\mathbf{s}$ is a homomorphic image of some $M_\mathbf{a}$ (see [**9**, 3.4], say), it follows that V is a homomorphic image of $\Omega_\mathbb{C}^{\otimes r}\otimes_{\mathcal{H}_\Delta(r)_\mathbb{C}}M_\mathbf{a}$, which is, by Lemma 4.6.1, isomorphic to $\Omega_\mathbb{C}(a_1)\otimes_\mathbb{C}\cdots\otimes_\mathbb{C}\Omega_\mathbb{C}(a_r)$. Hence, V is a homomorphic image of $\Omega_\mathbb{C}(a_1)\otimes_\mathbb{C}\cdots\otimes_\mathbb{C}\Omega_\mathbb{C}(a_r)$, proving (2). □

Remark 4.6.3. The algebras $\mathrm{U}_\mathbb{C}(\widehat{\mathfrak{gl}}_n)$, $\mathcal{S}_\Delta(n,r)_\mathbb{C}$, etc., under consideration are all defined over $\mathbb{C}$ with parameter z which is not a root of unity. As an $\mathcal{S}(n,r)_\mathbb{C}$-module, the finite dimensional tensor space $\Omega_{n,\mathbb{C}}^{\otimes r}$ is a semisimple module (i.e., is completely reducible). In the affine case, however, the infinite dimensional tensor space $\Omega_\mathbb{C}^{\otimes r}$ is not completely reducible. The fact that every simple $\mathcal{S}_\Delta(n,r)_\mathbb{C}$-module is a homomorphic image of $\Omega_\mathbb{C}^{\otimes r}$ does reflect a certain degree of the complete reducibility.

Using Corollary 4.6.2, we can prove the first key result for the classification theorem.

Proposition 4.6.4. *Every simple $\mathcal{S}_\Delta(n,r)_\mathbb{C}$-module is a polynomial representation of $\mathrm{U}_\mathbb{C}(\widehat{\mathfrak{gl}}_n)$.*

Proof. Let V be a simple $\mathcal{S}_\Delta(n,r)_\mathbb{C}$-module. Then V is finite dimensional by Theorem 4.1.6. Thus, V is a quotient module of $\Omega_\mathbb{C}(a_1)\otimes_\mathbb{C}\cdots\otimes_\mathbb{C}\Omega_\mathbb{C}(a_r)$ for some $\mathbf{a}\in(\mathbb{C}^*)^r$, by Corollary 4.6.2. Now, (4.6.0.1) implies that $\Omega_\mathbb{C}(a)$ is a polynomial representation of $\mathrm{U}_\mathbb{C}(\widehat{\mathfrak{gl}}_n)$. So, by [**28**, 4.3], the tensor product

$\Omega_{\mathbb{C}}(a_1) \otimes_{\mathbb{C}} \cdots \otimes_{\mathbb{C}} \Omega_{\mathbb{C}}(a_r)$ is a polynomial representation of $\mathrm{U}_{\mathbb{C}}(\widehat{\mathfrak{gl}}_n)$. Hence, V is also a polynomial representation of $\mathrm{U}_{\mathbb{C}}(\widehat{\mathfrak{gl}}_n)$. □

For $a \in \mathbb{C}^*$, the $\mathrm{U}_{\mathbb{C}}(\widehat{\mathfrak{gl}}_n)$-module Det_a defined in the following lemma plays the same role in the representation theory of $\mathrm{U}_{\mathbb{C}}(\widehat{\mathfrak{gl}}_n)$ as the quantum determinant in the representation theory of $U(n)_{\mathbb{C}} = U(\mathfrak{gl}_n)_{\mathbb{C}}$. One can easily see that by restriction Det_a is isomorphic to the quantum determinant for $U(n)_{\mathbb{C}}$.

Lemma 4.6.5. *Fix $a \in \mathbb{C}^*$. Let $\mathbf{a} = \mathbf{s} = (a, az^2, \ldots, az^{2(n-1)})$ regarded as a single segment, and let*

$$\mathrm{Det}_a := \Omega_{\mathbb{C}}^{\otimes n} \otimes_{\mathcal{H}_\Delta(n)_{\mathbb{C}}} V_a,$$

where $V_a = V_{\mathbf{s}}$ is the submodule of $M_{\mathbf{a}}$ generated by $\overline{y}_{(n)}$. Then $\dim \mathrm{Det}_a = 1$ *and*

$$\mathrm{Det}_a = \mathrm{span}_{\mathbb{C}}\{\omega_1 \otimes \cdots \otimes \omega_n \otimes \overline{y}_{(n)}\}.$$

Moreover, as a $\mathrm{U}_{\mathbb{C}}(\widehat{\mathfrak{gl}}_n)$-module, $\mathrm{Det}_a \cong L(\mathbf{Q})$, *where $\mathbf{Q} = (Q_1(u), \ldots, Q_n(u)) \in \mathcal{Q}(n)$ with $Q_i(u) = 1 - az^{2(n-i)}u$, for all $i = 1, 2, \ldots, n$.*

Proof. Recall the notation used in §4.3. We have $M_{\mathbf{a}} \cong \mathcal{H}(n)_{\mathbb{C}}$ and, by Theorem 4.3.2, $V_a = \mathcal{H}(n)_{\mathbb{C}}\overline{y}_{(n)} = \mathbb{C}\overline{y}_{(n)}$, since $T_i y_{(n)} = -y_{(n)}$, for all $1 \leqslant i \leqslant n-1$.

By Theorem 4.1.3, Det_a is a simple $\mathcal{S}_\Delta(n,n)_{\mathbb{C}}$-module. By Proposition 4.6.4, $\mathrm{Det}_a \cong L(\mathbf{Q})$, for some $\mathbf{Q} \in \mathcal{Q}(n)$. Since

$$\Omega_{n,\mathbb{C}}^{\otimes n} = \bigoplus_{\mathbf{i} \in I_\Delta(n,n)_0} \omega_{\mathbf{i}} \mathcal{H}(n)_{\mathbb{C}},$$

by Proposition 4.2.1,

$$\mathrm{Det}_a \cong \Omega_{n,\mathbb{C}}^{\otimes n} \otimes_{\mathcal{H}(n)_{\mathbb{C}}} V_a \cong \bigoplus_{\mathbf{i} \in I_\Delta(n,n)_0} \omega_{\mathbf{i}} \mathcal{H}(n)_{\mathbb{C}} \otimes_{\mathcal{H}(n)_{\mathbb{C}}} V_a.$$

Hence, $\mathrm{Det}_a = \mathrm{span}_{\mathbb{C}}\{\omega_{\mathbf{i}} \otimes \overline{y}_{(n)} \mid \mathbf{i} \in I_\Delta(n,n)_0\}$. If $i_k = i_{k+1}$ in an $\mathbf{i} \in I_\Delta(n,n)_0$ for some $1 \leqslant k \leqslant n-1$, then

$$z^2 \omega_{\mathbf{i}} \otimes \overline{y}_{(n)} = \omega_{\mathbf{i}} T_k \otimes \overline{y}_{(n)} = \omega_{\mathbf{i}} \otimes T_k \overline{y}_{(n)} = -\omega_{\mathbf{i}} \otimes \overline{y}_{(n)}.$$

This forces $\omega_{\mathbf{i}} \otimes \overline{y}_{(n)} = 0$ as z is not a root of unity. The only $\mathbf{i} \in I_\Delta(n,n)_0$ with $i_k \neq i_{k+1}$, for $1 \leqslant k \leqslant n-1$, is $\mathbf{i} = (1, 2, \ldots, n)$. Hence,

$$\mathrm{Det}_a = \mathbb{C}\,\omega_1 \otimes \cdots \otimes \omega_n \otimes \overline{y}_{(n)}.$$

Since $K_i w_0 = z w_0$, where $w_0 = \omega_1 \otimes \cdots \otimes \omega_n \otimes \overline{y}_{(n)}$, it follows that $\deg Q_i(u) = 1$, for all $1 \leqslant i \leqslant n$. On the other hand, since

$$P_i(u) = \frac{Q_i(uz^{i-1})}{Q_{i+1}(uz^{i+1})}$$

is a polynomial, we must have $P_i(u) = 1$, for all $1 \leqslant i \leqslant n-1$.

Suppose $Q_n(u) = 1 - bu$, for some $b \in \mathbb{C}^*$. Then $Q_i(u) = 1 - bz^{2(n-i)}u$ for all i. Thus, as in the proof of Theorem 4.4.2, Lemma 4.3.1 implies

$$\frac{tz^{\pm t}}{[t]_z} \sum_{1\leqslant i\leqslant n} \mathsf{g}_{i,\pm t} w_0 = \mathsf{z}_t^{\pm} w_0 = \sum_{1\leqslant k\leqslant n} (az^{2(k-1)})^{\pm t} w_0,$$

for all $t \geqslant 1$ and, hence,

$$\prod_{1\leqslant i\leqslant n} \left(1 - (bz^{2(n-i)}u)^{\pm 1}\right) w_0 = \prod_{1\leqslant i\leqslant n} Q_i^{\pm}(u) w_0$$
$$= \prod_{1\leqslant i\leqslant n} \mathcal{Q}_i^{\pm}(u) w_0 = \prod_{1\leqslant k\leqslant n} \left(1 - (az^{2(k-1)}u)^{\pm 1}\right) w_0.$$

Equating the coefficients of u forces $a = b$. □

Lemma 4.6.6. *Let V be a simple $\mathcal{S}_\Delta(n,k)_\mathbb{C}$-module and W be a simple $\mathcal{S}_\Delta(n,l)_\mathbb{C}$-module. Then $V \otimes W$ is an $\mathcal{S}_\Delta(n,k+l)_\mathbb{C}$-module.*

Proof. By Corollary 4.6.2, V is a quotient module of the $\mathrm{U}_\mathbb{C}(\widehat{\mathfrak{gl}}_n)$-module $\Omega_\mathbb{C}(a_1) \otimes_\mathbb{C} \cdots \otimes_\mathbb{C} \Omega_\mathbb{C}(a_k)$ for some $\mathbf{a} \in (\mathbb{C}^*)^k$ and W is a quotient module of $\Omega_\mathbb{C}(b_1) \otimes_\mathbb{C} \cdots \otimes_\mathbb{C} \Omega_\mathbb{C}(b_l)$ for some $\mathbf{b} \in (\mathbb{C}^*)^l$. Thus, $V \otimes W$ is a quotient module of the $\mathrm{U}_\mathbb{C}(\widehat{\mathfrak{gl}}_n)$-module

$$\Omega_\mathbb{C}(a_1) \otimes_\mathbb{C} \cdots \otimes_\mathbb{C} \Omega_\mathbb{C}(a_k) \otimes_\mathbb{C} \Omega_\mathbb{C}(b_1) \otimes_\mathbb{C} \cdots \otimes_\mathbb{C} \Omega_\mathbb{C}(b_l),$$

which is isomorphic to $\Omega_\mathbb{C}^{\otimes(k+l)} \otimes_{\mathcal{H}_\Delta(r)_\mathbb{C}} M_{(\mathbf{a},\mathbf{b})}$, by Lemma 4.6.1. Hence, as a quotient module of an $\mathcal{S}_\Delta(n,k+l)_\mathbb{C}$-module, $V \otimes W$ is an $\mathcal{S}_\Delta(n,k+l)_\mathbb{C}$-module. □

We remark that it is possible to embed $\mathcal{H}_\Delta(k)_\mathbb{C} \otimes \mathcal{H}_\Delta(l)_\mathbb{C}$ into $\mathcal{H}_\Delta(k+l)_\mathbb{C}$ as a subalgebra (see, e.g., [**9**, 3.2]) and, hence, $\mathcal{S}_\Delta(n,k+l)_\mathbb{C}$ is embedded as a subalgebra in $\mathcal{S}_\Delta(n,k)_\mathbb{C} \otimes \mathcal{S}_\Delta(n,l)_\mathbb{C}$. Thus, by restriction, the $\mathcal{S}_\Delta(n,k)_\mathbb{C} \otimes \mathcal{S}_\Delta(n,l)_\mathbb{C}$-module $V \otimes W$ is an $\mathcal{S}_\Delta(n,k+l)_\mathbb{C}$-module. See the map Ξ in §5.5 or [**56**, 1.2] for a geometric construction of the embedding.

For $1 \leqslant i \leqslant n-1$ and $a \in \mathbb{C}^*$, define $\mathbf{Q}_{i,a} \in \mathcal{Q}(n)$ by setting $Q_n(u) = 1$ and

$$\frac{Q_j(uz^{j-1})}{Q_{j+1}(uz^{j+1})} = (1-au)^{\delta_{i,j}},$$

for $1 \leqslant j \leqslant n-1$. In other words,

$$\mathbf{Q}_{i,a} = (1 - az^{i-1}u, \ldots, 1 - az^{-i+3}u, \underset{(i)}{1 - az^{-i+1}u}, 1, \ldots, 1).$$

Since $n > i$, by Theorem 4.4.2, the simple $\mathrm{U}_\mathbb{C}(\widehat{\mathfrak{gl}}_n)$-module

$$L_{i,a} := L(\mathbf{Q}_{i,a})$$

is also a simple $\mathcal{S}_\Delta(n,i)_\mathbb{C}$-module. The weight of a pseudo-highest weight vector of $L_{i,a}$ is $\lambda(i,a) = (1^i, 0^{n-i})$. Now we can prove the second key result for the classification theorem.

Proposition 4.6.7. *If* $\mathbf{Q} = (Q_i(u)) \in \mathcal{Q}(n)$ *with* $r = \sum_{1\leqslant i\leqslant n} \deg(Q_i(u))$, *then* $L(\mathbf{Q})$ *is (isomorphic to) a simple* $\mathcal{S}_\Delta(n,r)_\mathbb{C}$*-module.*

Proof. Let $\lambda = (\lambda_1, \dots, \lambda_n)$ with $\lambda_j = \deg(Q_j(u))$ and let

$$P_i(u) = \frac{Q_i(uz^{i-1})}{Q_{i+1}(uz^{i+1})}, \quad 1 \leqslant i \leqslant n-1.$$

Write $Q_n(u) = (1 - b_1 u) \cdots (1 - b_{\lambda_n} u)$ and

$$P_i(u) = \prod_{1\leqslant j\leqslant \mu_i} (1 - a_{i,j}u),$$

where $\mu_i = \lambda_i - \lambda_{i+1}$. Let

$$V = L_1 \otimes \cdots \otimes L_{n-1} \otimes \mathrm{Det}_{b_1} \otimes \cdots \otimes \mathrm{Det}_{b_{\lambda_n}},$$

where $L_i = L_{i,a_{i,1}} \otimes \cdots \otimes L_{i,a_{i,\mu_i}}$ for $1 \leqslant i \leqslant n-1$.

Let $w_{i,a_{i,k}}$ (resp., v_j) be a pseudo-highest weight vector of $L_{i,a_{i,k}}$ (resp., Det_{b_j}), and let

$$w_0 = w_1 \otimes w_2 \otimes \cdots \otimes w_{n-1} \text{ and } v_0 = v_1 \otimes \cdots \otimes v_n,$$

where

$$w_i = w_{i,a_{i,1}} \otimes w_{i,a_{i,2}} \otimes \cdots \otimes w_{i,a_{i,\mu_i}}, \text{ for } 1 \leqslant i \leqslant n-1.$$

Since $w_{i,a_{i,k}}$ has weight $(1^i, 0^{n-i})$ and

$$\mathcal{Q}_j^{\pm}(u) w_{i,a_{i,k}} = \begin{cases} (1 - (a_{i,k} z^{i-2j+1} u)^{\pm 1}) w_{i,a_{i,k}}, & \text{if } 1 \leqslant j \leqslant i; \\ w_{i,a_{i,k}}, & \text{if } i < j \leqslant n, \end{cases} \quad \text{and}$$

$$\mathcal{Q}_j^{\pm}(u) v_i = (1 - (b_i z^{2(n-j)} u)^{\pm 1}) v_i \text{ for } 1 \leqslant i \leqslant \lambda_n,$$

it follows from [**28**, Lem. 4.1] that

$$\mathcal{Q}_j^{\pm}(u)(w_0 \otimes v_0) = Q_j^{\pm}(u)(w_0 \otimes v_0).$$

Moreover, the weight of $w_0 \otimes v_0$ is

$$\lambda = (\mu_1, 0, \dots, 0) + (\mu_2, \mu_2, 0, \dots, 0) + (\mu_{n-1}, \dots, \mu_{n-1}, 0) + (\lambda_n, \dots, \lambda_n).$$

Let W be the submodule of V generated by $w_0 \otimes v_0$. Then W is a pseudo-highest weight module whose pseudo-highest weight vector is a common eigenvector of k_i and $\mathcal{Q}_{i,s}$ with eigenvalues z^{λ_i} and $Q_{i,s}$, respectively, where $Q_{i,s}$ are the coefficients of $Q_i^{\pm}(u)$. So the simple quotient module of W is isomorphic to $L(\mathbf{Q})$ (cf. the construction in [**28**, Lem. 4.8]). Since $\sum_{1\leqslant i\leqslant n-1} i\mu_i + n\lambda_n = \sum_{1\leqslant i\leqslant n}\lambda_i = r$, by Lemma 4.6.6, V is an $\mathcal{S}_\Delta(n,r)_\mathbb{C}$-module. Hence, $L(\mathbf{Q})$ has an $\mathcal{S}_\Delta(n,r)_\mathbb{C}$-module structure. □

Now using Propositions 4.6.4 and 4.6.7 we can prove the following classification theorem.

Theorem 4.6.8. *For any $n, r \geqslant 1$, the set $\{L(\mathbf{Q}) \mid \mathbf{Q} \in \mathcal{Q}(n)_r\}$ is a complete set of non-isomorphic simple $\mathcal{S}_\Delta(n,r)_\mathbb{C}$-modules.*

Proof. By Proposition 4.6.7, the set $\{L(\mathbf{Q}) \mid \mathbf{Q} \in \mathcal{Q}(n)_r\}$ consists of non-isomorphic simple $\mathcal{S}_\Delta(n,r)_\mathbb{C}$-modules. It remains to prove that every simple $\mathcal{S}_\Delta(n,r)_\mathbb{C}$-module is isomorphic to $L(\mathbf{Q})$ for some $\mathbf{Q} \in \mathcal{Q}(n)_r$.

Let V be a simple $\mathcal{S}_\Delta(n,r)_\mathbb{C}$-module. Then $V \cong L(\mathbf{Q})$ as a $\mathrm{U}_\mathbb{C}(\widehat{\mathfrak{gl}}_n)$-module for some $\mathbf{Q} \in \mathcal{Q}(n)$ by Proposition 4.6.4. Let $l = \sum_{1\leqslant i\leqslant n} \deg Q_i(u)$. Then, by Proposition 4.6.7, $L(\mathbf{Q})$ is an $\mathcal{S}_\Delta(n,l)_\mathbb{C}$-module. Thus, by restriction, V is a module for the q-Schur algebra $\mathcal{S}(n,r)_\mathbb{C}$ and V is also a module for the q-Schur algebra $\mathcal{S}(n,l)_\mathbb{C}$. Hence, $r = l$. □

Corollary 4.6.9. *Let V be a finite dimensional irreducible polynomial representation of $\mathrm{U}_\mathbb{C}(\widehat{\mathfrak{gl}}_n)$. Then V can be regarded as an $\mathcal{S}_\Delta(n,r)_\mathbb{C}$-module via $\xi'_{r,\mathbb{C}}$ if and only if V is of level r as a $U(n)_\mathbb{C}$-module.*

Proof. If V can be regarded as an $\mathcal{S}_\Delta(n,r)_\mathbb{C}$-module via $\xi'_{r,\mathbb{C}}$, then we may view V as an $\mathcal{S}(n,r)_\mathbb{C}$-module by restriction and, hence, V is of level r as a $U_\Delta(n)_\mathbb{C}$-module.

Conversely, suppose that V is of level r as a $U(n)_\mathbb{C}$-module and $V = L(\mathbf{Q})$ for some $\mathbf{Q} \in \mathcal{Q}(n)$. Then V is an $\mathcal{S}_\Delta(n,r')_\mathbb{C}$-module by Theorem 4.6.8, where $r' = \sum_{1\leqslant i\leqslant n} \deg Q_i(u)$. Hence, V is of level r' as a $U(n)_\mathbb{C}$-module. So $r = r'$ and V is an $\mathcal{S}_\Delta(n,r)_\mathbb{C}$-module. □

Remark 4.6.10. It is reasonable to make the following definition. A finite dimensional $\mathrm{U}_\mathbb{C}(\widehat{\mathfrak{gl}}_n)$-module is said to be of level r if it is an $\mathcal{S}_\Delta(n,r)_\mathbb{C}$-module via $\xi'_{r,\mathbb{C}}$. Thus, if V is a $\mathrm{U}_\mathbb{C}(\widehat{\mathfrak{gl}}_n)$-module of level r, then its composition factors are all homomorphic images of $\Omega_\mathbb{C}^{\otimes r}$. It would be interesting to know if the converse is also true.

It is natural to make a comparison between the Classification Theorems 4.3.4 and 4.6.8 and to raise the following problem.

Problem 4.6.11. Generalize the Identification Theorem 4.4.2 to the case where $n \leqslant r$.

4.7. Classification of simple $U_\Delta(n,r)_\mathbb{C}$-modules

The homomorphic image $U_\Delta(n,r)_\mathbb{C}$ of the extended affine quantum $\mathfrak{sl}_n$, $U_\Delta(n)_\mathbb{C}$, is a proper subalgebra of $\mathcal{S}_\Delta(n,r)_\mathbb{C}$ when $n \leqslant r$. In other words, by restriction, the surjective algebra homomorphism $\xi'_{r,\mathbb{C}} : \mathrm{U}_\mathbb{C}(\widehat{\mathfrak{gl}}_n) \to \mathcal{S}_\Delta(n,r)_\mathbb{C}$ induces a surjective algebra homomorphism $\xi'_{r,\mathbb{C}} : U_\Delta(n)_\mathbb{C} \to U_\Delta(n,r)_\mathbb{C}$. In this section, we will classify finite dimensional simple $U_\Delta(n,r)_\mathbb{C}$-modules.

Let $\mathbf{P} \in \mathcal{P}(n)$ and $\lambda \in \Lambda^+(n,r)$ be such that $\lambda_i - \lambda_{i+1} = \deg P_i(u)$, for $1 \leqslant i \leqslant n-1$. Define

$$\bar{M}(\mathbf{P},\lambda) = U_\Delta(n)_\mathbb{C}/\bar{I}(\mathbf{P},\lambda),$$

where $\bar{I}(\mathbf{P},\lambda)$ is the left ideal of $U_\Delta(n)_\mathbb{C}$ generated by $\mathrm{x}^+_{i,s}$, $\mathscr{P}_{i,s} - P_{i,s}$, and $\mathrm{k}_j - z^{\lambda_j}$ for $1 \leqslant i \leqslant n-1$, $s \in \mathbb{Z}$, and $1 \leqslant j \leqslant n$, where $P_{i,s}$ are defined using (4.2.1.7). The $U_\Delta(n)_\mathbb{C}$-module $\bar{M}(\mathbf{P},\lambda)$ has a unique simple quotient $U_\Delta(n)_\mathbb{C}$-module, which is denoted by $\bar{L}(\mathbf{P},\lambda)$.

Lemma 4.7.1. *For $\mathbf{Q} \in \mathcal{Q}(n)$, let $\lambda = (\lambda_1,\ldots,\lambda_n)$ with $\lambda_i = \deg(Q_i(u))$ for $1 \leqslant i \leqslant n$ and $\mathbf{P} = (P_1(u),\ldots,P_{n-1}(u)) \in \mathcal{P}(n)$ be such that*

$$P_j(u) = \frac{Q_j(uz^{j-1})}{Q_{j+1}(uz^{j+1})},$$

for $1 \leqslant j \leqslant n-1$. Then $\bar{L}(\mathbf{P},\lambda) \cong L(\mathbf{Q})|_{U_\Delta(n)_\mathbb{C}}$.

Proof. Let $w_0 \in L(\mathbf{Q})$ be a pseudo-highest weight vector. Since $\mathrm{U}_\mathbb{C}(\widehat{\mathfrak{gl}}_n)$ is generated by $U_\Delta(n)_\mathbb{C}$ and the central elements $\mathrm{z}^\pm_t$, for $t \geqslant 1$, every simple $\mathrm{U}_\mathbb{C}(\widehat{\mathfrak{gl}}_n)$-module is a simple $U_\Delta(n)_\mathbb{C}$-module by restriction. In particular, $L(\mathbf{Q})|_{U_\Delta(n)_\mathbb{C}}$ is simple. So we have $L(\mathbf{Q}) = U_\Delta(n)_\mathbb{C} w_0$. Hence, there is a surjective $U_\Delta(n)_\mathbb{C}$-module homomorphism $\varphi : \bar{M}(\mathbf{P},\lambda) \to L(\mathbf{Q})$ defined by sending $\bar{u}$ to uw_0, for $u \in U_\Delta(n)_\mathbb{C}$. Thus, $L(\mathbf{Q}) \cong \bar{M}(\mathbf{P},\lambda)/\mathrm{Ker}\,\varphi$ as $U_\Delta(n)_\mathbb{C}$-modules. Since $L(\mathbf{Q})|_{U_\Delta(n)_\mathbb{C}}$ is simple and $\bar{M}(\mathbf{P},\lambda)$ has a unique simple quotient $\bar{L}(\mathbf{P},\lambda)$, we have $\bar{L}(\mathbf{P},\lambda) \cong L(\mathbf{Q})$ as $U_\Delta(n)_\mathbb{C}$-modules. □

Corollary 4.7.2. *Let $\mathbf{P} \in \mathcal{P}(n)$ and $\lambda \in \Lambda^+(n,r)$ with $\lambda_i - \lambda_{i+1} = \deg P_i(u)$ for $1 \leqslant i \leqslant n-1$. Then $\bar{L}(\mathbf{P},\lambda)$ is a $U_\Delta(n,r)_\mathbb{C}$-module via $\xi'_{r,\mathbb{C}}$ and $\bar{L}(\mathbf{P},\lambda)|_{\mathrm{U}_\mathbb{C}(\widehat{\mathfrak{sl}}_n)} \cong \bar{L}(\mathbf{P})$.*

Proof. Let $Q_n(u) = 1 + u^{\lambda_n}$. Using the formula

$$P_j(u) = \frac{Q_j(uz^{j-1})}{Q_{j+1}(uz^{j+1})},$$

we define the polynomials $Q_i(u)$, for $1 \leqslant i \leqslant n-1$. Then we have $\mathbf{Q} = (Q_1(u),\ldots,Q_n(u)) \in \mathcal{Q}(n)$ and $\lambda_i = \deg Q_i(u)$, for $1 \leqslant i \leqslant n$. By

Theorem 4.6.8, $L(\mathbf{Q})$ is an $\mathcal{S}_\Delta(n,r)_\mathbb{C}$-module. So, by Lemma 4.7.1, $\bar{L}(\mathbf{P},\lambda) \cong L(\mathbf{Q})|_{U_\Delta(n)_\mathbb{C}}$ is a $U_\Delta(n,r)_\mathbb{C}$-module. Hence, $\bar{L}(\mathbf{P},\lambda)|_{\mathrm{U}_\mathbb{C}(\widehat{\mathfrak{sl}}_n)}$ is simple since the algebra homomorphism $\xi'_{r,\mathbb{C}} : \mathrm{U}_\mathbb{C}(\widehat{\mathfrak{sl}}_n) \to U_\Delta(n,r)_\mathbb{C}$ is surjective. □

Since $U_\Delta(n,r)_\mathbb{C}$ contains $\mathcal{S}(n,r)_\mathbb{C}$ as a subalgebra, it follows that, if $\lambda \in \Lambda^+(n,r)$, then $\bar{L}(\mathbf{P},\lambda)$ is an $\mathcal{S}(n,r)_\mathbb{C}$-module and

$$\bar{L}(\mathbf{P},\lambda) = \bigoplus_{\substack{\mu\trianglelefteq\lambda\\ \mu\in\Lambda(n,r)}} \bar{L}(\mathbf{P},\lambda)_\mu, \tag{4.7.2.1}$$

where $\bar{L}(\mathbf{P},\lambda)_\mu$ denotes the weight space of $\bar{L}(\mathbf{P},\lambda)$ as a $U(n)_\mathbb{C}$-module.

Lemma 4.7.3. *For partitions λ,λ^* and $\mathbf{P},\mathbf{P}^* \in \mathcal{P}(n)$, if $\bar{L}(\mathbf{P},\lambda) \cong \bar{L}(\mathbf{P}^*,\lambda^*)$, then $\mathbf{P} = \mathbf{P}^*$ and $\lambda = \lambda^*$. In particular, for dominant polynomials $\mathbf{Q},\mathbf{Q}^* \in \mathcal{Q}(n)$, $L(\mathbf{Q})|_{U_\Delta(n)_\mathbb{C}} \cong L(\mathbf{Q}^*)|_{U_\Delta(n)_\mathbb{C}}$ if and only if $\deg Q_i(u) = \deg Q_i^*(u)$ and $Q_j(uz^{j-1})/Q_{j+1}(uz^{j+1}) = Q_j^*(uz^{j-1})/Q_{j+1}^*(uz^{j+1})$, for all $1 \leqslant i \leqslant n$ and $1 \leqslant j \leqslant n-1$.*

Proof. By (4.7.2.1) we have $\lambda = \lambda^*$. Since $\bar{L}(\mathbf{P},\lambda) \cong \bar{L}(\mathbf{P}^*,\lambda^*)$, it follows from Corollary 4.7.2 that $\bar{L}(\mathbf{P}) \cong \bar{L}(\mathbf{P},\lambda)|_{\mathrm{U}_\mathbb{C}(\widehat{\mathfrak{sl}}_n)} \cong \bar{L}(\mathbf{P}^*,\lambda^*)|_{\mathrm{U}_\mathbb{C}(\widehat{\mathfrak{sl}}_n)} \cong \bar{L}(\mathbf{P}^*)$. Therefore, $\mathbf{P} = \mathbf{P}^*$. □

Lemma 4.7.4. *Let V be a finite dimensional simple $U_\Delta(n,r)_\mathbb{C}$-module. Then there exist $\mathbf{P} = (P_1(u),\ldots,P_{n-1}(u)) \in \mathcal{P}(n)$ and $\lambda \in \Lambda^+(n,r)$ with $\lambda_i - \lambda_{i+1} = \deg P_i(u)$, for all $1 \leqslant i \leqslant n-1$, such that $V \cong \bar{L}(\mathbf{P},\lambda)$.*

Proof. Since $\xi'_{r,\mathbb{C}} : \mathrm{U}_\mathbb{C}(\widehat{\mathfrak{sl}}_n) \to U_\Delta(n,r)_\mathbb{C}$ is surjective, V is a simple $\mathrm{U}_\mathbb{C}(\widehat{\mathfrak{sl}}_n)$-module. Let w_0 be a pseudo-highest weight vector satisfying

$$\mathrm{x}^+_{i,s} w_0 = 0,\ \mathscr{P}_{i,s} w_0 = P_{i,s} w_0, \text{ and } \widetilde{\mathrm{k}}_i w_0 = z^{\mu_i} w_0,$$

for all $1 \leqslant i \leqslant n-1$ and $s \in \mathbb{Z}$, where $\mu_i = \deg P_i(u)$. Using the idempotent decomposition $1 = \sum_{\nu\in\Lambda_\Delta(n,r)} \mathfrak{l}_\nu$, $\sum_{\nu\in\Lambda_\Delta(n,r)} \mathfrak{l}_\nu w_0 = w_0 \neq 0$ implies that there exists $\lambda \in \Lambda_\Delta(n,r)$ such that $\mathfrak{l}_\lambda w_0 \neq 0$.

It is clear that $\mathfrak{l}_\lambda w_0$ is also a pseudo-highest weight vector satisfying

$$\mathrm{x}^+_{i,s}\mathfrak{l}_\lambda w_0 = 0,\ \mathscr{P}_{i,s}\mathfrak{l}_\lambda w_0 = P_{i,s}\mathfrak{l}_\lambda w_0, \text{ and } \widetilde{\mathrm{k}}_i\mathfrak{l}_\lambda w_0 = z^{\mu_i}\mathfrak{l}_\lambda w_0.$$

On the other hand, $\mathrm{k}_i\mathfrak{l}_\lambda w_0 = z^{\lambda_i}\mathfrak{l}_\lambda w_0$, for $1 \leqslant i \leqslant n$. Thus, $\widetilde{\mathrm{k}}_i\mathfrak{l}_\lambda w_0 = z^{\lambda_i-\lambda_{i+1}}\mathfrak{l}_\lambda w_0$. So $\lambda_i - \lambda_{i+1} = \mu_i$, for $1 \leqslant i \leqslant n-1$. Hence, there is a surjective $U_\Delta(n)_\mathbb{C}$-module homomorphism $\varphi : \bar{M}(\mathbf{P},\lambda) \to V$ defined by sending $\bar{u}$ to $u\mathfrak{l}_\lambda w_0$, for all $u \in U_\Delta(n)_\mathbb{C}$. This surjection induces a $U_\Delta(n)_\mathbb{C}$-module isomorphism $V \cong \bar{L}(\mathbf{P},\lambda)$. □

Altogether this gives the following classification theorem.

Theorem 4.7.5. *The set*

$$\{L(\mathbf{P},\lambda)\mid \mathbf{P}\in\mathcal{P}(n),\ \lambda\in\Lambda^+(n,r),\ \lambda_i-\lambda_{i+1}=\deg P_i(u),\ \forall 1\leqslant i<n\}$$

is a complete set of non-isomorphic finite dimensional simple $U_\Delta(n,r)_\mathbb{C}$-modules.

Define an equivalence relation $\sim$ on $\mathcal{Q}(n)$ be setting, for $\mathbf{Q},\mathbf{Q}'\in\mathcal{Q}(n)$,

$$\mathbf{Q}\sim\mathbf{Q}' \iff \deg Q_i(u)=\deg Q_i'(u),\ 1\leqslant i\leqslant n,\ \text{and}$$
$$\frac{Q_j(uz^{j-1})}{Q_{j+1}(uz^{j+1})}=\frac{Q_j'(uz^{j-1})}{Q_{j+1}'(uz^{j+1})},\ 1\leqslant j\leqslant n-1.$$

Corollary 4.7.6. *Let $\Pi_r=\mathcal{Q}(n)_r/\!\sim$ denote the set of equivalence classes and choose a representative $\mathbf{Q}_\pi\in\pi$ for every $\pi\in\Pi_r$. Then the set*

$$\{L(\mathbf{Q}_\pi)|_{U_\Delta(n)_\mathbb{C}}:\pi\in\Pi_r\}$$

is a complete set of non-isomorphic finite dimensional simple $U_\Delta(n,r)_\mathbb{C}$-modules. Moreover, if $n>r$, then $\Pi_r=\mathcal{Q}(n)_r$.

Proof. The last assertion follows from the fact that, if $n>r$, then $U_\Delta(n,r)_\mathbb{C}=\mathcal{S}_\Delta(n,r)_\mathbb{C}$. □

As seen in Theorem 4.2.3, there is a rougher equivalence relation $\sim'$ on $\mathcal{Q}(n)$ defined by setting for $\mathbf{Q},\mathbf{Q}'\in\mathcal{Q}(n)$,

$$\mathbf{Q}\sim'\mathbf{Q}' \iff \frac{Q_j(uz^{j-1})}{Q_{j+1}(uz^{j+1})}=\frac{Q_j'(uz^{j-1})}{Q_{j+1}'(uz^{j+1})},\ \text{for } 1\leqslant j\leqslant n-1,$$

such that the equivalence classes are in one-to-one correspondence to simple $\mathrm{U}_\mathbb{C}(\widehat{\mathfrak{sl}}_n)$-modules.

5

The presentation and realization problems

As seen in Chapters 2 and 3, the double Ringel–Hall algebra $\mathfrak{D}_\Delta(n)$ is presented by generators and relations, while the affine quantum Schur algebra $\boldsymbol{\mathcal{S}}_\Delta(n,r)$ is defined as an endomorphism algebra which is a vector space with an explicitly defined multiplication. Now the algebra epimorphism from $\mathfrak{D}_\Delta(n)$ to $\boldsymbol{\mathcal{S}}_\Delta(n,r)$ raises two natural questions: how to present affine quantum Schur algebras $\boldsymbol{\mathcal{S}}_\Delta(n,r)$ in terms of generators and relations, and how to realize the double Ringel–Hall algebra $\mathfrak{D}_\Delta(n)$ in terms of a vector spacc together with an explicitly defined multiplication. In this chapter and the next, we will tackle these problems.

Since $\mathfrak{D}_\Delta(n) \cong \mathbf{U}_\Delta(n) \otimes \mathbf{Z}_\Delta(n)$ by Remark 2.3.6(2), it follows that $\boldsymbol{\mathcal{S}}_\Delta(n,r) = \mathbf{U}_\Delta(n,r)\mathbf{Z}_\Delta(n,r)$, where $\mathbf{U}_\Delta(n,r)$ (resp., $\mathbf{Z}_\Delta(n,r)$) is the homomorphic image of the quantum group $\mathbf{U}_\Delta(n)$, the extended quantum affine $\mathfrak{sl}_n$, (resp., the central subalgebra $\mathbf{Z}_\Delta(n)$) under the map ξ_r. We first review in §5.1 a presentation of McGerty for $\mathbf{U}_\Delta(n,r)$. This is a proper subalgebra of $\boldsymbol{\mathcal{S}}_\Delta(n,r)$ if $n \leqslant r$. As a natural affine analogue of the presentation given by Doty–Giaquinto [**18**], we will modify McGerty's presentation to obtain a Drinfeld–Jimbo type presentation for $\mathbf{U}_\Delta(n,r)$ (Theorem 5.1.3). We then determine the structure of the central subalgebra $\mathbf{Z}_\Delta(n,r)$ of $\boldsymbol{\mathcal{S}}_\Delta(n,r)$ (Proposition 5.2.3). However, it is almost impossible to combine the two to give a presentation for $\boldsymbol{\mathcal{S}}_\Delta(n,r)$. In §5.3, we will use the multiplication formulas given in §3.4 to derive some extra relations for an extra generator required for presenting $\boldsymbol{\mathcal{S}}_\Delta(r,r)$ for all $r \geqslant 1$ (Theorem 5.3.5). In particular, we will then easily see why the Hopf algebra $\widehat{\mathbf{U}}$ considered in [**35**, 3.1.1][1] maps onto affine quantum Schur algebras when $n \geqslant r$; see Remark 5.3.2. Strictly speaking, $\widehat{\mathbf{U}}$ cannot be regarded as a quantum enveloping algebra since it does not have a triangular decomposition.

[1] This algebra $\widehat{\mathbf{U}}$ is denoted by $U(\widehat{gl}_n)$ in [**35**].

From §5.4 onwards, we will discuss the realization problem. We first formulate a realization conjecture in §5.4, as suggested in [**24**, 5.2(2)], and its classical ($v = 1$) version. In the last section, we show that Lusztig's transfer maps are not compatible with the map ξ_r for the double Ringel–Hall algebra $'\mathfrak{D}_\Delta(n)$ considered in Remark 2.3.6(3). This justifies why we cannot have a realization in terms of an inverse limit of the transfer maps. We will then establish the conjecture for the classical case in the next chapter.

5.1. McGerty's presentation for $\mathbf{U}_\Delta(n, r)$

The presentation problem for $\boldsymbol{\mathcal{S}}_\Delta(n, r)$ when $n > r$ is relatively easy. In this case, $\boldsymbol{\mathcal{S}}_\Delta(n, r) = \mathbf{U}_\Delta(n, r)$ is a homomorphic image of $\mathbf{U}_\Delta(n)$. By using McGerty's presentation for $\mathbf{U}_\Delta(n, r)$, we obtain a new presentation for $\mathbf{U}_\Delta(n, r)$ similar to that for quantum Schur algebras given in [**18**] (cf. [**26**]). In particular, this gives Doty–Green's result [**19**] for $\boldsymbol{\mathcal{S}}_\Delta(n, r)$ with $n > r$ (removing the condition $n \geqslant 3$ required there).

Let $\xi_r : \mathfrak{D}_\Delta(n) \twoheadrightarrow \boldsymbol{\mathcal{S}}_\Delta(n, r)$ be the surjective homomorphism defined in (3.5.5.4). For each $r \geqslant 1$, define

$$\mathbf{U}_\Delta(n, r) := \xi_r(\mathbf{U}_\Delta(n)).$$

Clearly, $\mathbf{U}_\Delta(n, r)$ is generated by the elements

$$\mathfrak{e}_i := \xi_r(E_i) = E^\Delta_{i,i+1}(\mathbf{0}, r),\ \mathfrak{f}_i := \xi_r(F_i) = E^\Delta_{i+1,i}(\mathbf{0}, r),\ \text{and}$$
$$\mathfrak{k}_i := \xi_r(K_i) = 0(\boldsymbol{e}^\Delta_i, r),$$

for all $i \in I$.

Let $C = (c_{i,j})$ denote the generalized Cartan matrix of type $\widetilde{A}_{n-1}$ as in (1.3.2.1). Recall the elements $\mathfrak{l}_\lambda$, $\lambda \in \Lambda_\Delta(n, r)$, defined in Proposition 3.7.4(2). The following result is taken from [**58**, Prop. 6.4 & Lem. 6.6].

Theorem 5.1.1. *As a $\mathbb{Q}(v)$-algebra, $\mathbf{U}_\Delta(n, r)$ is generated by*

$$\mathfrak{e}_i,\ \mathfrak{f}_i,\ \mathfrak{l}_\lambda \qquad (i \in I,\ \lambda \in \Lambda_\Delta(n, r))$$

subject to the following relations

(1) $\mathfrak{l}_\lambda \mathfrak{l}_\mu = \delta_{\lambda,\mu}\mathfrak{l}_\lambda$, $\sum_{\lambda \in \Lambda_\Delta(n,r)} \mathfrak{l}_\lambda = 1$;

(2) $$\mathfrak{e}_i \mathfrak{l}_\lambda = \begin{cases} \mathfrak{l}_{\lambda+\boldsymbol{e}^\Delta_i-\boldsymbol{e}^\Delta_{i+1}}\mathfrak{e}_i, & \text{if } \lambda + \boldsymbol{e}^\Delta_i - \boldsymbol{e}^\Delta_{i+1} \in \Lambda_\Delta(n, r); \\ 0, & \text{otherwise}; \end{cases}$$

(3) $$\mathfrak{f}_i \mathfrak{l}_\lambda = \begin{cases} \mathfrak{l}_{\lambda-\boldsymbol{e}^\Delta_i+\boldsymbol{e}^\Delta_{i+1}}\mathfrak{f}_i, & \text{if } \lambda - \boldsymbol{e}^\Delta_i + \boldsymbol{e}^\Delta_{i+1} \in \Lambda_\Delta(n, r); \\ 0, & \text{otherwise}; \end{cases}$$

(4) $\mathfrak{e}_i\mathfrak{f}_j - \mathfrak{f}_j\mathfrak{e}_i = \delta_{i,j}\sum_{\lambda\in\Lambda_\Delta(n,r)}[\lambda_i - \lambda_{i+1}]\mathfrak{l}_\lambda$;

(5) $\displaystyle\sum_{a+b=1-c_{i,j}}(-1)^a\begin{bmatrix}1-c_{i,j}\\ a\end{bmatrix}\mathfrak{e}_i^a\mathfrak{e}_j\mathfrak{e}_i^b = 0$ *for* $i\neq j$;

(6) $\displaystyle\sum_{a+b=1-c_{i,j}}(-1)^a\begin{bmatrix}1-c_{i,j}\\ a\end{bmatrix}\mathfrak{f}_i^a\mathfrak{f}_j\mathfrak{f}_i^b = 0$ *for* $i\neq j$.

This theorem is the affine version of Theorem 3.4 in [**18**]. Naturally, one expects the affinization of Theorem 3.1 in [**18**] for a Drinfeld–Jimbo type presentation. In fact, this is an easy consequence of the following result for the Laurent polynomial algebra $\mathbf{U}^0 := \mathbb{Q}(v)[K_1^{\pm1},\ldots,K_n^{\pm1}] = \mathbf{U}_\Delta(n)^0$, whose proof can be found in [**18**, Prop. 8.2, 8.3] and [**26**, 4.5, 4.6] in the context of quantum $\mathfrak{gl}_n$ and quantum Schur algebras (though the result itself was not explicitly stated there).

For any $\mu\in\mathbb{N}^n$ and $1\leqslant i\leqslant n$, let

$$\mathfrak{L}_\mu = \prod_{i=1}^n\begin{bmatrix}K_i;0\\ \mu_i\end{bmatrix},\quad \boldsymbol{v}^\mu = (v^{\mu_1},\ldots,v^{\mu_n}),\quad\text{and}\qquad (5.1.1.1)$$
$$[K_i;r+1]^! = (K_i-1)(K_i-v)\cdots(K_i-v^r).$$

If we regard $\mathfrak{L}_\mu$ as a function from $\mathbf{U}^0$ to $\mathbb{Q}(v)$, then $\mathfrak{L}_\lambda(\boldsymbol{v}^\mu) = \delta_{\lambda,\mu}$, for all $\lambda,\mu\in\Lambda(n,r)$.

Lemma 5.1.2. *The ideals $\langle I_r\rangle$ and $\langle J_r\rangle$ of $\mathbf{U}^0$ generated by the sets*

$$I_r = \{1-\Sigma_{\lambda\in\Lambda(n,r)}\mathfrak{L}_\lambda\}\cup\{\mathfrak{L}_\lambda\mathfrak{L}_\mu - \delta_{\lambda,\mu}\mathfrak{L}_\lambda \mid \lambda,\mu\in\Lambda(n,r)\}$$
$$\cup\{K_i\mathfrak{L}_\lambda - v^{\lambda_i}\mathfrak{L}_\lambda \mid 1\leqslant i\leqslant n,\ \lambda\in\Lambda(n,r)\}\ \textit{and}$$
$$J_r = \{\kappa := K_1\cdots K_n - v^r, [K_i;r+1]^! \mid 1\leqslant i\leqslant n\},$$

are the same.

Proof. For completeness, we provide here a direct proof. Consider the algebra epimorphism

$$\varphi:\mathbf{U}^0\longrightarrow\bigoplus_{\mu\in\Lambda(n,r)}\mathbf{U}^0/\langle K_1-v^{\mu_1},\ldots,K_n-v^{\mu_n}\rangle,\ f\longmapsto(f(\boldsymbol{v}^\mu))_{\mu\in\Lambda(n,r)}.$$

It is clear from the relations $\mathfrak{L}_\lambda(\boldsymbol{v}^\mu) = \delta_{\lambda,\mu}$ that $\langle I_r\rangle = \operatorname{Ker}\varphi$. Applying the Chinese Remainder Theorem yields[2]

[2] The ideal J appearing in the proof of [**12**, Lem. 13.36] should be

$$\bigcap_{0\leqslant\mu_1,\ldots,\mu_n\leqslant r}\langle x_1-v^{\mu_1},\ldots,x_n-v^{\mu_n}\rangle,$$

while the quotient algebra R/J should have dimension $(r+1)^n$.

$$\begin{aligned}
\langle I_r\rangle &= \bigcap_{\mu\in\Lambda(n,r)} \langle K_1 - v^{\mu_1}, \ldots, K_n - v^{\mu_n}\rangle \\
&= \bigcap_{0\leqslant \mu_1,\ldots,\mu_n\leqslant r} (\langle\kappa\rangle + \langle K_1 - v^{\mu_1}, \ldots, K_n - v^{\mu_n}\rangle) \\
&= \langle\kappa\rangle + \bigcap_{0\leqslant \mu_1,\ldots,\mu_n\leqslant r} \langle K_1 - v^{\mu_1}, \ldots, K_n - v^{\mu_n}\rangle \\
&= \langle\kappa\rangle + \langle [K_1; r+1]^!, \ldots, [K_n; r+1]^!\rangle \\
&= \langle J_r\rangle,
\end{aligned}$$

since $\kappa \in \langle K_1 - v^{\mu_1}, \ldots, K_n - v^{\mu_n}\rangle \iff \mu_1 + \cdots + \mu_n = r$. □

The above lemma together with Theorem 5.1.1 gives the following result, which, as mentioned above, was proved in [**19**] under the assumption that $n \geqslant 3$ and $n > r$.

Theorem 5.1.3. *The algebra* $\mathbf{U}_\triangle(n, r)$ *is generated by the elements*

$$\mathfrak{e}_i,\ \mathfrak{f}_i,\ \mathfrak{k}_i\ (i \in I = \mathbb{Z}/n\mathbb{Z})$$

subject to the relations:

(QS1) $\mathfrak{k}_i\mathfrak{k}_j = \mathfrak{k}_j\mathfrak{k}_i$;

(QS2) $\mathfrak{k}_i\mathfrak{e}_j = v^{\delta_{i,j}-\delta_{i,j+1}}\mathfrak{e}_j\mathfrak{k}_i,\ \mathfrak{k}_i\mathfrak{f}_j = v^{-\delta_{i,j}+\delta_{i,j+1}}\mathfrak{f}_j\mathfrak{k}_i$;

(QS3) $\mathfrak{e}_i\mathfrak{f}_j - \mathfrak{f}_j\mathfrak{e}_i = \delta_{i,j}\frac{\widetilde{\mathfrak{k}}_i - \widetilde{\mathfrak{k}}_i^{-1}}{v - v^{-1}}$, *where* $\widetilde{\mathfrak{k}}_i = \mathfrak{k}_i\mathfrak{k}_{i+1}^{-1}$;

(QS4) $\displaystyle\sum_{a+b=1-c_{i,j}} (-1)^a \begin{bmatrix} 1 - c_{i,j} \\ a \end{bmatrix} \mathfrak{e}_i^a\mathfrak{e}_j\mathfrak{e}_i^b = 0$, *for* $i \neq j$;

(QS5) $\displaystyle\sum_{a+b=1-c_{i,j}} (-1)^a \begin{bmatrix} 1 - c_{i,j} \\ a \end{bmatrix} \mathfrak{f}_i^a\mathfrak{f}_j\mathfrak{f}_i^b = 0$, *for* $i \neq j$;

(QS6) $[\mathfrak{k}_i; r+1]^! = 0,\ \mathfrak{k}_1\cdots\mathfrak{k}_n = v^r$.

Proof. First, by the lemma, the ideal $\mathcal{I}_r$ of $\mathbf{U}_\triangle(n)$ generated by I_r is the same as the ideal $\mathcal{J}_r$ generated by J_r. Second, if $\xi_r^* : \mathbf{U}_\triangle(n) \to \boldsymbol{\mathcal{S}}_\triangle(n, r)$ denotes the restriction of ξ_r to $\mathbf{U}_\triangle(n)$, then it is clear that $\mathcal{I}_r \subseteq \operatorname{Ker}\xi_r^*$ (see the proof of Proposition 3.7.4). Thus, we obtain an epimorphism

$$\varphi : \mathbf{U}_\triangle(n)/\mathcal{I}_r \longrightarrow \mathbf{U}_\triangle(n)/\operatorname{Ker}\xi_r^* \cong \mathbf{U}_\triangle(n, r)$$

satisfying $\varphi(E_i + \mathcal{I}_r) = \mathfrak{e}_i$, $\varphi(F_i + \mathcal{I}_r) = \mathfrak{f}_i$, and $\varphi(\mathfrak{L}_\lambda + \mathcal{I}_r) = \mathfrak{l}_\lambda$ for $i \in I$ and $\lambda \in \Lambda_\triangle(n, r)$.

On the other hand, it is straightforward to check that all relations given in Theorem 5.1.1 hold in $\mathbf{U}_\triangle(n)/\mathcal{I}_r$ (see, e.g., the proof of [**12**, Lem 13.40]). Thus, applying Theorem 5.1.1 yields a natural algebra homomorphism

$$\psi : \mathbf{U}_\Delta(n,r) \longrightarrow \mathbf{U}_\Delta(n)/\mathcal{I}_r$$

satisfying $\mathfrak{e}_i \mapsto E_i + \mathcal{I}_r$, $\mathfrak{f}_i \mapsto F_i + \mathcal{I}_r$, and $\mathfrak{l}_\lambda \mapsto \mathfrak{L}_\lambda + \mathcal{I}_r$. Therefore, φ has to be an isomorphism, forcing $\operatorname{Ker} \xi_r^* = \mathcal{I}_r = \mathcal{J}_r$. □

Remarks 5.1.4. (1) It is possible to replace the relations in (QS6) by the relations

$$[\mathfrak{k}_1; \mu_1]^![\mathfrak{k}_2; \mu_2]^! \cdots [\mathfrak{k}_n; \mu_n]^! = 0, \text{ for all } \mu \in \mathbb{N}_\Delta^n \text{ with } \sigma(\mu) = r+1,$$

and replace $\mathfrak{k}_n$ used in (QS1)–(QS5) by $\mathfrak{k}_n := v^r \mathfrak{k}_1^{-1} \cdots \mathfrak{k}_{n-1}^{-1}$. The new presentation uses only $3n-1$ generators. For more details, see [**26**] or [**12**, §13.10]. It is interesting to point out that the homomorphic image $\mathbf{U}(\infty, r)$ of $\mathbf{U}(\mathfrak{gl}_\infty)$, which is a proper subalgebra of the infinite quantum Schur algebra $\boldsymbol{\mathcal{S}}(\infty, r)$, for all $r \geqslant 1$, has only a presentation of this type; see [**24**, 4.7,5.4].

(2) If $\widetilde{\mathcal{I}}_r$ denotes the ideal of $\mathfrak{D}_\Delta(n)$ generated by I_r, then

$$\mathfrak{D}_\Delta(n)/\widetilde{\mathcal{I}}_r \cong \mathbf{U}_\Delta(n,r) \otimes \mathbf{Z}_\Delta(n).$$

Thus, adding (QS6) to relations (QGL1)–(QGL8) in Theorem 2.3.1 gives a presentation for this algebra.

The relations (QS1)–(QS6) in Theorem 5.1.3 will form part of the relations in a presentation for $\boldsymbol{\mathcal{S}}_\Delta(r,r)$, $r \geqslant 1$; see §5.3.

5.2. Structure of affine quantum Schur algebras

When $n \leqslant r$, $\mathbf{U}_\Delta(n,r)$ is a proper subalgebra of $\boldsymbol{\mathcal{S}}_\Delta(n,r)$. The next two sections are devoted to the study of the structure of affine quantum Schur algebras $\boldsymbol{\mathcal{S}}_\Delta(n,r)$ in this case. We will first see in this section a general structure of $\boldsymbol{\mathcal{S}}_\Delta(n,r)$ inherited from $\mathfrak{D}_\Delta(n)$ and give an explicit presentation for $\boldsymbol{\mathcal{S}}_\Delta(r,r)$ in §5.3.

We first endow $\boldsymbol{\mathcal{S}}_\Delta(n,r)$ with a $\mathbb{Z}$-grading through the surjective algebra homomorphism

$$\xi_r : \mathfrak{D}_\Delta(n) \longrightarrow \boldsymbol{\mathcal{S}}_\Delta(n,r) = \operatorname{End}_{\mathcal{H}_\Delta(r)}(\boldsymbol{\Omega}^{\otimes r}).$$

If we assign to each u_A^+ (resp., u_A^-, K_i) the degree $\mathfrak{d}(A) = \dim M(A)$ (resp., $-\mathfrak{d}(A)$, 0), then $\mathfrak{D}_\Delta(n)$ admits a $\mathbb{Z}$-grading $\mathfrak{D}_\Delta(n) = \oplus_{m\in\mathbb{Z}} \mathfrak{D}_\Delta(n)_m$. By definition, we have, for each $x \in \mathfrak{D}_\Delta(n)_m$ and $\omega_{\mathbf{i}} = \omega_{i_1} \otimes \cdots \otimes \omega_{i_r} \in \boldsymbol{\Omega}^{\otimes r}$,

$$x \cdot \omega_{\mathbf{i}} = \sum_{p=1}^{l} a_p \omega_{\mathbf{j}^{(p)}},$$

where $a_p \in \mathbb{Q}(v)$ and $\mathbf{j}^{(p)} = (j_{1,p}, \ldots, j_{r,p}) \in \mathbb{Z}^r$ satisfy $\sum_{s=1}^r j_{s,p} = \sum_{s=1}^r i_s - m$, for all $1 \leqslant p \leqslant l$. Thus, letting $\mathcal{S}_\Delta(n,r)_m = \xi_r(\mathfrak{D}_\Delta(n)_m)$ gives a decomposition

$$\mathcal{S}_\Delta(n,r) = \bigoplus_{m\in\mathbb{Z}} \mathcal{S}_\Delta(n,r)_m$$

of $\mathcal{S}_\Delta(n,r)$, i.e., $\mathcal{S}_\Delta(n,r)$ is $\mathbb{Z}$-graded, too.

By Remark 2.3.6(2), there is a central subalgebra $\mathbf{Z}_\Delta(n) = \mathbb{Q}(v)[\mathsf{z}_m^+, \mathsf{z}_m^-]_{m\geqslant 1}$ of $\mathfrak{D}_\Delta(n)$ such that $\mathfrak{D}_\Delta(n) = \mathbf{U}_\Delta(n) \otimes_{\mathbb{Q}(v)} \mathbf{Z}_\Delta(n)$. This gives another subalgebra of $\mathcal{S}_\Delta(n,r)$

$$\mathbf{Z}_\Delta(n,r) := \xi_r(\mathbf{Z}_\Delta(n))$$

such that $\mathcal{S}_\Delta(n,r) = \mathbf{U}_\Delta(n,r)\mathbf{Z}_\Delta(n,r)$. In other words, ξ_r induces a surjective algebra homomorphism

$$\mathbf{U}_\Delta(n,r) \otimes \mathbf{Z}_\Delta(n) \longrightarrow \mathcal{S}_\Delta(n,r),\ x \otimes y \longmapsto x\xi_r(y).$$

Clearly, $\mathbf{Z}_\Delta(n,r)$ is contained in the center of $\mathcal{S}_\Delta(n,r)$.

By [**56**, Th. 7.10 & 8.4] (see also Corollary 3.5.9), we have the following result.

Lemma 5.2.1. *The equality* $\mathbf{U}_\Delta(n,r) = \mathcal{S}_\Delta(n,r)$ *holds if and only if* $n > r$. *In other words,* $\mathbf{Z}_\Delta(n,r) \subseteq \mathbf{U}_\Delta(n,r)$ *if and only if* $n > r$.

Moreover, for each $m \geqslant 1$, $\xi_r(\mathsf{z}_m^+) \in \mathcal{S}_\Delta(n,r)_{mn}$ and $\xi_r(\mathsf{z}_m^-) \in \mathcal{S}_\Delta(n,r)_{-mn}$, and the $\mathbb{Z}$-grading of $\mathcal{S}_\Delta(n,r)$ induces a $\mathbb{Z}$-grading

$$\mathbf{U}_\Delta(n,r) = \bigoplus_{m\in\mathbb{Z}} \mathbf{U}_\Delta(n,r)_m$$

of $\mathbf{U}_\Delta(n,r)$, where $\mathbf{U}_\Delta(n,r)_m = \mathbf{U}_\Delta(n,r) \cap \mathcal{S}_\Delta(n,r)_m$.

We are now going to determine the structure of both $\mathbf{Z}_\Delta(n,r)$ and $\mathbf{U}_\Delta(n,r)$. For all $1 \leqslant s \leqslant r$, define commuting $\mathbb{Q}(v)$-linear maps

$$\phi_s : \boldsymbol{\Omega}^{\otimes r} \longrightarrow \boldsymbol{\Omega}^{\otimes r},\ \omega_{\mathbf{i}} \longmapsto \omega_{\mathbf{i}-ne_s} = \omega_{i_1} \otimes \cdots \otimes \omega_{i_s-n} \otimes \cdots \otimes \omega_{i_r}$$

and set, for each $m \geqslant 1$,

$$\mathfrak{p}_m = \sum_{s=1}^r \phi_s^m \ \text{ and }\ \mathfrak{q}_m = \sum_{s=1}^r \phi_s^{-m}.$$

By (3.5.4.1), $\mathfrak{p}_m = \xi_r(\mathsf{z}_m^+)$ and $\mathfrak{q}_m = \xi_r(\mathsf{z}_m^-)$. Thus, they both lie in $\mathcal{S}_\Delta(n,r)$. Moreover,

$$\mathbf{Z}_\Delta(n,r) = \mathbb{Q}(v)[\mathfrak{p}_m, \mathfrak{q}_m]_{m\geqslant 1}.$$

Remark 5.2.2. Recall from (3.5.5.4) the algebra homomorphism

$$\xi_r^\vee : \mathcal{H}_\Delta(r) \longrightarrow \mathrm{End}_{\mathbb{Q}(v)}(\mathbf{\Omega}^{\otimes r})^{\mathrm{op}}.$$

It is easy to see from the definition that, for each $m \geqslant 1$,

$$\mathfrak{p}_m = \xi_r^\vee(X_1^m + \cdots + X_r^m) \text{ and } \mathfrak{q}_m = \xi_r^\vee(X_1^{-m} + \cdots + X_r^{-m}).$$

It is well known that, for each $m \in \mathbb{Z}$, the element $X_1^m + \cdots + X_r^m$ is central in $\mathcal{H}_\Delta(r)$.

Now let $\sigma_1, \ldots, \sigma_r$ (resp., $\tau_1, \ldots, \tau_r$) denote the elementary symmetric polynomials in $\phi_1, \ldots, \phi_r$ (resp., $\phi_1^{-1}, \ldots, \phi_r^{-1}$), i.e., for $1 \leqslant s \leqslant r$,

$$\sigma_s = \sum_{1 \leqslant t_1 < \cdots < t_s \leqslant r} \phi_{t_1} \cdots \phi_{t_s} \text{ (resp., } \tau_s = \sum_{1 \leqslant t_1 < \cdots < t_s \leqslant r} \phi_{t_1}^{-1} \cdots \phi_{t_s}^{-1}).$$

Then $\sigma_s, \tau_s \in \mathbf{Z}_\Delta(n, r)$ and

$$\mathbf{Z}_\Delta(n, r) = \mathbb{Q}(v)[\sigma_1, \ldots, \sigma_r, \tau_1, \ldots, \tau_r].$$

Since

$$\tau_r = \sigma_r^{-1} \text{ and } \tau_s = \sigma_{r-s}\tau_r \text{ for each } 1 \leqslant s < r,$$

this implies that

$$\mathbf{Z}_\Delta(n, r) = \mathbb{Q}(v)[\sigma_1, \ldots, \sigma_r, \sigma_r^{-1}].$$

Proposition 5.2.3. *The set*

$$\mathcal{X} := \{\sigma_1^{\lambda_1} \cdots \sigma_{r-1}^{\lambda_{r-1}} \sigma_r^{\lambda_r} \mid \lambda_1, \ldots, \lambda_{r-1} \in \mathbb{N}, \lambda_r \in \mathbb{Z}\}$$

forms a $\mathbb{Q}(v)$-basis of $\mathbf{Z}_\Delta(n, r)$. In other words, $\mathbf{Z}_\Delta(n, r)$ is a (Laurent) polynomial algebra in $\sigma_1, \ldots, \sigma_r, \sigma_r^{-1}$.

Proof. It is obvious that $\mathbf{Z}_\Delta(n, r)$ is spanned by $\mathcal{X}$. It remains to show that $\mathcal{X}$ is linearly independent.

For $\mathbf{i} = (i_1, \ldots, i_r), \mathbf{j} = (j_1, \ldots, j_r) \in \mathbb{Z}^r$, we define the lexicographic order $\mathbf{i} \leqslant_{\mathrm{lex}} \mathbf{j}$ if $\mathbf{i} = \mathbf{j}$ or there exists $1 < s \leqslant r$ such that

$$i_r = j_r, \ldots, i_s = j_s, i_{s-1} < j_{s-1}.$$

Clearly, this gives a linear ordering on $\mathbb{Z}^r$.

For each $\lambda = (\lambda_1, \ldots, \lambda_{r-1}, \lambda_r) \in \mathbb{N}^{r-1} \times \mathbb{Z}$, define

$$\sigma^\lambda = \sigma_1^{\lambda_1} \cdots \sigma_{r-1}^{\lambda_{r-1}} \sigma_r^{\lambda_r}.$$

Then, by definition, for $\mathbf{i} = (i_1, \ldots, i_r) \in \mathbb{Z}^r$, we obtain that[3]

$$\sigma^{\lambda}(\omega_{\mathbf{i}}) = \omega_{\mathbf{i}*\lambda} + \sum_{\mathbf{i}*\lambda <_{\text{lex}} \mathbf{j}} b_{\mathbf{j}} \omega_{\mathbf{j}},$$

where all but finitely many $b_{\mathbf{j}} \in \mathbb{N}$ are zero and

$$\mathbf{i} * \lambda = \big(i_1 - \lambda_r n, i_2 - (\lambda_{r-1} + \lambda_r)n, \ldots, i_r - (\lambda_1 + \cdots + \lambda_r)n\big).$$

Note that, for $\lambda, \mu \in \mathbb{N}^{r-1} \times \mathbb{Z}$,

$$\mathbf{i} * \lambda = \mathbf{i} * \mu \Longleftrightarrow \lambda = \mu.$$

Now suppose

$$\sum_{t=1}^{m} a_t \sigma^{\lambda^{(t)}} = 0,$$

where $a_t \in \mathbb{Q}(v)$ and $\lambda^{(t)} \in \mathbb{N}^{r-1} \times \mathbb{Z}$, for $1 \leqslant t \leqslant m$. Fix an $\mathbf{i} \in \mathbb{Z}^r$. Without loss of generality, we may suppose $\mathbf{i} * \lambda^{(1)} <_{\text{lex}} \mathbf{i} * \lambda^{(t)}$, for all $2 \leqslant t \leqslant m$. Then

$$0 = \sum_{t=1}^{m} a_t \sigma^{\lambda^{(t)}}(\omega_{\mathbf{i}}) = a_1 \omega_{\mathbf{i}*\lambda^{(1)}} + x',$$

where x' is a linear combination of $\omega_{\mathbf{j}}$ with $\mathbf{i} * \lambda^{(1)} <_{\text{lex}} \mathbf{j}$. Hence, $a_1 = 0$. Inductively, we deduce that $a_t = 0$, for all $1 \leqslant t \leqslant m$. This finishes the proof. □

Remark 5.2.4. It has been proved in [**80**] that the center of the affine Schur algebra of type A is isomorphic to the polynomial algebra $\mathbb{Q}[\sigma_1, \ldots, \sigma_r, \sigma_r^{-1}]$. It is natural to conjecture that $\mathbf{Z}_\triangle(n, r) = \mathbb{Q}(v)[\sigma_1, \ldots, \sigma_r, \sigma_r^{-1}]$ is exactly the center of $\boldsymbol{\mathcal{S}}_\triangle(n, r)$.

Recall that $\mathfrak{e}_i = \xi_r(E_i)$, $\mathfrak{f}_i = \xi_r(F_i)$, and $\mathfrak{k}_i = \xi_r(K_i)$, for each $1 \leqslant i \leqslant n$. Let $\boldsymbol{\mathcal{S}}_\triangle(n, r)^0$ denote the subalgebra of $\boldsymbol{\mathcal{S}}_\triangle(n, r)$ generated by $\mathfrak{k}_i$, for $1 \leqslant i \leqslant n$. Then, by [**26**, Cor. 4.7(1)] (see also [**12**, Lem. 3.29]), $\boldsymbol{\mathcal{S}}_\triangle(n, r)^0$ has a basis

$$\{\mathfrak{k}_{\mathbf{j}} = \mathfrak{k}_1^{j_1} \cdots \mathfrak{k}_{n-1}^{j_{n-1}} \mid \mathbf{j} = (j_1, \ldots, j_{n-1}) \in (\mathbb{N}^{n-1})_{\leqslant r}\},$$

where $(\mathbb{N}^{n-1})_{\leqslant r} = \{\mathbf{j} = (j_1, \ldots, j_{n-1}) \in \mathbb{N}^{n-1} \mid j_1 + \cdots + j_{n-1} \leqslant r\}$. Let, further, $\widehat{\mathbf{Z}}_\triangle(n, r)$ be the subalgebra of $\boldsymbol{\mathcal{S}}_\triangle(n, r)$ generated by $\mathbf{Z}_\triangle(n, r)$ and $\boldsymbol{\mathcal{S}}_\triangle(n, r)^0$.

[3] The linear maps σ_i and σ^λ should not be confused with the sum function σ defined on $M_{\triangle,n}(\mathbb{Z})$ and $\mathbb{Z}_\triangle^n$ in §1.1.

Proposition 5.2.5. *The multiplication map*

$$\mathbf{Z}_\Delta(n,r)\otimes_{\mathbb{Q}(v)}\boldsymbol{\mathcal{S}}_\Delta(n,r)^0\longrightarrow\widehat{\mathbf{Z}}_\Delta(n,r)$$

is a $\mathbb{Q}(v)$*-algebra isomorphism.*

Proof. It remains to show that the set

$$\{\sigma^\lambda\mathfrak{k}_{\mathbf{j}}\mid\lambda\in\mathbb{N}^{n-1}\times\mathbb{Z},\mathbf{j}\in(\mathbb{N}^{n-1})_{\leqslant r}\}$$

is linearly independent. By definition, for each $\mathbf{i}\in\mathbb{Z}^r$, $\mathfrak{k}_{\mathbf{j}}(\omega_{\mathbf{i}})=v^{f(\mathbf{i},\mathbf{j})}\omega_{\mathbf{i}}$, where $f(\mathbf{i},\mathbf{j})\in\mathbb{Z}$ is determined by $\mathbf{i}$ and $\mathbf{j}$. Thus, by the proof of Proposition 5.2.3, we infer that, for $\lambda\in\mathbb{N}^{n-1}\times\mathbb{Z}$, $\mathbf{j}\in(\mathbb{N}^{n-1})_{\leqslant r}$, and $\mathbf{i}\in\mathbb{Z}^r$,

$$\sigma^\lambda\mathfrak{k}_{\mathbf{j}}(\omega_{\mathbf{i}})=v^{f(\mathbf{i},\mathbf{j})}\omega_{\mathbf{i}*\lambda}+\sum_{\mathbf{i}*\lambda<_{\mathrm{lex}}\mathbf{j}}c_{\mathbf{j}}\omega_{\mathbf{j}},$$

where all but finitely many $c_{\mathbf{j}}\in\mathbb{Q}(v)$ are zero.

Since $\mathbf{i}*\lambda=\mathbf{i}*\mu$ if and only if $\lambda=\mu$, it suffices to show that, for each fixed λ, $\{\sigma^\lambda\mathfrak{k}_{\mathbf{j}}\mid\mathbf{j}\in(\mathbb{N}^{n-1})_{\leqslant r}\}$ is a linearly independent set. This follows from the fact that $\{\mathfrak{k}_{\mathbf{j}}\mid\mathbf{j}\in(\mathbb{N}^{n-1})_{\leqslant r}\}$ is a linearly independent set. □

By the discussion above, the center of the positive part $\boldsymbol{\mathcal{S}}_\Delta(n,r)^+$ of $\boldsymbol{\mathcal{S}}_\Delta(n,r)$ contains the polynomial algebra $\mathbb{Q}(v)[\sigma_1,\dots,\sigma_r]:=\mathbf{Z}_\Delta(n,r)^+$, and the center of the negative part $\boldsymbol{\mathcal{S}}_\Delta(n,r)^-$ contains the polynomial algebra $\mathbb{Q}(v)[\tau_1,\dots,\tau_r]:=\mathbf{Z}_\Delta(n,r)^-$. Moreover,

$$\boldsymbol{\mathcal{S}}_\Delta(n,r)^+=\mathbf{U}_\Delta(n,r)^+\cdot\mathbf{Z}_\Delta(n,r)^+\quad\text{and}\quad\boldsymbol{\mathcal{S}}_\Delta(n,r)^-=\mathbf{U}_\Delta(n,r)^-\cdot\mathbf{Z}_\Delta(n,r)^-.$$

For each $m\geqslant 0$, set

$$\boldsymbol{\mathcal{S}}_\Delta(n,r)^\pm_m=\boldsymbol{\mathcal{S}}_\Delta(n,r)^\pm\cap\boldsymbol{\mathcal{S}}_\Delta(n,r)_{\pm m}\quad\text{and}$$
$$\mathbf{U}_\Delta(n,r)^\pm_m=\mathbf{U}_\Delta(n,r)^\pm\cap\mathbf{U}_\Delta(n,r)_{\pm m}.$$

Then we obtain an $\mathbb{N}$-grading on $\boldsymbol{\mathcal{S}}_\Delta(n,r)^\pm$ and $\mathbf{U}_\Delta(n,r)^\pm$:

$$\boldsymbol{\mathcal{S}}_\Delta(n,r)^\pm=\bigoplus_{m\geqslant 0}\boldsymbol{\mathcal{S}}_\Delta(n,r)^\pm_m\quad\text{and}\quad\mathbf{U}_\Delta(n,r)^\pm=\bigoplus_{m\geqslant 0}\mathbf{U}_\Delta(n,r)^\pm_m.$$

Note that, for $m\geqslant 1$, $\mathfrak{p}_m\in\boldsymbol{\mathcal{S}}_\Delta(n,r)^+_{mn}$ and $\mathfrak{q}_m\in\boldsymbol{\mathcal{S}}_\Delta(n,r)^-_{mn}$ and, for $1\leqslant s\leqslant r$, $\sigma_s\in\boldsymbol{\mathcal{S}}_\Delta(n,r)^+_{sn}$ and $\tau_s\in\boldsymbol{\mathcal{S}}_\Delta(n,r)^-_{sn}$.

Proposition 5.2.6. *For each* $m\geqslant 0$,

$$\begin{aligned}&\dim\boldsymbol{\mathcal{S}}_\Delta(n,r)^\pm_m/\mathbf{U}_\Delta(n,r)^\pm_m\\&=|\{A\in\Theta^+_\Delta(n)\mid A\text{ is periodic with }\sigma(A)\leqslant r,\mathfrak{d}(A)=m\}|.\end{aligned}$$

Proof. We only prove the assertion for the "+" case; the "−" case is similar.

By [**13**, §8], the composition subalgebra $\mathfrak{C}_\Delta(n)^+$ of $\mathfrak{D}_\Delta(n)^+$ has a basis[4]

$$\{E_A^+ \mid A \in \Theta_\Delta^+(n) \text{ is aperiodic}\},$$

where $E_A^+ = \widetilde{u}_A^+ + \sum_{B<_{\mathrm{dg}}A} \eta_B^A \widetilde{u}_B^+$ with all B periodic and $\eta_B^A \in \mathbb{Q}(v)$. Furthermore, by [**11**, §6], for $A, B \in \Theta_\Delta^+(n)$,

$$B \leqslant_{\mathrm{dg}} A \Longrightarrow \sigma(B) \geqslant \sigma(A).$$

Thus, if $A \in \Theta_\Delta^+(n)$ is aperiodic with $\sigma(A) > r$, then by (3.4.0.2),

$$\xi_r(E_A^+) = A(\mathbf{0}, r) + \sum_{B<_{\mathrm{dg}}A} \eta_B^A B(\mathbf{0}, r) = 0.$$

Hence, for each $m \geqslant 0$, the set

$$\mathscr{X}_m := \{\xi_r(E_A^+) \mid A \in \Theta_\Delta^+(n) \text{ is aperiodic with } \sigma(A) \leqslant r, \mathfrak{d}(A) = m\}$$

is a spanning set for $\mathbf{U}_\Delta(n, r)_m^+$. By Proposition 3.7.4(1), the set

$$\{A(\mathbf{0}, r) \mid A \in \Theta_\Delta^+(n), \sigma(A) \leqslant r, \mathfrak{d}(A) = m\}$$

is a basis of $\boldsymbol{\mathcal{S}}_\Delta(n, r)_m^+$. Hence, $\mathscr{X}_m$ is a linearly independent set and, thus, is a basis of $\mathbf{U}_\Delta(n, r)_m^+$. This gives the desired assertion. □

Corollary 5.2.7. *For each* $r \geqslant 2$,

$$\boldsymbol{\mathcal{S}}_\Delta(r, r)^+ = \mathbf{U}_\Delta(r, r)^+ \oplus \bigoplus_{t\geqslant 1} \mathbb{Q}(v)\sigma_1^t \quad \text{and} \quad \boldsymbol{\mathcal{S}}_\Delta(r, r)^- = \mathbf{U}_\Delta(r, r)^- \oplus \bigoplus_{t\geqslant 1} \mathbb{Q}(v)\tau_1^t.$$

Proof. By Proposition 5.2.6,

$$\dim \boldsymbol{\mathcal{S}}_\Delta(r, r)_m^+ / \mathbf{U}_\Delta(r, r)_m^+ = \begin{cases} 1, & \text{if } m \neq 0 \text{ and } m \equiv 0 \bmod r, \\ 0, & \text{otherwise.} \end{cases}$$

Since $\sigma_1 = \mathfrak{p}_1 = \xi_r(\mathsf{z}_1^+)$, it follows from Theorem 3.8.1(2) that $\boldsymbol{\mathcal{S}}_\Delta(r, r)^+$ is generated by $\mathfrak{e}_i$ $(1 \leqslant i \leqslant r)$ and σ_1. This implies the first decomposition. The second one can be proved similarly. □

5.3. Presentation of $\boldsymbol{\mathcal{S}}_\Delta(r, r)$

In this section we give a presentation for $\boldsymbol{\mathcal{S}}_\Delta(r, r)$ by describing explicitly a complement of $\mathbf{U}_\Delta(r, r)$ in $\boldsymbol{\mathcal{S}}_\Delta(r, r)$.

[4] The basis was used to give an elementary construction for the canonical basis of $\mathfrak{C}_\Delta(n)^\pm$ in [**13**].

We first consider the general case and recall the surjective algebra homomorphism $\xi_r : \mathfrak{D}_\Delta(n) \to \mathcal{S}_\Delta(n, r)$. For each $\mathbf{a} = (a_i) \in \mathbb{N}^n$, we write in $\mathfrak{D}_\Delta(n)$,

$$u_{\mathbf{a}}^{\pm} = u_{[S_{\mathbf{a}}]}^{\pm} \text{ and } \widetilde{u}_{\mathbf{a}}^{\pm} = v^{\sum_i a_i(a_i-1)} u_{\mathbf{a}}^{\pm} \quad (\text{cf. §1.4}).$$

By (1.4.2.1), if $\mathbf{a}$ is not sincere, say $a_i = 0$, then

$$\begin{aligned}\widetilde{u}_{\mathbf{a}}^{\pm} &= v^{\sum_j a_j(a_j-1)} u_{\mathbf{a}}^{\pm} \\ &= (u_{i-1}^{\pm})^{(a_{i-1})} \cdots (u_1^{\pm})^{(a_1)} (u_n^{\pm})^{(a_n)} \cdots (u_{i+1}^{\pm})^{(a_{i+1})} \in \mathbf{U}_\Delta(n).\end{aligned}$$

Thus, if $\mathbf{a}$ is not sincere, then both $\mathfrak{e}_{\mathbf{a}} := \xi_r(\widetilde{u}_{\mathbf{a}}^{+})$ and $\mathfrak{f}_{\mathbf{a}} := \xi_r(\widetilde{u}_{\mathbf{a}}^{-})$ lie in $\mathbf{U}_\Delta(n, r)$.

Now consider the following element ρ in $\mathrm{End}_{\mathbb{Q}(v)}(\mathbf{\Omega}^{\otimes r})$:

$$\rho : \mathbf{\Omega}^{\otimes r} \longrightarrow \mathbf{\Omega}^{\otimes r}, \ \omega_{\mathbf{i}} \longmapsto \omega_{\mathbf{i}-e_1-\cdots-e_r},$$

i.e., $\rho(\omega_{i_1} \otimes \cdots \otimes \omega_{i_r}) = \omega_{i_1-1} \otimes \cdots \otimes \omega_{i_r-1}$. It is clear that $\rho^r = \sigma_r \in \mathcal{S}_\Delta(n, r)$.

Lemma 5.3.1. *Suppose $n \geqslant r$. Then*

$$\rho = \sum_{\mathbf{a} \in \mathbb{N}^n, \sigma(\mathbf{a}) = r} \mathfrak{e}_{\mathbf{a}} \text{ and } \rho^{-1} = \sum_{\mathbf{a} \in \mathbb{N}^n, \sigma(\mathbf{a}) = r} \mathfrak{f}_{\mathbf{a}}.$$

In particular, if $n > r$, then $\rho, \rho^{-1} \in \mathbf{U}_\Delta(n, r)$.

Proof. For each $\mathbf{a} \in \mathbb{N}^n$, let $Y_{\mathbf{a}}$ denote the set of the sequences $\mathbf{j} = (j_1, \ldots, j_r)$ satisfying the condition that, for each $1 \leqslant i \leqslant n$,

$$a_i = |\{1 \leqslant s \leqslant r \mid j_s = i\}|.$$

By Corollary 3.5.8, for each $\omega_{i_1} \otimes \cdots \otimes \omega_{i_r} \in \mathbf{\Omega}^{\otimes r}$,

$$\widetilde{u}_{\mathbf{a}}^{+} \cdot (\omega_{i_1} \otimes \cdots \otimes \omega_{i_r}) = \sum_{\mathbf{j} \in Y_{\mathbf{a}}} \delta_{\overline{j_1}, \overline{i_1 - 1}} \cdots \delta_{\overline{j_r}, \overline{i_r - 1}} \omega_{i_1-1} \otimes \cdots \otimes \omega_{i_r-1},$$

where, for $m \in \mathbb{Z}$, $\overline{m}$ denotes its residue class in $\mathbb{Z}/n\mathbb{Z}$. Therefore,

$$\Big(\sum_{\mathbf{a} \in \mathbb{N}^n, \sigma(\mathbf{a}) = r} \widetilde{u}_{\mathbf{a}}^{+} \Big) \cdot (\omega_{i_1} \otimes \cdots \otimes \omega_{i_r}) = \omega_{i_1-1} \otimes \cdots \otimes \omega_{i_r-1},$$

that is, the first equality holds. The second one can be proved similarly. □

The above lemma implies that

$$\mathfrak{e}'_\delta := \rho - \mathfrak{e}_\delta \text{ and } \mathfrak{f}'_\delta := \rho^{-1} - \mathfrak{f}_\delta \tag{5.3.1.1}$$

are in $\mathbf{U}_\Delta(r,r)$, where $\delta = (1,\dots,1)$. More precisely,

$$\mathfrak{e}'_\delta = \sum_{i=1}^{n} \sum_{\substack{\mathbf{a}\in\mathbb{N}^n \\ \sigma(\mathbf{a})=r, a_i=0}} (\mathfrak{e}_{i-1})^{(a_{i-1})}\cdots(\mathfrak{e}_1)^{(a_1)}(\mathfrak{e}_n)^{(a_n)}\cdots(\mathfrak{e}_{i+1})^{(a_{i+1})} \quad \text{and}$$

$$\mathfrak{f}'_\delta = \sum_{i=1}^{n} \sum_{\substack{\mathbf{a}\in\mathbb{N}^n \\ \sigma(\mathbf{a})=r, a_i=0}} (\mathfrak{f}_{i-1})^{(a_{i-1})}\cdots(\mathfrak{f}_1)^{(a_1)}(\mathfrak{f}_n)^{(a_n)}\cdots(\mathfrak{f}_{i+1})^{(a_{i+1})}. \tag{5.3.1.2}$$

From now on, we assume $n = r$. By §1.4 and Theorem 3.8.1(2), $\boldsymbol{\mathcal{S}}_\Delta(r,r)$ is generated by $\mathfrak{e}_i, \mathfrak{f}_i, \mathfrak{k}_i$ $(1 \leqslant i \leqslant n = r)$, and $\mathfrak{e}_\delta, \mathfrak{f}_\delta$. It follows from (5.3.1.1) and (5.3.1.2) that $\boldsymbol{\mathcal{S}}_\Delta(r,r)$ is also generated by the $\mathfrak{e}_i, \mathfrak{f}_i, \mathfrak{k}_i$, and ρ, ρ^{-1}. Since

$$\boldsymbol{\mathcal{S}}_\Delta(r,r)_r^+ = \mathbf{U}_\Delta(r,r)_r^+ \oplus \mathbb{Q}(v)\rho = \mathbf{U}_\Delta(r,r)_r^+ \oplus \mathbb{Q}(v)\sigma_1 \quad \text{and}$$
$$\boldsymbol{\mathcal{S}}_\Delta(r,r)_r^- = \mathbf{U}_\Delta(r,r)_r^- \oplus \mathbb{Q}(v)\rho^{-1} = \mathbf{U}_\Delta(r,r)_r^- \oplus \mathbb{Q}(v)\tau_1,$$

it follows from Corollary 5.2.7 that

$$\boldsymbol{\mathcal{S}}_\Delta(r,r)^+ = \mathbf{U}_\Delta(r,r)^+ \oplus \bigoplus_{t\geqslant 1}\mathbb{Q}(v)\rho^t \quad \text{and}$$
$$\boldsymbol{\mathcal{S}}_\Delta(r,r)^- = \mathbf{U}_\Delta(r,r)^- \oplus \bigoplus_{t\geqslant 1}\mathbb{Q}(v)\rho^{-t}. \tag{5.3.1.3}$$

Remark 5.3.2. As a consequence of the above discussion, we obtain [**35**, Th. 3.4.8] which states that (for $n \geqslant r$ with $n \geqslant 3$) there is a surjective algebra homomorphism $\alpha_r : \widehat{\mathbf{U}} \to \boldsymbol{\mathcal{S}}_\Delta(n,r)$ taking R to ρ^{-1}, where $\widehat{\mathbf{U}}$ is a Hopf algebra obtained from $\mathbf{U}_\Delta(n)$ by adding primitive elements R, R^{-1} with $RR^{-1} = R^{-1}R = 1$. It seems to us that $\widehat{\mathbf{U}}$ is not a quantum group in a strict sense since it does not admit a triangular decomposition.

Proposition 5.3.3. *For each* $1 \leqslant i \leqslant r$, *we have in* $\boldsymbol{\mathcal{S}}_\Delta(r,r)$,

(1) $\mathfrak{k}_i\mathfrak{e}_\delta = v\mathfrak{e}_\delta$;
(2) $\mathfrak{e}_i\mathfrak{e}_\delta = \frac{1}{v+v^{-1}}\mathfrak{e}_i^2\mathfrak{e}_{i-1}\cdots\mathfrak{e}_1\mathfrak{e}_r\cdots\mathfrak{e}_{i+1}$;
(3) $\mathfrak{f}_i\mathfrak{e}_\delta = \frac{1}{1-v^{-2}}\mathfrak{e}_{i-1}\cdots\mathfrak{e}_1\mathfrak{e}_r\cdots\mathfrak{e}_{i+1}\mathfrak{k}_i^{-1}(\mathfrak{k}_{i+1} - \mathfrak{k}_{i+1}^{-1})$;
(4) $\mathfrak{k}_i\mathfrak{f}_\delta = v\mathfrak{f}_\delta$;
(5) $\mathfrak{f}_i\mathfrak{f}_\delta = \frac{1}{v+v^{-1}}\mathfrak{f}_i^2\mathfrak{f}_{i+1}\cdots\mathfrak{f}_r\mathfrak{f}_1\cdots\mathfrak{f}_{i-1}$;
(6) $\mathfrak{e}_i\mathfrak{f}_\delta = \frac{1}{1-v^{-2}}\mathfrak{f}_{i+1}\cdots\mathfrak{f}_r\mathfrak{f}_1\cdots\mathfrak{f}_{i-1}\mathfrak{k}_{i+1}^{-1}(\mathfrak{k}_i - \mathfrak{k}_i^{-1})$.

Proof. All these relations can be deduced from the multiplication formulas in Theorem 3.4.2. However, we provide here a direct proof for (1) and (2). The relations (4) and (5) can be proved in a similar manner.

By definition, we have in $\mathfrak{D}_\Delta(r)$,

$$u_i^+u_\delta^+ = u_{[M]}^+ + (v^2+1)u_{\delta+e_i}^+,$$

where $M = S_1 \oplus \cdots \oplus S_{i-1} \oplus S_i[2] \oplus S_{i+2} \oplus \cdots \oplus S_r$. On the other hand,

$$\begin{aligned}(u_i^+)^2u_{i-1}^+\cdots u_1^+u_r^+\cdots u_{i+1}^+ &= v(v^2+1)u_{[2S_i]}^+u_{[S_1\oplus\cdots S_{i-1}\oplus S_{i+1}\oplus\cdots\oplus S_r]}^+\\ &= (v+v^{-1})(u_{[M]}^+ + u_{\delta+e_i}^+).\end{aligned}$$

Since $S_{\delta+e_i}$ is semisimple of dimension $r+1$, it follows that $\xi_r(u_{\delta+e_i}^+) = 0$. Hence,

$$\begin{aligned}\mathfrak{e}_i\mathfrak{e}_\delta = \xi_r(u_i^+u_\delta^+) = \xi_r(u_{[M]}^+) &= \frac{1}{v+v^{-1}}\xi_r((u_i^+)^2u_{i-1}^+\cdots u_1^+u_r^+\cdots u_{i+1}^+)\\ &= \frac{1}{v+v^{-1}}\mathfrak{e}_i^2\mathfrak{e}_{i-1}\cdots\mathfrak{e}_1\mathfrak{e}_r\cdots\mathfrak{e}_{i+1},\end{aligned}$$

which gives the relation (2).

The relation (1) follows from the fact that for each $\omega_{i_1}\otimes\cdots\otimes\omega_{i_r}\in\boldsymbol{\Omega}^{\otimes r}$,

$$\begin{aligned}u_\delta\cdot(\omega_{i_1}\otimes\cdots\otimes\omega_{i_r}) &= \sum_{\mathbf{j}\in Y_\delta}\delta_{\overline{j_1},\overline{i_1-1}}\cdots\delta_{\overline{j_r},\overline{i_r-1}}\omega_{i_1-1}\otimes\cdots\otimes\omega_{i_r-1}\\ &= \begin{cases}\omega_{i_1-1}\otimes\cdots\otimes\omega_{i_r-1}, & \text{if } \bar{i}_1,\ldots,\bar{i}_r \text{ are pairwise distinct;}\\ 0, & \text{otherwise.}\end{cases}\end{aligned}$$

Note that, if $\bar{i}_1,\ldots,\bar{i}_r$ are pairwise distinct, then

$$K_i\cdot(\omega_{i_1}\otimes\cdots\otimes\omega_{i_r}) = v\omega_{i_1}\otimes\cdots\otimes\omega_{i_r}.$$

□

The above proposition together with $\rho = \mathfrak{e}_\delta' + \mathfrak{e}_\delta$ and $\rho^{-1} = \mathfrak{f}_\delta' + \mathfrak{f}_\delta$ gives the following result.

Corollary 5.3.4. *For each $1 \leqslant i \leqslant r$, we have in $\boldsymbol{S}_\Delta(r,r)$,*

(QS1′) $(\mathfrak{k}_i - v)\rho = (\mathfrak{k}_i - v)\mathfrak{e}_\delta'$;
(QS2′) $\mathfrak{e}_i\rho = \mathfrak{e}_i\mathfrak{e}_\delta' + \frac{1}{v+v^{-1}}\mathfrak{e}_i^2\mathfrak{e}_{i-1}\cdots\mathfrak{e}_1\mathfrak{e}_r\cdots\mathfrak{e}_{i+1}$;
(QS3′) $\mathfrak{f}_i\rho = \mathfrak{f}_i\mathfrak{e}_\delta' + \frac{1}{1-v^{-2}}\mathfrak{e}_{i-1}\cdots\mathfrak{e}_1\mathfrak{e}_r\cdots\mathfrak{e}_{i+1}\mathfrak{k}_i^{-1}(\mathfrak{k}_{i+1} - \mathfrak{k}_{i+1}^{-1})$;
(QS4′) $(\mathfrak{k}_i - v)\rho^{-1} = (\mathfrak{k}_i - v)\mathfrak{f}_\delta'$;
(QS5′) $\mathfrak{f}_i\rho^{-1} = \mathfrak{f}_i\mathfrak{f}_\delta' + \frac{1}{v+v^{-1}}\mathfrak{f}_i^2\mathfrak{f}_{i+1}\cdots\mathfrak{f}_r\mathfrak{f}_1\cdots\mathfrak{f}_{i-1}$;
(QS6′) $\mathfrak{e}_i\rho^{-1} = \mathfrak{e}_i\mathfrak{f}_\delta' + \frac{1}{1-v^{-2}}\mathfrak{f}_{i+1}\cdots\mathfrak{f}_r\mathfrak{f}_1\cdots\mathfrak{f}_{i-1}\mathfrak{k}_{i+1}^{-1}(\mathfrak{k}_i - \mathfrak{k}_i^{-1})$,

where $\mathfrak{e}_\delta'$ and $\mathfrak{f}_\delta'$ are given in (5.3.1.2).

Theorem 5.3.5. *The* $\mathbb{Q}(v)$*-algebra* $\boldsymbol{\mathcal{S}}_\vartriangle(r,r)$ *is generated by* $\mathfrak{e}_i, \mathfrak{f}_i, \mathfrak{k}_i^{\pm 1}$ $(1 \leqslant i \leqslant r)$ *and* $\rho^{\pm 1}$ *subject to the relations*

(QS0′) $\quad \rho\rho^{-1} = \rho^{-1}\rho = 1,\ \rho\mathfrak{e}_i\rho^{-1} = \mathfrak{e}_{i-1},\ \rho\mathfrak{f}_i\rho^{-1} = \mathfrak{f}_{i-1},\ \rho\mathfrak{k}_i\rho^{-1} = \mathfrak{k}_{i-1}$

together with the relations (QS1′)–(QS6′) *and the relations* (QS1)–(QS6) *in Theorem 5.1.3. In particular,*

$$\boldsymbol{\mathcal{S}}_\vartriangle(r,r) = \mathbf{U}_\vartriangle(r,r) \oplus \bigoplus_{0 \neq m \in \mathbb{Z}} \mathbb{Q}(v)\rho^m.$$

Proof. Let $\boldsymbol{\mathcal{S}}'$ be the $\mathbb{Q}(v)$-algebra generated by $\mathfrak{x}_i, \mathfrak{y}_i, \mathfrak{z}_i^{\pm 1}$ $(1 \leqslant i \leqslant r)$ and $\eta^{\pm 1}$ with the relations (QS1)–(QS6) and (QS0′)–(QS6′) (here we replace $\mathfrak{e}_i, \mathfrak{f}_i, \mathfrak{k}_i^{\pm 1}$, and $\rho^{\pm 1}$ by $\mathfrak{x}_i, \mathfrak{y}_i, \mathfrak{z}_i^{\pm 1}$, and $\eta^{\pm 1}$, respectively). Thus, there is a surjective $\mathbb{Q}(v)$-algebra homomorphism

$$\begin{aligned} \Upsilon : \boldsymbol{\mathcal{S}}' &\longrightarrow \boldsymbol{\mathcal{S}}_\vartriangle(r,r), \\ \mathfrak{x}_i &\longmapsto \mathfrak{e}_i,\ \mathfrak{y}_i \longmapsto \mathfrak{f}_i,\ \mathfrak{z}_i^{\pm} \longmapsto \mathfrak{k}_i^{\pm},\ \eta^{\pm 1} \longmapsto \rho^{\pm 1}. \end{aligned}$$

Let $\mathbf{U}'$ be the $\mathbb{Q}(v)$-subalgebra of $\boldsymbol{\mathcal{S}}'$ generated by $\mathfrak{x}_i, \mathfrak{y}_i$, and $\mathfrak{z}_i^{\pm 1}$ $(1 \leqslant i \leqslant r)$. Then Υ induces a surjective homomorphism $\Upsilon_1 : \mathbf{U}' \to \mathbf{U}_\vartriangle(r,r)$. Since the relations (QS1)–(QS6) are the defining relations for $\mathbf{U}_\vartriangle(r,r)$, there is also a natural surjective homomorphism $\Phi : \mathbf{U}_\vartriangle(r,r) \to \mathbf{U}'$. Clearly, the compositions $\Upsilon_1\Phi$ and $\Phi\Upsilon_1$ are identity maps. Thus, both Υ_1 and Φ are isomorphisms.

It is clear that for each $0 \neq t \in \mathbb{Z}$, ρ^t lies in $\boldsymbol{\mathcal{S}}_\vartriangle(r,r)_{rt}$. We claim that ρ^t does not lie in $\mathbf{U}_\vartriangle(r,r)_{rt}$. Otherwise, applying relations (QS0′)–(QS6′) would give that

$$\rho = \rho^t\rho^{-t+1} \in \big(\mathbf{U}_\vartriangle(r,r)_{rt}\big)\rho^{-t+1} \subseteq \mathbf{U}_\vartriangle(r,r)_r.$$

This implies $\boldsymbol{\mathcal{S}}_\vartriangle(r,r) = \mathbf{U}_\vartriangle(r,r)$, which contradicts Lemma 5.2.1. Hence, $\rho^t \notin \mathbf{U}_\vartriangle(r,r)_{rt}$ for all $0 \neq t \in \mathbb{Z}$.

For each $m \in \mathbb{Z}$, choose a $\mathbb{Q}(v)$-basis $\mathcal{B}_m$ of $\mathbf{U}_\vartriangle(r,r)_m$. Then the set $\mathcal{B} := \cup_{m\in\mathbb{Z}}\mathcal{B}_m$ forms a basis of $\mathbf{U}_\vartriangle(r,r)$. Moreover, by the discussion above, the set $\mathcal{B} \cup \{\rho^t \mid 0 \neq t \in \mathbb{Z}\}$ is linearly independent in $\boldsymbol{\mathcal{S}}_\vartriangle(r,r)$. On the other hand, by the definition of $\boldsymbol{\mathcal{S}}'$, the set

$$\{\Phi(c) \mid c \in \mathcal{B}\} \cup \{\eta^m \mid 0 \neq m \in \mathbb{Z}\}$$

is a spanning set of $\boldsymbol{\mathcal{S}}'$. Since $\Upsilon(\Phi(c)) = c$ and $\Upsilon(\eta^m) = \rho^m$, it follows that the above spanning set is a basis of $\boldsymbol{\mathcal{S}}'$. Therefore, Υ is an isomorphism. This finishes the proof. □

Remark 5.3.6. We can also present $\boldsymbol{\mathcal{S}}_\vartriangle(r,r)$ by using generators $\mathfrak{e}_i, \mathfrak{f}_i, \mathfrak{k}_i^{\pm 1}$ $(1 \leqslant i \leqslant r)$, $\mathfrak{e}_\delta$ and $\mathfrak{f}_\delta$, but the relation between $\mathfrak{e}_\delta$ and $\mathfrak{f}_\delta$ is not clear. Furthermore, if we use generators σ_1 and τ_1 instead of ρ and ρ^{-1}, the relations would be much more complicated; see the example below. From the relations obtained in the case $n = r$, it seems very hard to get the relations for the general case $n < r$.

Example 5.3.7. We consider the special case $n = r = 2$. In this case, $\mathbf{U}_\Delta(2,2) = \xi_2(\mathbf{U}_\Delta(2))$ is generated by $\mathfrak{e}_i, \mathfrak{f}_i, \mathfrak{k}_i$ for $i = 1, 2$, and $\mathbf{Z}_\Delta(2,2) = \mathbb{Q}(v)[\sigma_1, \sigma_2, \sigma_2^{-1}]$. Moreover, $\rho \in \mathrm{End}_{\mathbb{Q}(v)}(\boldsymbol{\Omega}^{\otimes 2})$ is defined by $\rho(\omega_s \otimes \omega_t) = \omega_{s-1} \otimes \omega_{t-1}$ and satisfies $\rho^2 = \sigma_2$. A direct calculation shows that

$$\rho = \frac{1}{v+v^{-1}}(\mathfrak{e}_1^2 + \mathfrak{e}_1\mathfrak{e}_2 + \mathfrak{e}_2\mathfrak{e}_1 + \mathfrak{e}_2^2 - \sigma_1).$$

By (5.3.1.3), we obtain that

$$\boldsymbol{\mathcal{S}}_\Delta(2,2)^+ = \mathbf{U}_\Delta(2,2)^+ \oplus \bigoplus_{m\geqslant 1} \mathbb{Q}(v)\rho^m = \mathbf{U}_\Delta(2,2)^+ \oplus \bigoplus_{m\geqslant 1} \mathbb{Q}(v)\sigma_1^m.$$

Similarly,

$$\rho^{-1} = \frac{1}{v+v^{-1}}(\mathfrak{f}_1^2 + \mathfrak{f}_1\mathfrak{f}_2 + \mathfrak{f}_2\mathfrak{f}_1 + \mathfrak{f}_2^2 - \tau_1)$$

and

$$\boldsymbol{\mathcal{S}}_\Delta(2,2)^- = \mathbf{U}_\Delta(2,2)^- \oplus \bigoplus_{m\geqslant 1} \mathbb{Q}(v)\rho^{-m} = \mathbf{U}_\Delta(2,2)^- \oplus \bigoplus_{m\geqslant 1} \mathbb{Q}(v)\tau_1^m.$$

Hence, $\sigma_1\tau_1 \in \mathbf{U}_\Delta(2,2)$ and

$$\boldsymbol{\mathcal{S}}_\Delta(2,2) = \mathbf{U}_\Delta(2,2) \oplus \bigoplus_{0\neq m\in\mathbb{Z}} \mathbb{Q}(v)\rho^m = \mathbf{U}_\Delta(2,2) \oplus \bigoplus_{m\geqslant 1} \big(\mathbb{Q}(v)\sigma_1^m \oplus \mathbb{Q}(v)\tau_1^m\big).$$

Furthermore, the following relations hold in $\boldsymbol{\mathcal{S}}_\Delta(2,2)$:

$$\begin{aligned}
&\sigma_1\mathfrak{e}_1 = \mathfrak{e}_1\mathfrak{e}_2\mathfrak{e}_1,\ \sigma_1\mathfrak{e}_2 = \mathfrak{e}_2\mathfrak{e}_1\mathfrak{e}_2,\\
&\tau_1\mathfrak{f}_1 = \mathfrak{f}_1\mathfrak{f}_2\mathfrak{f}_1,\ \tau_1\mathfrak{f}_2 = \mathfrak{f}_2\mathfrak{f}_1\mathfrak{f}_2,\\
&\sigma_1\mathfrak{f}_1 = \mathfrak{e}_1\mathfrak{e}_2\mathfrak{f}_1 + \mathfrak{f}_1\mathfrak{e}_2\mathfrak{e}_1,\ \sigma_1\mathfrak{f}_2 = \mathfrak{e}_2\mathfrak{e}_1\mathfrak{f}_2 + \mathfrak{f}_2\mathfrak{e}_1\mathfrak{e}_2,\\
&\tau_1\mathfrak{e}_1 = \mathfrak{f}_1\mathfrak{f}_2\mathfrak{e}_1 + \mathfrak{e}_1\mathfrak{f}_2\mathfrak{f}_1,\ \tau_1\mathfrak{e}_2 = \mathfrak{f}_2\mathfrak{f}_1\mathfrak{e}_2 + \mathfrak{e}_2\mathfrak{f}_1\mathfrak{f}_2,\\
&\sigma_1(\mathfrak{k}_i - v) = (\mathfrak{e}_1\mathfrak{e}_2 + \mathfrak{e}_2\mathfrak{e}_1)(\mathfrak{k}_i - v),\ i = 1, 2,\ \text{and}\\
&\tau_1(\mathfrak{k}_i - v) = (\mathfrak{f}_1\mathfrak{f}_2 + \mathfrak{f}_2\mathfrak{f}_1)(\mathfrak{k}_i - v),\ i = 1, 2.
\end{aligned} \tag{5.3.7.1}$$

Remark 5.3.8. Recall from §2.3 that $\mathfrak{D}_\Delta(n)$ is the $\mathcal{Z}$-subalgebra of $\boldsymbol{\mathfrak{D}}_\Delta(n)$ generated by $K_i^{\pm 1}$, $\left[{K_i;0 \atop t}\right]$, z_s^+, z_s^-, $(u_i^+)^{(m)}$, and $(u_i^-)^{(m)}$ for $i \in I$ and $s, t, m \geqslant 1$. By Proposition 3.6.1 and Corollary 3.7.5, all $\xi_r(\mathsf{z}_s^\pm)$ lie in $\mathcal{S}_\Delta(n,r)$. Thus, the $\mathbb{Q}(v)$-algebra homomorphism $\xi_r : \boldsymbol{\mathfrak{D}}_\Delta(n) \to \boldsymbol{\mathcal{S}}_\Delta(n,r)$ induces a $\mathcal{Z}$-algebra homomorphism $\xi_{r,\mathcal{Z}} : \mathfrak{D}_\Delta(n) \to \mathcal{S}_\Delta(n,r)$. By Corollary 3.8.2, $\xi_{r,\mathcal{Z}}$

is surjective in case $n > r$. However, it is in general not surjective as shown below.

Let $n = r = 2$. By Theorem 5.3.5,

$$\boldsymbol{\mathcal{S}}_\Delta(2,2) = \mathbf{U}_\Delta(2,2) \oplus \bigoplus_{0 \neq m \in \mathbb{Z}} \mathbb{Q}(v)\rho^m,$$

where $\rho = \sum_{\mathbf{a}\in\mathbb{N}^2, \sigma(\mathbf{a})=2} \mathfrak{e}_\mathbf{a}$ and $\rho^{-1} = \sum_{\mathbf{a}\in\mathbb{N}^2, \sigma(\mathbf{a})=2} \mathfrak{f}_\mathbf{a}$; see Lemma 5.3.1. By definition, $\mathfrak{e}_\mathbf{a} = \xi_2(\widetilde{u}_\mathbf{a}^+)$ and $\mathfrak{f}_\mathbf{a} = \xi_2(\widetilde{u}_\mathbf{a}^-)$. It follows that ρ and ρ^{-1} lie in $\mathcal{S}_\Delta(2,2)$. Using an argument similar to the proof of Corollary 3.8.2, we can show that $\mathcal{S}_\Delta(2,2)$ can be generated by $\mathfrak{e}_i^{(m)}, \mathfrak{f}_i^{(m)}$ $(i = 1,2, m \geqslant 1)$, $\mathfrak{l}_\lambda$ $(\lambda \in \Lambda_\Delta(2,2))$, ρ, and ρ^{-1}. Thus, we obtain that

$$\mathcal{S}_\Delta(2,2) = U_\Delta(2,2) \oplus \bigoplus_{0 \neq m \in \mathbb{Z}} \mathcal{Z}\rho^m,$$

where $U_\Delta(2,2)$ is the $\mathcal{Z}$-subalgebra of $\boldsymbol{\mathcal{S}}_\Delta(2,2)$ generated by $\mathfrak{e}_i^{(m)}, \mathfrak{f}_i^{(m)}, \mathfrak{l}_\lambda$ for $i = 1,2$, $m \geqslant 1$, and $\lambda \in \Lambda_\Delta(2,2)$.

On the other hand, the image $\operatorname{Im}\xi$ for $\xi := \xi_{2,\mathcal{Z}} : \mathfrak{D}_\Delta(2) \to \mathcal{S}_\Delta(2,2)$ is the $\mathcal{Z}$-subalgebra of $\mathcal{S}_\Delta(2,2)$ generated by $\mathfrak{k}_i^{\pm1}$, $\left[\begin{smallmatrix}\mathfrak{k}_i;0\\ t\end{smallmatrix}\right]$, $\xi(\mathsf{z}_s^+)$, $\xi(\mathsf{z}_s^-)$, $\mathfrak{e}_i^{(m)}$, and $\mathfrak{f}_i^{(m)}$ for $i = 1,2$ and $s,t,m \geqslant 1$. Since $\sigma_1 = \xi(\mathsf{z}_1^+)$ and $\tau_1 = \xi(\mathsf{z}_1^-)$, we have by the example above that

$$\xi(\mathsf{z}_1^+) = -(v+v^{-1})\rho + (\mathfrak{e}_1^2 + \mathfrak{e}_1\mathfrak{e}_2 + \mathfrak{e}_2\mathfrak{e}_1 + \mathfrak{e}_2^2)$$

and

$$\xi(\mathsf{z}_1^-) = -(v+v^{-1})\rho^{-1} + (\mathfrak{f}_1^2 + \mathfrak{f}_1\mathfrak{f}_2 + \mathfrak{f}_2\mathfrak{f}_1 + \mathfrak{f}_2^2).$$

Furthermore, for each $s \geqslant 1$,

$$\xi(\mathsf{z}_s^\pm) = \begin{cases} (\xi(\mathsf{z}_1^\pm))^s - \sum_{t=1}^{(s-1)/2} \binom{s}{t}\xi(\mathsf{z}_{s-2t}^\pm)\rho^{\pm 2t}, & \text{if } s \text{ is odd;} \\ (\xi(\mathsf{z}_1^\pm))^s - \sum_{t=1}^{s/2} \binom{s}{t}\xi(\mathsf{z}_{s-2t}^\pm)\rho^{\pm 2t}, & \text{if } s \text{ is even.} \end{cases}$$

By (5.3.7.1), all the elements

$$\mathfrak{e}_i\rho^{\pm1},\ \mathfrak{f}_i\rho^{\pm1},\ \rho^{\pm1}\mathfrak{e}_i,\ \rho^{\pm1}\mathfrak{e}_i$$

lie in $\mathbf{U}_\Delta(2,2)$ for $i = 1,2$. Thus, an inductive argument implies that for each $s \geqslant 1$,

$$\xi(\mathsf{z}_s^\pm) \equiv (-1)^s(v^s + v^{-s})\rho^{\pm s} \mod U_\Delta(2,2).$$

From (5.3.7.1) it also follows that

$$\mathfrak{k}_i\rho^{\pm1} \equiv v\rho^{\pm1} \equiv \rho^{\pm1}\mathfrak{k}_i \mod \mathbf{U}_\Delta(2,2) \text{ for } i = 1,2.$$

We conclude that neither of ρ and ρ^{-1} lies in $\operatorname{Im}\xi$. Therefore, the $\mathcal{Z}$-algebra homomorphism $\xi = \xi_{2,\mathcal{Z}} : \mathfrak{D}_\Delta(2) \to \mathcal{S}_\Delta(2,2)$ is not surjective.

5.4. The realization conjecture

We now look at the realization problem for quantum affine $\mathfrak{gl}_n$. In the non-affine case, Beilinson–Lusztig–MacPherson [**4**] provided a construction for quantum $\mathfrak{gl}_n$ via quantum Schur algebras. In order to generalize the BLM approach to the affine case, a modified BLM approach has been introduced in [**24**]. On the one hand, this approach produces a realization of quantum $\mathfrak{gl}_n$ which serves as a "cut-down" version of the original BLM realization. On the other hand, most of the constructions in this approach can be generalized to the affine case. The conjecture proposed below is a natural outcome from this consideration; see [**24**, 5.2].

Let $\mathcal{K}_\triangle(n)$ be the $\mathcal{Z}$-algebra which has $\mathcal{Z}$-basis $\{[A]\}_{A\in\Theta_\triangle(n)}$ and multiplication defined by $[A]\cdot[B]=0$ if $\mathrm{co}(A)\neq\mathrm{ro}(B)$, and $[A]\cdot[B]$ as given in $\mathcal{S}_\triangle(n,r)$ if $\mathrm{co}(A)=\mathrm{ro}(B)$ and $r=\sigma(A)$, where $\mathrm{co}(A)$ (resp., $\mathrm{ro}(A)$) is the column (resp., row) sum vector associated with A (see §1.1). This algebra has no identity but infinitely many idempotents $[\mathrm{diag}(\lambda)]$ for all $\lambda\in\mathbb{N}^n_\triangle$. Moreover, $\mathcal{K}_\triangle(n)\cong\oplus_{r\geqslant 0}\mathcal{S}_\triangle(n,r)$. (Note that $\mathcal{S}_\triangle(n,0)=\mathcal{Z}$ with a basis labeled by the zero matrix.)

Let $\boldsymbol{\mathcal{K}}_\triangle(n)=\mathcal{K}_\triangle(n)_{\mathbb{Q}(v)}$ and let $\widehat{\boldsymbol{\mathcal{K}}}_\triangle(n)$ be the vector space of all formal (possibly infinite) $\mathbb{Q}(v)$-linear combinations $\sum_{A\in\Theta_\triangle(n)}\beta_A[A]$ which have the following properties:

$$\begin{aligned}&\forall\,\mathbf{x}\in\mathbb{N}^n_\triangle,\text{ the sets }\{A\in\Theta_\triangle(n)\mid\beta_A\neq 0,\ \mathrm{ro}(A)=\mathbf{x}\}\\&\text{and }\{A\in\Theta_\triangle(n)\mid\beta_A\neq 0,\ \mathrm{co}(A)=\mathbf{x}\}\text{ are finite.}\end{aligned}\tag{5.4.0.1}$$

In other words, for $\lambda,\mu\in\mathbb{N}^n_\triangle$, both

$$\sum_{A\in\Theta_\triangle(n)}\beta_A[\mathrm{diag}(\lambda)]\cdot[A]\ \text{ and }\ \sum_{A\in\Theta_\triangle(n)}\beta_A[A]\cdot[\mathrm{diag}(\mu)]$$

are finite sums. Thus, there is a well-defined multiplication on $\widehat{\boldsymbol{\mathcal{K}}}_\triangle(n)$ by setting

$$\Big(\sum_{A\in\Theta_\triangle(n)}\alpha_A[A]\Big)\cdot\Big(\sum_{B\in\Theta_\triangle(n)}\beta_B[B]\Big):=\sum_{A,B\in\Theta_\triangle(n)}\alpha_A\beta_B[A][B].$$

This defines an associative algebra structure on $\widehat{\boldsymbol{\mathcal{K}}}_\triangle(n)$. This algebra has an identity element $\sum_{\lambda\in\mathbb{N}^n_\triangle}[\mathrm{diag}(\lambda)]$, the sum of all $[D]$ with D a diagonal matrix in $\Theta_\triangle(n)$, and contains $\boldsymbol{\mathcal{K}}_\triangle(n)$ as a natural subalgebra without identity. Note that $\widehat{\boldsymbol{\mathcal{K}}}_\triangle(n)$ is isomorphic to the direct product algebra $\prod_{r\geqslant 0}\boldsymbol{\mathcal{S}}_\triangle(n,r)$, and that the anti-involutions τ_r given in (3.1.3.4) induce the algebra anti-involution

$$\tau:=\sum_r\tau_r:\widehat{\boldsymbol{\mathcal{K}}}_\triangle(n)\longrightarrow\widehat{\boldsymbol{\mathcal{K}}}_\triangle(n),\qquad\sum_{A\in\Theta_\triangle(n)}\beta_A[A]\longmapsto\sum_{A\in\Theta_\triangle(n)}\beta_A[{}^tA].\tag{5.4.0.2}$$

For $A \in \Theta^{\pm}_{\triangle}(n)$ and $\mathbf{j} \in \mathbb{Z}^n_{\triangle}$, define $A(\mathbf{j}) \in \widehat{\mathcal{K}}_{\triangle}(n)$ by

$$A(\mathbf{j}) = \sum_{\lambda \in \mathbb{N}^n_{\triangle}} v^{\lambda \centerdot \mathbf{j}}[A + \mathrm{diag}(\lambda)],$$

where $\lambda \centerdot \mathbf{j} = \sum_{1 \leqslant i \leqslant n} \lambda_i j_i$. Clearly, $A(\mathbf{j}) = \sum_{r \geqslant 0} A(\mathbf{j}, r)$.

Let $\mathfrak{A}_{\triangle}(n)$ be the subspace of $\widehat{\mathcal{K}}_{\triangle}(n)$ spanned by $A(\mathbf{j})$ for all $A \in \Theta^{\pm}_{\triangle}(n)$ and $\mathbf{j} \in \mathbb{Z}^n_{\triangle}$. Since the structure constants (with respect to the BLM basis) appearing in the multiplication formulas given in Theorem 3.4.2 are independent of r, we immediately obtain the following similar formulas in $\mathfrak{A}_{\triangle}(n)$. Here, again for the simplicity of the statement, we extend the definition of $A(\mathbf{j})$ to all the matrices in $M_{n,\triangle}(\mathbb{Z})$ by setting $A(\mathbf{j}) = 0$ if some off-diagonal entries of A are negative.

Theorem 5.4.1. *Maintain the notation used in Theorem 3.4.2. The following multiplication formulas hold in $\mathfrak{A}_{\triangle}(n)$:*

$$\begin{aligned} 0(\mathbf{j})A(\mathbf{j}') &= v^{\mathbf{j} \centerdot \mathrm{ro}(A)} A(\mathbf{j} + \mathbf{j}'), \\ A(\mathbf{j}')0(\mathbf{j}) &= v^{\mathbf{j} \centerdot \mathrm{co}(A)} A(\mathbf{j} + \mathbf{j}'), \end{aligned} \tag{5.4.1.1}$$

where 0 *stands for the zero matrix,*

$$\begin{aligned} E^{\triangle}_{h,h+1}(\mathbf{0})A(\mathbf{j}) = &\sum_{i<h;\, a_{h+1,i} \geqslant 1} v^{f(i)} \overline{\left[\!\!\left[\begin{matrix} a_{h,i}+1 \\ 1 \end{matrix} \right]\!\!\right]} (A + E^{\triangle}_{h,i} - E^{\triangle}_{h+1,i})(\mathbf{j} + \boldsymbol{\alpha}^{\triangle}_h) \\ &+ \sum_{i>h+1;\, a_{h+1,i} \geqslant 1} v^{f(i)} \overline{\left[\!\!\left[\begin{matrix} a_{h,i}+1 \\ 1 \end{matrix} \right]\!\!\right]} (A + E^{\triangle}_{h,i} - E^{\triangle}_{h+1,i})(\mathbf{j}) \\ &+ v^{f(h)-j_h-1} \frac{(A - E^{\triangle}_{h+1,h})(\mathbf{j} + \boldsymbol{\alpha}^{\triangle}_h) - (A - E^{\triangle}_{h+1,h})(\mathbf{j} + \boldsymbol{\beta}^{\triangle}_h)}{1 - v^{-2}} \\ &+ v^{f(h+1)+j_{h+1}} \overline{\left[\!\!\left[\begin{matrix} a_{h,h+1}+1 \\ 1 \end{matrix} \right]\!\!\right]} (A + E^{\triangle}_{h,h+1})(\mathbf{j}), \end{aligned} \tag{5.4.1.2}$$

and

$$\begin{aligned} E^{\triangle}_{h+1,h}(\mathbf{0})A(\mathbf{j}) = &\sum_{i<h;\, a_{h,i} \geqslant 1} v^{f'(i)} \overline{\left[\!\!\left[\begin{matrix} a_{h+1,i}+1 \\ 1 \end{matrix} \right]\!\!\right]} (A - E^{\triangle}_{h,i} + E^{\triangle}_{h+1,i})(\mathbf{j}) \\ &+ \sum_{i>h+1;\, a_{h,i} \geqslant 1} v^{f'(i)} \overline{\left[\!\!\left[\begin{matrix} a_{h+1,i}+1 \\ 1 \end{matrix} \right]\!\!\right]} (A - E^{\triangle}_{h,i} + E^{\triangle}_{h+1,i})(\mathbf{j} - \boldsymbol{\alpha}^{\triangle}_h) \\ &+ v^{f'(h+1)-j_{h+1}-1} \frac{(A - E^{\triangle}_{h,h+1})(\mathbf{j} - \boldsymbol{\alpha}^{\triangle}_h) - (A - E^{\triangle}_{h,h+1})(\mathbf{j} + \boldsymbol{\beta}^{\triangle}_h)}{1 - v^{-2}} \\ &+ v^{f'(h)+j_h} \overline{\left[\!\!\left[\begin{matrix} a_{h+1,h}+1 \\ 1 \end{matrix} \right]\!\!\right]} (A + E^{\triangle}_{h+1,h})(\mathbf{j}). \end{aligned} \tag{5.4.1.3}$$

There is a parallel construction in the non-affine case as discussed in [**24**, §8]. By removing the sub/superscripts △, we obtain the corresponding objects and multiplication formulas in this (non-affine) case. Since the quantum $\mathfrak{gl}_n$, denoted by $\mathbf{U}(n)$, is generated by E_i, F_i, and $K_j^{\pm 1}$ $(1 \leqslant i \leqslant n-1, 1 \leqslant j \leqslant n)$, the corresponding multiplication formulas define an algebra isomorphism $\mathbf{U}(n) \to \mathfrak{A}(n)$ sending E_h to $E_{h,h+1}(\mathbf{0})$, F_h to $E_{h+1,h}(\mathbf{0})$, and K_j to $0(\boldsymbol{e}_j)$. In this way, we obtain a realization of the quantum $\mathfrak{gl}_n$. This is the so-called modified approach introduced in [**24**]. In this approach, the algebra $\mathcal{K}(n)$ as a direct sum of all quantum Schur algebras $\boldsymbol{\mathcal{S}}(n, r)$ is just a homomorphic image of the BLM algebra $\mathbf{K}$ constructed in [**4**], which is isomorphic to the modified quantum group $\dot{\mathbf{U}}(n)$. However, since we avoid using the stabilization property (see [**4**, §4]) required in the construction of $\mathbf{K}$, it can be generalized to obtain the affine construction above.

The $\mathbb{Q}(v)$-space $\mathfrak{A}_\triangle(n)$ is a natural candidate for a realization of $\mathfrak{D}_\triangle(n) \cong \mathbf{U}(\widehat{\mathfrak{gl}}_n)$. Since $\mathfrak{D}_\triangle(n)$ has generators other than simple generators, the formulas in Theorem 5.4.1 are not sufficient to show that $\mathfrak{A}_\triangle(n)$ is a subalgebra. However, as seen in [**24**, 5.4 &7.5], these formulas are sufficient to embed the subalgebras $\mathbf{U}_\triangle(n)$, $\mathfrak{H}_\triangle(n)^{\geqslant 0}$ and $\mathfrak{H}_\triangle(n)^{\leqslant 0}$ into $\mathfrak{A}_\triangle(n)$. Thus, it is natural to formulate the following conjecture.

Conjecture 5.4.2. ([**24**, 5.5(2)]) The $\mathbb{Q}(v)$-space $\mathfrak{A}_\triangle(n)$ is a subalgebra of $\widehat{\mathcal{K}}_\triangle(n)$ which is isomorphic to $\mathfrak{D}_\triangle(n)$ or the quantum loop algebra $\mathbf{U}(\widehat{\mathfrak{gl}}_n)$.

As seen in §1.4, there are three types of extra generators for $\mathfrak{D}_\triangle(n)$. We expect to derive more multiplication formulas between these extra generators and the BLM basis elements in $\boldsymbol{\mathcal{S}}_\triangle(n, r)$ (see, e.g., Problem 6.4.2) and, hence, to prove the conjecture. As a first test, we will establish the conjecture for the classical $(v = 1)$ case in the next chapter. We end this section with a formulation of the conjecture in this case.

Let $\mathcal{Z} \to \mathbb{Q}$ be the specialization by sending v to 1, and let $\widehat{\mathcal{K}}_\triangle(n)_\mathbb{Q}$ be the vector space of all formal (possibly infinite) $\mathbb{Q}$-linear combinations $\sum_{A\in\Theta_\triangle(n)} \beta_A[A]_1$ satisfying (5.4.0.1). Here $[A]_1$ denotes the image of $[A]$ in $\mathcal{S}_\triangle(n, r)_\mathbb{Q}$. For any $r > 0$, $A \in \Theta_\triangle^\pm(n)$, and $\mathbf{j} \in \mathbb{N}_\triangle^n$, define in $\mathcal{S}_\triangle(n, r)_\mathbb{Q}$ (cf. [**30**, (3.0.3)])

$$A[\mathbf{j}, r] = \begin{cases} \sum_{\lambda\in\Lambda_\triangle(n,r-\sigma(A))} \lambda^{\mathbf{j}}[A + \mathrm{diag}(\lambda)]_1, & \text{if } \sigma(A) \leqslant r; \\ 0, & \text{otherwise}, \end{cases} \tag{5.4.2.1}$$

where $\lambda^{\mathbf{j}} = \prod_{i=1}^n \lambda_i^{j_i}$. (These elements play a role similar to the elements $A(\mathbf{j}, r)$ for affine quantum Schur algebras. See §6.2 below for more discussion of the elements.) Let

$$A[\mathbf{j}] = \sum_{r=0}^{\infty} A[\mathbf{j}, r] \in \widehat{\mathcal{K}}_\triangle(n)_\mathbb{Q}. \tag{5.4.2.2}$$

Then the classical version of Conjecture 5.4.2 claims that the $\mathbb{Q}$-span $\mathfrak{A}_{\triangle,\mathbb{Q}}[n]$ of all $A[\mathbf{j}]$ is a subalgebra of $\widehat{\mathcal{K}}_\triangle(n)_\mathbb{Q}$ which is isomorphic to the algebra

$$\overline{\mathfrak{D}_\triangle(n)_\mathbb{Q}} := \mathfrak{D}_\triangle(n)_\mathbb{Q}/\langle K_i - 1 \mid 1 \leqslant i \leqslant n\rangle,$$

where $\mathfrak{D}_\triangle(n)$ is the integral form defined in Definition 2.4.4. We will prove in §6.1 that the specialized algebra $\overline{\mathfrak{D}_\triangle(n)_\mathbb{Q}}$ is isomorphic to the universal enveloping algebra $\mathcal{U}(\widehat{\mathfrak{gl}}_n)$ over $\mathbb{Q}$ introduced in §1.1.

5.5. Lusztig's transfer maps on semisimple generators

In [**57**] Lusztig defined an algebra homomorphism $\phi_{r,r-n} : \boldsymbol{\mathcal{S}}_\triangle(n, r) \longrightarrow \boldsymbol{\mathcal{S}}_\triangle(n, r-n)$, for $r \geqslant n$, called the *transfer map*. In this section, we describe the images of $A(\mathbf{0}, r)$ and ${}^tA(\mathbf{0}, r)$ under $\phi_{r,r-n}$ for all $A = A_\lambda = \sum_{i=1}^n \lambda_i E^\triangle_{i,i+1} \in \Theta_\triangle(n)$.

Throughout this section, for a prime power q, $\mathcal{S}_\triangle(n, r)_\mathbb{C} = \mathcal{S}_\triangle(n, r) \otimes_\mathcal{Z} \mathbb{C}$ always denotes the specialization of $\mathcal{S}_\triangle(n, r)$ at $v = \sqrt{q}$. In other words, $\mathcal{S}_\triangle(n, r)_\mathbb{C}$ is a $\mathbb{C}$-vector space with a basis $\{e_A = e_A \otimes 1 \mid A \in \Theta_\triangle(n, r)\}$; see Definition 3.1.1.

We first recall from [**57**, §1] the definition of $\phi_{r,r-n}$.

As in §3.1, let V be a free $\mathbb{F}[\varepsilon, \varepsilon^{-1}]$-module of rank r and let $\mathscr{F}_\triangle = \mathscr{F}_{\triangle n}(V)$ be the set of all cyclic flags $\mathbf{L} = (L_i)_{i\in\mathbb{Z}}$, where each L_i is a lattice in V such that $L_{i-1} \subseteq L_i$ and $L_{i-n} = \varepsilon L_i$ for all $i \in \mathbb{Z}$. If $\mathbb{F}$ is the finite field with q elements, then we use the notation $\mathscr{Y} = \mathscr{F}_\triangle(q)$ for $\mathscr{F}_\triangle$, and $\mathcal{S}_\triangle(n, r)_\mathbb{C}$ is identified with $\mathbb{C}_G(\mathscr{Y} \times \mathscr{Y})$, where $G = G_V$ is the group of automorphisms of the $\mathbb{F}[\varepsilon, \varepsilon^{-1}]$-module V. Given $\mathbf{L} = (L_i), \widetilde{\mathbf{L}} = (\widetilde{L}_i) \in \mathscr{F}_\triangle$ with $\widetilde{\mathbf{L}} \subseteq \mathbf{L}$, i.e., $\widetilde{L}_i \subseteq L_i$ for all $i \in \mathbb{Z}$, $\mathbf{L}/\widetilde{\mathbf{L}}$ can be viewed as a nilpotent representation of the cyclic quiver $\triangle(n)$; see §3.6.

As in (1.2.0.1), for each $\lambda = (\lambda_i)_{i\in\mathbb{Z}} \in \mathbb{N}^n_\triangle$, let $S_\lambda = \oplus_{1\leqslant i\leqslant n}\lambda_i S_i$, where S_i denotes the simple representation of $\triangle(n)$ corresponding to the vertex i, and set $A_\lambda = \sum_{1\leqslant i\leqslant n} \lambda_i E^\triangle_{i,i+1} \in \Theta^+_\triangle(n)$.

For $A \in \Theta^\pm_\triangle(n)$ and $\mathbf{j} \in \mathbb{Z}^n_\triangle$, let $A(\mathbf{j}, r) \in \mathcal{S}_\triangle(n, r)$ be the BLM basis elements defined as in (3.4.0.2). We also view $A(\mathbf{j}, r)$ as an element in $\mathcal{S}_\triangle(n, r)_\mathbb{C}$ via specializing v to $\sqrt{q}$.

By Corollary 3.6.2, for all $\lambda \in \mathbb{N}^n_\triangle$ and $\mathbf{L} = (L_i), \widetilde{\mathbf{L}} = (\widetilde{L}_i) \in \mathscr{F}_\triangle$, we have

$$A_\lambda(\mathbf{0}, r)(\mathbf{L}, \widetilde{\mathbf{L}}) = \begin{cases} v^{-c(\mathbf{L},\widetilde{\mathbf{L}})}, & \text{if } \widetilde{\mathbf{L}} \subseteq \mathbf{L} \text{ and } \mathbf{L}/\widetilde{\mathbf{L}} \cong S_\lambda; \\ 0, & \text{otherwise,} \end{cases} \tag{5.5.0.1}$$

where

$$c(\mathbf{L},\widetilde{\mathbf{L}}) = \sum_{1\leqslant i\leqslant n} \dim(L_{i+1}/\widetilde{L}_{i+1})\big(\dim(\widetilde{L}_{i+1}/\widetilde{L}_i) - \dim(L_i/\widetilde{L}_i)\big)$$
$$= \sum_{1\leqslant i\leqslant n} \dim(L_i/\widetilde{L}_i)\big(\dim(L_i/L_{i-1}) - \dim(L_i/\widetilde{L}_i)\big).$$

Moreover,

$$0(\lambda,r)(\mathbf{L},\widetilde{\mathbf{L}}) = \begin{cases} v^{\lambda\centerdot\mu}, & \text{if } \widetilde{\mathbf{L}} = \mathbf{L};\\ 0, & \text{otherwise,}\end{cases} \tag{5.5.0.2}$$

where $\mu = (\mu_i) \in \mathbb{N}^n_\Delta$ with $\mu_i = \dim L_i/L_{i-1}$.

Now let $r, r', r'' \geqslant 0$ satisfy $r = r' + r''$. Let V'' be a direct summand of V of rank r''. Set $V' = V/V''$. Define $\pi' : \mathscr{F}_{\Delta,n}(V) \to \mathscr{F}_{\Delta,n}(V')$, $\mathbf{L} \mapsto \mathbf{L}'$ (resp., $\pi'' : \mathscr{F}_{\Delta,n}(V) \to \mathscr{F}_{\Delta,n}(V'')$, $\mathbf{L} \mapsto \mathbf{L}''$) by setting

$$L'_i = (L_i + V'')/V'' \quad (\text{resp., } L''_i = L_i \cap V''),$$

for all $i \in \mathbb{Z}$. Thus, for each $i \in \mathbb{Z}$, there is an exact sequence

$$0 \longrightarrow L''_i/L''_{i-1} \longrightarrow L_i/L_{i-1} \longrightarrow L'_i/L'_{i-1} \longrightarrow 0$$

of $\mathbb{F}$-vector spaces.

Let $\mathbb{F}$ be the finite field with q elements. Following [**57**, 1.2], let Ξ : $\mathcal{S}_\Delta(n,r)_\mathbb{C} \to \mathcal{S}_\Delta(n,r')_\mathbb{C} \otimes \mathcal{S}_\Delta(n,r'')_\mathbb{C}$ be the map defined by[5]

$$\Xi(f)(\mathbf{L}',\widetilde{\mathbf{L}}',\mathbf{L}'',\widetilde{\mathbf{L}}'') = \sum_{\widetilde{\mathbf{L}}\in\mathscr{F}_{\Delta,n}(V)} f(\mathbf{L},\widetilde{\mathbf{L}})$$

for $f \in \mathcal{S}_\Delta(n,r)_\mathbb{C}$, where $\mathbf{L}', \widetilde{\mathbf{L}}' \in \mathscr{F}_{\Delta,n}(V')$, $\mathbf{L}'', \widetilde{\mathbf{L}}'' \in \mathscr{F}_{\Delta,n}(V'')$, and $\mathbf{L}$ is a fixed element in $\mathscr{F}_{\Delta,n}(V)$ satisfying $\pi'(\mathbf{L}) = \mathbf{L}'$ and $\pi''(\mathbf{L}) = \mathbf{L}''$, and the sum is taken over all $\widetilde{\mathbf{L}} \in \mathscr{F}_{\Delta,n}(V)$ such that $\pi'(\widetilde{\mathbf{L}}) = \widetilde{\mathbf{L}}'$ and $\pi''(\widetilde{\mathbf{L}}) = \widetilde{\mathbf{L}}''$.

For two lattices L, L' in V, define

$$(L : L') = \dim_\mathbb{F}(L/\widetilde{L}) - \dim_\mathbb{F}(L'/\widetilde{L}),$$

where $\widetilde{L}$ is a lattice contained in $L \cap L'$. For $\mathbf{L} = (L_i), \mathbf{L}' = (L'_i) \in \mathscr{F}_{\Delta,n}(V)$, set

$$(\mathbf{L} : \mathbf{L}') = \sum_{i=1}^{n} (L_i : L'_i).$$

We finally define a $\mathbb{C}$-linear isomorphism $\xi : \mathcal{S}_\Delta(n,r)_\mathbb{C} \to \mathcal{S}_\Delta(n,r)_\mathbb{C}$ by

$$\xi(f)(\mathbf{L},\mathbf{L}') = q^{(\mathbf{L}:\mathbf{L}')/2} f(\mathbf{L},\mathbf{L}')$$

[5] The map Ξ is denoted by Δ in [**57**].

for $f \in \mathcal{S}_\Delta(n,r)_\mathbb{C}$ and $\mathbf{L}, \mathbf{L}' \in \mathscr{F}_{\Delta,n}(V)$. Then ξ is an algebra isomorphism; see [**57**, 1.7].

As before, let $\delta = (\delta_i) \in \mathbb{N}_\Delta^n$ with all $\delta_i = 1$. For each $A = (a_{i,j}) \in \Theta_\Delta(n,n)$, if $\mathrm{ro}(A) \neq \delta$ or $\mathrm{co}(A) \neq \delta$, then we set $\mathrm{sgn}_A = 0$. Suppose now $\mathrm{ro}(A) = \mathrm{co}(A) = \delta$. Then there is a unique permutation $w : \mathbb{Z} \to \mathbb{Z}$ such that $a_{i,j} = \delta_{w(j),i}$. In this case, we set $\mathrm{sgn}_A = (-1)^{\mathrm{Inv}(w)}$, where

$$\mathrm{Inv}(w) = |\{(i,j) \in \mathbb{Z}^2 \mid 1 \leqslant i \leqslant n, i < j, w(i) > w(j)\}|$$

(cf. §3.2). By [**57**, 1.8], there is an algebra homomorphism $\chi : \mathcal{S}_\Delta(n,n)_\mathbb{C} \to \mathbb{C}$ such that $\chi(e_A) = \mathrm{sgn}_A$. In particular, for $1 \leqslant i \leqslant n$ and $\lambda \in \mathbb{Z}_\Delta^n$,

$$\chi(E^\Delta_{i,i+1}(\mathbf{0},n)) = 0, \ \ \chi(E^\Delta_{i+1,i}(\mathbf{0},n)) = 0, \ \ \text{and} \ \ \chi(0(\lambda,n)) = q^{(\lambda\centerdot\delta)/2}.$$

Suppose $r \geqslant n$ and let $\phi^q_{r,r-n}$ denote the composition

$$\begin{aligned}\mathcal{S}_\Delta(n,r)_\mathbb{C} \xrightarrow{\ \Xi\ } \mathcal{S}_\Delta(n,r-n)_\mathbb{C} \otimes \mathcal{S}_\Delta(n,n)_\mathbb{C} &\xrightarrow{\xi\otimes\chi} \mathcal{S}_\Delta(n,r-n)_\mathbb{C} \otimes \mathbb{C} \\ &= \mathcal{S}_\Delta(n,r-n)_\mathbb{C}.\end{aligned}$$

Lusztig [**57**] showed that, for each pair $A \in \Theta_\Delta(n,r)$ and $B \in \Theta_\Delta(n,r-n)$, there is a uniquely determined polynomial $f_{A,B}(v,v^{-1}) \in \mathcal{Z} = \mathbb{Z}[v,v^{-1}]$ such that for each finite field $\mathbb{F}$ of q elements,

$$\phi^q_{r,r-n}(q^{-d_A/2}e_A) = \sum_{B\in\Theta_\Delta(n,r-n)} f_{A,B}(q^{1/2},q^{-1/2})q^{-d_B/2}e_B.$$

This gives a $\mathbb{Q}(v)$-algebra homomorphism $\phi_{r,r-n} : \boldsymbol{\mathcal{S}}_\Delta(n,r) \to \boldsymbol{\mathcal{S}}_\Delta(n,r-n)$ defined by setting

$$\phi_{r,r-n}([A]) = \sum_{B\in\Theta_\Delta(n,r-n)} f_{A,B}(v,v^{-1})[B],$$

which is called the *transfer map*. Moreover, $\phi_{r,r-n}$ takes

$$\begin{aligned}E^\Delta_{i,i+1}(\mathbf{0},r) &\longmapsto E^\Delta_{i,i+1}(\mathbf{0},r-n),\\ E^\Delta_{i+1,i}(\mathbf{0},r) &\longmapsto E^\Delta_{i+1,i}(\mathbf{0},r-n), \ \ \text{and}\\ 0(\lambda,r) &\longmapsto v^{\lambda\centerdot\delta}0(\lambda,r-n),\end{aligned}$$

for $i \in \mathbb{Z}$ and $\lambda \in \mathbb{Z}^n_\Delta$.

By the definition, for each $i \in I$, we have

$$\phi_{r,r-n}\xi_r(K_i) = v\,0(\boldsymbol{e}^\Delta_i, n-r) \neq 0(\boldsymbol{e}^\Delta_i, n-r) = \xi_{r-n}(K_i).$$

Thus, $\phi_{r,r-n}\xi_r \neq \xi_{r-n}$. However, if we view $\mathbf{U}(\widehat{\mathfrak{sl}}_n)$ as a subalgebra of $\mathfrak{D}_\Delta(n)$ generated by E_i, F_i, $\widetilde{K}_i$ for $i \in I$ and denote the restriction of ξ_r (resp., ξ_{r-n}) to $\mathbf{U}(\widehat{\mathfrak{sl}}_n)$ still by ξ_r (resp., ξ_{r-n}), then $\phi_{r,r-n}\xi_r = \xi_{r-n}$, i.e., the following triangle commutes

$$\begin{array}{ccc} & \xrightarrow{\ \xi_r\ } & \boldsymbol{\mathcal{S}}_\Delta(n,r) \\ \mathbf{U}(\widehat{\mathfrak{sl}}_n) & & \Big\downarrow \phi_{r,r-n} \\ & \xrightarrow{\ \xi_{r-n}\ } & \boldsymbol{\mathcal{S}}_\Delta(n,r-n) \end{array}$$

As a result, there is an induced algebra homomorphism $\mathbf{U}(\widehat{\mathfrak{sl}}_n) \to \varprojlim \boldsymbol{\mathcal{S}}_\Delta(n, n+m)$; see [**57**, 3.4]. As seen above (a fact pointed out by Lusztig in [**57**]), if $\mathbf{U}(\widehat{\mathfrak{sl}}_n)$ is replaced by $\mathbf{U}_\Delta(n)$, then the diagram above is not commutative. Hence, the homomorphism $\mathbf{U}(\widehat{\mathfrak{sl}}_n) \to \varprojlim \boldsymbol{\mathcal{S}}_\Delta(n, n+m)$ cannot be extended to $\mathbf{U}_\Delta(n)$, nor to $\mathfrak{D}_\Delta(n)$. However, it is natural to ask if this homomorphism can be extended to the double Ringel–Hall algebra $'\mathfrak{D}_\Delta(n)$ introduced in Remark 2.3.6(3), which has the same 0-part as $\mathbf{U}(\widehat{\mathfrak{sl}}_n)$. We now show below that this is not the case.

It is known from §3.7 that $\boldsymbol{\mathcal{S}}_\Delta(n,r)$ is generated by the $A_\lambda(\mathbf{0},r)$, ${}^tA_\lambda(\mathbf{0},r)$ and $0(\boldsymbol{e}_i^\Delta, r)$ for $\lambda \in \mathbb{N}_\Delta^n$ and $i \in I$. In the following we describe the images of $A_\lambda(\mathbf{0},r)$ and ${}^tA_\lambda(\mathbf{0},r)$ under $\phi_{r,r-n}$. As in §2.1, for $\lambda = (\lambda_i) \in \mathbb{Z}_\Delta^n$, define $\tau\lambda \in \mathbb{Z}_\Delta^n$ by setting $(\tau\lambda)_i = \lambda_{i-1}$, for all $i \in \mathbb{Z}$.

Proposition 5.5.1. *Keep the notation above. Let $\mathbb{F}$ be the finite field with q elements. For each $\lambda = (\lambda_i) \in \mathbb{N}_\Delta^n$, we have*

$$\Xi(A_\lambda(\mathbf{0},r)) = \sum_{\substack{\mu,\nu\in\mathbb{N}_\Delta^n \\ \mu+\nu=\lambda}} q^{(\mu\centerdot\tau\nu-\tau\mu\centerdot\nu)/2} A_\mu(\mathbf{0},r')0(\nu,r') \otimes A_\nu(\mathbf{0},r'')0(-\mu,r'').$$

Dually, we have

$$\Xi({}^tA_\lambda(\mathbf{0},r)) = \sum_{\substack{\mu,\nu\in\mathbb{N}_\Delta^n \\ \mu+\nu=\lambda}} q^{(\mu\centerdot\tau\nu-\tau\mu\centerdot\nu)/2}\, {}^tA_\mu(\mathbf{0},r')0(\tau\nu,r') \otimes {}^tA_\nu(\mathbf{0},r'')0(-\tau\mu,r'').$$

Proof. We only prove the first formula. The second one can be proved similarly.

For $\mu+\nu=\lambda$ with $\mu,\nu \in \mathbb{N}_\Delta^n$, we write

$$\Psi_{\mu,\nu} = A_\mu(\mathbf{0},r')0(\nu,r') \otimes A_\nu(\mathbf{0},r'')0(-\mu,r'').$$

Thus, it suffices to show that for all fixed $\mathbf{L}', \widetilde{\mathbf{L}}' \in \mathscr{F}_{\Delta,n}(V')$ and $\mathbf{L}'', \widetilde{\mathbf{L}}'' \in \mathscr{F}_{\Delta,n}(V'')$,

$$\Xi(A_\lambda(\mathbf{0},r))(\mathbf{L}', \widetilde{\mathbf{L}}', \mathbf{L}'', \widetilde{\mathbf{L}}'') = \sum_{\substack{\mu,\nu\in\mathbb{N}_\Delta^n \\ \mu+\nu=\lambda}} q^{(\mu\centerdot\tau\nu-\tau\mu\centerdot\nu)/2} \Psi_{\mu,\nu}(\mathbf{L}', \widetilde{\mathbf{L}}', \mathbf{L}'', \widetilde{\mathbf{L}}'').$$

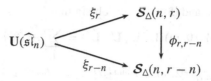

By the definition and (5.5.0.1), we obtain that

$$\begin{aligned}\Xi(A_\lambda(\mathbf{0},r))(\mathbf{L}',\widetilde{\mathbf{L}}',\mathbf{L}'',\widetilde{\mathbf{L}}'') &= \sum_{\widetilde{\mathbf{L}}} A_\lambda(\mathbf{0},r)(\mathbf{L},\widetilde{\mathbf{L}})\\ &= q^{(-\sum_{1\leqslant i\leqslant n}\lambda_i(\dim L_i/L_{i-1}-\lambda_i))/2}\ell,\end{aligned}$$

where ℓ denotes the cardinality of the set

$$\mathscr{L} := \{\widetilde{\mathbf{L}} \in \mathscr{F}_{\Delta n}(V) \mid \widetilde{\mathbf{L}} \subseteq \mathbf{L},\ \mathbf{L}/\widetilde{\mathbf{L}} \cong S_\lambda,\ \pi'(\widetilde{\mathbf{L}}) = \widetilde{\mathbf{L}}',\ \pi''(\widetilde{\mathbf{L}}) = \widetilde{\mathbf{L}}''\}.$$

Now let $\widetilde{\mathbf{L}} \in \mathscr{L}$. For each $i \in \mathbb{Z}$, consider the projection

$$\theta_i : L_i \longrightarrow L_i' = (L_i + V'')/V''.$$

Thus, $\theta_i^{-1}(\widetilde{L}_i) = \widetilde{L}_i + L_i \cap V'' = \widetilde{L}_i + L_i''$ and $L_i/\theta_i^{-1}(\widetilde{L}_i) \cong L_i'/\widetilde{L}_i'$. This implies that

$$\dim \theta_i^{-1}(\widetilde{L}_i)/\widetilde{L}_i = \dim L_i/\widetilde{L}_i - \dim L_i'/\widetilde{L}_i' = \dim L_i''/\widetilde{L}_i''.$$

The semisimplicity of $\mathbf{L}/\widetilde{\mathbf{L}} \cong S_\lambda$ shows $L_{i-1} \subseteq \widetilde{L}_i$. Hence, we have the inclusions

$$L_{i-1} + \widetilde{L}_i'' \subseteq \widetilde{L}_i \subseteq \theta_i^{-1}(\widetilde{L}_i) = \widetilde{L}_i + L_i'' \subseteq L_i.$$

Then

$$\theta_i^{-1}(\widetilde{L}_i)/(L_{i-1}+\widetilde{L}_i'') = \widetilde{L}_i/(L_{i-1}+\widetilde{L}_i'') + (L_{i-1}+L_i'')/(L_{i-1}+\widetilde{L}_i'') \text{ and}$$

$$\dim \theta_i^{-1}(\widetilde{L}_i)/(L_{i-1}+\widetilde{L}_i'') = \dim \widetilde{L}_i/(L_{i-1}+\widetilde{L}_i'') + \dim L_i''/\widetilde{L}_i''.$$

The inequality $\dim(L_{i-1}+L_i'')/(L_{i-1}+\widetilde{L}_i'') \leqslant \dim L_i''/\widetilde{L}_i''$ gives that

$$\dim(L_{i-1}+L_i'')/(L_{i-1}+\widetilde{L}_i'') = \dim L_i''/\widetilde{L}_i'' \text{ and}$$

$$\theta_i^{-1}(\widetilde{L}_i)/(L_{i-1}+\widetilde{L}_i'') = \widetilde{L}_i/(L_{i-1}+\widetilde{L}_i'') \oplus (L_{i-1}+L_i'')/(L_{i-1}+\widetilde{L}_i'').$$

Thus, $\widetilde{L}_i/(L_{i-1}+\widetilde{L}_i'')$ is a complement of $(L_{i-1}+L_i'')/(L_{i-1}+\widetilde{L}_i'')$.

Consequently,

$$\ell = \prod_{1\leqslant i\leqslant n} \ell_i,$$

where ℓ_i is the number of subspaces in $\theta_i^{-1}(\widetilde{L}_i)/(L_{i-1}+\widetilde{L}_i'')$ which are complementary to $(L_{i-1}+L_i'')/(L_{i-1}+\widetilde{L}_i'')$. Further,

$$\begin{aligned}&\dim \theta_i^{-1}(\widetilde{L}_i)/(L_{i-1}+\widetilde{L}_i'') = \dim L_i/(L_{i-1}+\widetilde{L}_i'') - \dim L_i/\theta_i^{-1}(\widetilde{L}_i)\\ =\ &\dim L_i/(L_{i-1}+L_i'') + \dim(L_{i-1}+L_i'')/(L_{i-1}+\widetilde{L}_i'') - \dim L_i'/\widetilde{L}_i'\\ =\ &\dim L_i/L_{i-1} - \dim(L_{i-1}+L_i'')/L_{i-1} + \dim L_i''/\widetilde{L}_i'' - \dim L_i'/\widetilde{L}_i'\\ =\ &\dim L_i'/L_{i-1}' + \dim L_i''/\widetilde{L}_i'' - \dim L_i'/\widetilde{L}_i'.\end{aligned}$$

Therefore,

$$\ell_i = q^{\dim L''_i/\widetilde{L}''_i(\dim L'_i/L'_{i-1} - \dim L'_i/\widetilde{L}'_i)}.$$

We finally get that

$$\Xi(A_\lambda(\mathbf{0}, r))(\mathbf{L}', \widetilde{\mathbf{L}}', \mathbf{L}'', \widetilde{\mathbf{L}}'') = q^{a/2},$$

where

$$a = \sum_{1\leqslant i\leqslant n} \big(-\lambda_i(\dim L_i/L_{i-1} - \lambda_i) + 2\dim L''_i/\widetilde{L}''_i(\dim L'_i/L'_{i-1} - \dim L'_i/\widetilde{L}'_i)\big).$$

On the other hand, by (5.5.0.1) and (5.5.0.2),

$$\begin{aligned}
&\Psi_{\mu,\nu}(\mathbf{L}', \widetilde{\mathbf{L}}', \mathbf{L}'', \widetilde{\mathbf{L}}'') \neq 0\\
\Longleftrightarrow\ &\widetilde{\mathbf{L}}' \subseteq \mathbf{L}',\ \widetilde{\mathbf{L}}'' \subseteq \mathbf{L}'',\ \mathbf{L}'/\widetilde{\mathbf{L}}' \cong S_\mu,\ \mathbf{L}''/\widetilde{\mathbf{L}}'' \cong S_\nu\\
\Longleftrightarrow\ &\widetilde{L}'_i \subseteq L'_i,\ \widetilde{L}''_i \subseteq L''_i,\ L'_i/\widetilde{L}'_i \cong \mathbb{F}^{\mu_i},\ L''_i/\widetilde{L}''_i \cong \mathbb{F}^{\nu_i},\ \forall i \in \mathbb{Z}.
\end{aligned}$$

Moreover, if this is the case, then

$$\begin{aligned}
&\Psi_{\mu,\nu}(\mathbf{L}', \widetilde{\mathbf{L}}', \mathbf{L}'', \widetilde{\mathbf{L}}'')\\
=&A_\mu(\mathbf{0}, r')(\mathbf{L}', \widetilde{\mathbf{L}}')\mathrm{O}(\nu, r')(\widetilde{\mathbf{L}}', \widetilde{\mathbf{L}}')A_\nu(\mathbf{0}, r'')(\mathbf{L}'', \widetilde{\mathbf{L}}'')\mathrm{O}(-\mu, r'')(\widetilde{\mathbf{L}}'', \widetilde{\mathbf{L}}'')\\
=&q^{b/2},
\end{aligned}$$

where

$$b = -c(\mathbf{L}', \widetilde{\mathbf{L}}') - c(\mathbf{L}'', \widetilde{\mathbf{L}}'') + \sum_{1\leqslant i\leqslant n} (\nu_i \dim \widetilde{L}'_i/\widetilde{L}'_{i-1} - \mu_i \dim \widetilde{L}''_i/\widetilde{L}''_{i-1}).$$

Since

$$\begin{aligned}
c(\mathbf{L}', \widetilde{\mathbf{L}}') &= \textstyle\sum_{1\leqslant i\leqslant n} \mu_i(\dim L'_i/L'_{i-1} - \mu_i),\\
c(\mathbf{L}'', \widetilde{\mathbf{L}}'') &= \textstyle\sum_{1\leqslant i\leqslant n} \nu_i(\dim L''_i/L''_{i-1} - \nu_i),\\
\dim \widetilde{L}'_i/\widetilde{L}'_{i-1} &= \dim L'_i/L'_{i-1} + \mu_{i-1} - \mu_i, \text{ and}\\
\dim \widetilde{L}''_i/\widetilde{L}''_{i-1} &= \dim L''_i/L''_{i-1} + \nu_{i-1} - \nu_i,
\end{aligned}$$

it follows that

$$\begin{aligned}
b = &\sum_{1\leqslant i\leqslant n} \big(-\mu_i \dim L'_i/L'_{i-1} - \nu_i \dim L''_i/L''_{i-1} + \mu_i^2 + \nu_i^2\\
&\qquad + \nu_i L'_i/L'_{i-1} - \mu_i \dim L''_i/L''_{i-1} + \nu_i\mu_{i-1} - \mu_i\nu_{i-1}\big)\\
=&a + \tau\mu \centerdot \nu - \mu \centerdot \tau\nu.
\end{aligned}$$

Note that $\lambda_i = \mu_i + \nu_i$ and $\dim L_i/L_{i-1} = \dim L'_i/L'_{i-1} + \dim L''_i/L''_{i-1}$, for all $i \in \mathbb{Z}$. Therefore,

$$\Xi(A_\lambda(\mathbf{0}, r))(\mathbf{L}', \widetilde{\mathbf{L}}', \mathbf{L}'', \widetilde{\mathbf{L}}'') = \sum_{\mu+\nu=\lambda} q^{(\mu\centerdot\tau\nu - \tau\mu\centerdot\nu)/2}\Psi_{\mu,\nu}(\mathbf{L}', \widetilde{\mathbf{L}}', \mathbf{L}'', \widetilde{\mathbf{L}}''),$$

which finishes the proof. □

Corollary 5.5.2. *Suppose $r \geqslant n$. If $\lambda = (\lambda_i) \in \mathbb{N}^n_\Delta$ satisfies $\lambda_i \geqslant 1$ for all $i \in \mathbb{Z}$, then*

$$\phi_{r,r-n}(A_\lambda(\mathbf{0}, r)) = A_\lambda(\mathbf{0}, r-n) + A_{\lambda-\delta}(\mathbf{0}, r-n) \cdot 0(\delta, r-n)$$

and

$$\phi_{r,r-n}({}^tA_\lambda(\mathbf{0}, r)) = {}^tA_\lambda(\mathbf{0}, r-n) + {}^tA_{\lambda-\delta}(\mathbf{0}, r-n) \cdot 0(\delta, r-n).$$

Proof. Write $r' = r - n$. Fix a finite field $\mathbb{F}$ with q elements. Then applying the above proposition gives

$$\Xi(A_\lambda(\mathbf{0}, r)) = \sum_{\substack{\mu,\nu\in\mathbb{N}^n_\Delta \\ \mu+\nu=\lambda}} q^{(\mu\centerdot\tau\nu - \tau\mu\centerdot\nu)/2} A_\mu(\mathbf{0}, r')0(\nu, r') \otimes A_\nu(\mathbf{0}, n)0(-\mu, n).$$

If $\sigma(\nu) > n$, then $A_\nu(\mathbf{0}, n) = 0$. If $1 \leqslant \sigma(\nu) \leqslant n$ and $\nu \neq \delta$, then $\chi(A_\nu(\mathbf{0}, n)) = 0$. Thus,

$$\begin{aligned}
&\phi^q_{r,r'}(A_\lambda(\mathbf{0}, r)) = (\xi \otimes \chi)\Xi(A_\lambda(\mathbf{0}, r)) \\
=&\xi(A_\lambda(\mathbf{0}, r'))\chi(0(-\lambda, n)) + \xi(A_{\lambda-\delta}(\mathbf{0}, r')0(\delta, r'))\chi(A_\delta(\mathbf{0}, n)0(-\lambda+\delta, n)) \\
=&A_\lambda(\mathbf{0}, r') + A_{\lambda-\delta}(\mathbf{0}, r')0(\delta, r').
\end{aligned}$$

The second formula can be proved similarly. □

By the above corollary, if $\lambda = (\lambda_i) \in \mathbb{N}^n_\Delta$ satisfies $\lambda_i \geqslant 1$, for all $i \in \mathbb{Z}$ and $\sigma(\lambda) \leqslant r$, then

$$\phi_{r,r-n}\xi_r(A_\lambda(\mathbf{0}, r)) \neq \xi_{r-n}(A_\lambda(\mathbf{0}, r)) \text{ and}$$
$$\phi_{r,r-n}\xi_r({}^tA_\lambda(\mathbf{0}, r)) \neq \xi_{r-n}({}^tA_\lambda(\mathbf{0}, r)).$$

This shows that the transfer maps $\phi_{r,r-n}$ are not compatible with the homomorphisms $\xi_r : {}'\mathfrak{D}_\Delta(n) \to \boldsymbol{\mathcal{S}}_\Delta(n, r)$.

6
The classical ($v = 1$) case

Let $\mathcal{U}(\widehat{\mathfrak{gl}}_n)$ be the universal enveloping algebra of the loop algebra $\widehat{\mathfrak{gl}}_n(\mathbb{Q})$ as mentioned in §1.1. We will establish, on the one hand, a surjective homomorphism from $\mathcal{U}(\widehat{\mathfrak{gl}}_n)$ to the affine Schur algebra $\mathcal{S}_\Delta(n,r)_\mathbb{Q}$ via the natural action of $\widehat{\mathfrak{gl}}_n(\mathbb{Q})$ on the $\mathbb{Q}$-space $\Omega_\mathbb{Q}$, and prove, on the other hand, that $\mathcal{U}(\widehat{\mathfrak{gl}}_n)$ is isomorphic to the specialization $\overline{\mathfrak{D}_\Delta(n)_\mathbb{Q}}$ of the integral double Ringel–Hall algebra $\mathfrak{D}_\Delta(n)$ at $v = 1$ and $K_i = 1$. In this way, we obtain a surjective algebra homomorphism $\eta_r : \overline{\mathfrak{D}_\Delta(n)_\mathbb{Q}} \to \mathcal{S}_\Delta(n,r)_\mathbb{Q}$ which is regarded as the classical ($v = 1$) version of the surjective homomorphism $\xi_r : \mathfrak{D}_\Delta(n) \to \mathcal{S}_\Delta(n,r)$. We then prove the classical version of Conjecture 5.4.2 via the homomorphisms η_r. A crucial step to establish the conjecture in this case is the extension of the multiplication formulas given in §3.4 to formulas between homogeneous indecomposable generators and arbitrary BLM basis elements. This is done in §6.2. The conjecture in the classical case is proved in §6.3.

In order to distinguish the specializations at non-roots-of-unity $v = z \in \mathbb{C}$ considered in previous chapters from the specialization at $v = 1$, we will particularly consider the specialization $\mathcal{Z} \to \mathbb{Q}$ by sending v to 1 throughout the chapter. Thus, $\mathcal{H}_\Delta(r)_\mathbb{Q}$ identifies the group algebra $\mathbb{Q}\mathfrak{S}_{\Delta,r}$, $\Omega_\mathbb{Q}$ is the $\mathbb{Q}$-space with basis $\{\omega_i\}_{i\in\mathbb{Z}}$, and $\mathcal{S}_\Delta(n,r)_\mathbb{Q}$ identifies the classical affine Schur algebra. In other words, $\mathcal{S}_\Delta(n,r)_\mathbb{Q} \cong \mathrm{End}_{\mathbb{Q}\mathfrak{S}_{\Delta,r}}(\Omega_\mathbb{Q}^{\otimes r})$.

6.1. The universal enveloping algebra $\mathcal{U}(\widehat{\mathfrak{gl}}_n)$

Recall from §1.1 that the loop algebra $\widehat{\mathfrak{gl}}_n(\mathbb{Q})$ has a basis $\{E^\Delta_{i,j}\}_{1\leqslant i\leqslant n, j\in\mathbb{Z}}$. Thus, the natural action of these basis elements on $\Omega_\mathbb{Q}$ defined by

$$E^\Delta_{i,j}\omega_k = \begin{cases} \omega_{i+tn}, & \text{if } k = j + tn; \\ 0, & \text{otherwise} \end{cases} \tag{6.1.0.1}$$

gives rise to a $\mathcal{U}(\widehat{\mathfrak{gl}}_n)$-module structure on $\Omega_{\mathbb{Q}}$ and, hence, on the r-fold tensor product $\Omega_{\mathbb{Q}}^{\otimes r}$. Thus, we obtain an algebra homomorphism

$$\eta_r : \mathcal{U}(\widehat{\mathfrak{gl}}_n) \longrightarrow \mathcal{S}_\Delta(n, r)_{\mathbb{Q}}. \tag{6.1.0.2}$$

The fact that η_r is surjective has already been established in [**79**, Th. 6.6(i)] by a coordinate algebra approach. We now present a different proof by identifying the above action with the Hall algebra action as discussed in §§3.5–3.6. Thus, a more explicit description for η_r is obtained.

First, we interpret $\mathcal{U}(\widehat{\mathfrak{gl}}_n)$ as the specialization $\overline{\mathfrak{D}_\Delta(n)_{\mathbb{Q}}}$ of the integral form $\mathfrak{D}_\Delta(n)$ given in Definition 2.4.4 at $v = 1$ and all $K_i = 1$. Let $U_\Delta(n)$ be the $\mathcal{Z}$-subalgebra of $\mathfrak{D}_\Delta(n)$ generated by $K_i^{\pm 1}$, $\left[{K_i;0 \atop t} \right]$, $(u_i^+)^{(m)}$ and $(u_i^-)^{(m)}$, for $1 \leqslant i \leqslant n$ and $t, m \geqslant 1$. Then $\mathfrak{D}_\Delta(n) \cong U_\Delta(n) \otimes_{\mathcal{Z}} \mathcal{Z}[\mathsf{z}_m^{\pm}]_{m \geqslant 1}$; see (2.4.4.2).

By specializing v to 1, $\mathbb{Q}$ is regarded as a $\mathcal{Z}$-module. Consider the $\mathbb{Q}$-algebra

$$\begin{aligned}\overline{\mathfrak{D}_\Delta(n)_{\mathbb{Q}}} &:= \mathfrak{D}_\Delta(n)_{\mathbb{Q}} / \langle K_i - 1 \mid 1 \leqslant i \leqslant n \rangle \\ &\cong (\mathfrak{D}_\Delta(n) \otimes_{\mathcal{Z}} \mathbb{Q}[v, v^{-1}]) / \langle v - 1, K_i - 1 \mid 1 \leqslant i \leqslant n \rangle.\end{aligned} \tag{6.1.0.3}$$

If $x \in \mathfrak{D}_\Delta(n)$, then $\bar{x}$ denotes the image in $\overline{\mathfrak{D}_\Delta(n)_{\mathbb{Q}}}$. Since $[m]_{v=1} = m \geqslant 1$, $\overline{\mathfrak{D}_\Delta(n)_{\mathbb{Q}}}$ is generated by $\overline{u_i^{\pm}}$ $(1 \leqslant i \leqslant n)$, $\overline{\mathsf{z}_m^{\pm}}$ $(m \geqslant 1)$, $\overline{\left[{K_i;0 \atop t} \right]}$ $(t \geqslant 0)$. Note that $\overline{\mathfrak{D}_\Delta(n)_{\mathbb{Q}}}$ inherits a Hopf algebra structure from $\mathfrak{D}_\Delta(n)_{\mathbb{Q}}$.

Let $\widehat{\mathfrak{sl}}_n(\mathbb{Q}) = \mathfrak{sl}_n(\mathbb{Q}) \otimes \mathbb{Q}[t, t^{-1}]$ and set $\widehat{\mathfrak{gl}}_n^{\mathrm{L}} = \widehat{\mathfrak{gl}}_n^{\mathrm{L}}(\mathbb{Q}) = \widehat{\mathfrak{sl}}_n(\mathbb{Q}) \oplus \mathbb{Q}E^\Delta$, where $E^\Delta = \sum_{1 \leqslant i \leqslant n} E_{i,i}^\Delta$ is in the center of $\widehat{\mathfrak{gl}}_n^{\mathrm{L}}(\mathbb{Q})$. Then the set

$$\begin{aligned} X := &\{E_{i,j+ln}^\Delta \mid 1 \leqslant i, j \leqslant n, i \neq j, l \in \mathbb{Z}\} \cup \{E_{i,i}^\Delta \mid 1 \leqslant i \leqslant n\} \\ &\cup \{E_{i,i+ln}^\Delta - E_{i+1,i+1+ln}^\Delta \mid 1 \leqslant i \leqslant n-1, l \in \mathbb{Z}, l \neq 0\}\end{aligned}$$

forms a $\mathbb{Q}$-basis for $\widehat{\mathfrak{gl}}_n^{\mathrm{L}}$. Let

$$\mathsf{z}_m = \sum_{1 \leqslant h \leqslant n} E_{h,h+mn}^\Delta \quad \text{for } m \neq 0.$$

Then $X \cup \{\mathsf{z}_m \mid m \in \mathbb{Z}, m \neq 0\}$ forms a $\mathbb{Q}$-basis for $\widehat{\mathfrak{gl}}_n = \widehat{\mathfrak{gl}}_n(\mathbb{Q})$. By the PBW theorem for universal enveloping algebras,

$$\mathcal{U}(\widehat{\mathfrak{gl}}_n) \cong \mathcal{U}(\widehat{\mathfrak{gl}}_n^{\mathrm{L}}) \otimes \mathbb{Q}[\mathsf{z}_m]_{m \in \mathbb{Z} \setminus \{0\}}, \tag{6.1.0.4}$$

where $\mathcal{U}(\widehat{\mathfrak{gl}}_n^{\mathrm{L}})$ is the enveloping algebra of $\widehat{\mathfrak{gl}}_n^{\mathrm{L}}$. Note that the z_m are central elements in $\mathcal{U}(\widehat{\mathfrak{gl}}_n)$.

Let $\binom{E_{i,i}^\Delta}{t} := \frac{E_{i,i}^\Delta (E_{i,i}^\Delta - 1) \cdots (E_{i,i}^\Delta - t + 1)}{t!} \in \mathcal{U}(\widehat{\mathfrak{gl}}_n)$. Let $\mathfrak{D}_\Delta(n)^{\pm} = U_\Delta(n)^{\pm} \otimes \mathcal{Z}[\mathsf{z}_m^{\pm}]_{m \geqslant 1}$ be the $\pm$-part of $\mathfrak{D}_\Delta(n)$ (see (2.4.4.1)) and let $\mathfrak{D}_\Delta(n)_{\mathbb{Q}}^{\pm} = \mathfrak{D}_\Delta(n)^{\pm} \otimes \mathbb{Q}$.

Theorem 6.1.1. *There is a Hopf algebra isomorphism* $\phi : \overline{\mathfrak{D}_\Delta(n)_\mathbb{Q}} \to \mathcal{U}(\widehat{\mathfrak{gl}}_n)$ *defined by sending* $\overline{\left[{K_i;0 \atop t} \right]}$ *to* $\binom{E^\Delta_{i,i}}{t}$, $\overline{u_i^+}$ *to* $E^\Delta_{i,i+1}$, $\overline{u_i^-}$ *to* $E^\Delta_{i+1,i}$, *and* $\overline{\mathsf{z}_s^\pm}$ *to* $\mathsf{z}_{\pm s}$ *for* $1 \leqslant i \leqslant n$ *and* $s, t \geqslant 1$. *In particular,* $\mathfrak{D}_\Delta(n)^\pm_\mathbb{Q} \cong \overline{\mathfrak{D}_\Delta(n)^\pm_\mathbb{Q}}$, *the homomorphic image of* $\mathfrak{D}_\Delta(n)^\pm$ *in* $\overline{\mathfrak{D}_\Delta(n)_\mathbb{Q}}$.

Thus, the bar on the elements of $\overline{\mathfrak{D}_\Delta(n)^\pm_\mathbb{Q}}$ can be dropped.

Proof. By [**52**, 6.7] and [**54**], specializing v to 1 induces an algebra isomorphism

$$\phi_1 : \overline{U_\Delta(n)_\mathbb{Q}} := U_\Delta(n)_\mathbb{Q}/\langle K_i - 1 \mid 1 \leqslant i \leqslant n\rangle \longrightarrow \mathcal{U}(\widehat{\mathfrak{gl}}^{\mathrm{L}}_n)$$

defined by taking $\overline{\left[{K_i;0 \atop t} \right]} \mapsto \binom{E^\Delta_{i,i}}{t}$, $\overline{u_i^+} \mapsto E^\Delta_{i,i+1}$, $\overline{u_i^-} \mapsto E^\Delta_{i+1,i}$, for all $1 \leqslant i \leqslant n$ and $t \geqslant 1$.

Since $\mathfrak{D}_\Delta(n) \cong U_\Delta(n) \otimes \mathcal{Z}[\mathsf{z}_m^\pm]_{m\geqslant 1}$ is an algebra isomorphism, it follows that $\overline{\mathfrak{D}_\Delta(n)_\mathbb{Q}} \cong \overline{U_\Delta(n)_\mathbb{Q}} \otimes \mathbb{Q}[\mathsf{z}_m^\pm]_{m\geqslant 1}$. Thus, the isomorphism ϕ_1 together with (6.1.0.4) gives the required algebra isomorphism

$$\phi = \phi_1 \otimes \phi_2 : \overline{\mathfrak{D}_\Delta(n)_\mathbb{Q}} \longrightarrow \mathcal{U}(\widehat{\mathfrak{gl}}_n),$$

where $\phi_2 : \mathbb{Q}[\mathsf{z}_m^\pm]_{m\geqslant 1} \to \mathbb{Q}[\mathsf{z}_m]_{m\in\mathbb{Z}\setminus\{0\}}$ is the isomorphism sending $\mathsf{z}_m^\pm$ to $\mathsf{z}_{\pm m}$ for all $m \geqslant 1$.

The compatibility of Hopf structures is clear since $\overline{u_i^\pm}$ and $\mathsf{z}_m^\pm$ are primitive elements.

Finally, choose a $\mathcal{Z}$-basis $\{y_j^\pm \mid j \in J\}$ of $U_\Delta(n)^\pm$. Then

$$Y^\pm = \left\{ y_j^\pm \prod_{i=1}^{k} (\mathsf{z}_i^\pm)^{a_i} \;\middle|\; j \in J,\ k \geqslant 1,\ a_i \in \mathbb{N},\ \forall i \right\} \tag{6.1.1.1}$$

is a basis for $\mathfrak{D}_\Delta(n)^\pm$. So base change gives a basis $Y^\pm_\mathbb{Q} = \{y = y \otimes 1\}_{y\in Y^\pm}$ for $\mathfrak{D}_\Delta(n)^\pm_\mathbb{Q}$. Let $\overline{Y^\pm_\mathbb{Q}}$ be the image of $Y^\pm_\mathbb{Q}$ in $\overline{\mathfrak{D}_\Delta(n)^\pm_\mathbb{Q}}$. Since $\phi(\overline{Y^\pm_\mathbb{Q}})$, as part of a basis for $\mathcal{U}(\widehat{\mathfrak{gl}}_n)$, is linearly independent, it follows that the bar map $\mathfrak{D}_\Delta(n)^\pm_\mathbb{Q} \to \overline{\mathfrak{D}_\Delta(n)^\pm_\mathbb{Q}}$ is injective, proving the last assertion. □

The integral form $\mathfrak{D}_\Delta(n)$ given in (2.4.4.1) acts on the free $\mathcal{Z}$-module Ω, thanks to (3.5.4.1). This induces an action of $\mathfrak{D}_\Delta(n)_\mathbb{Q}$ on $\Omega_\mathbb{Q}$ with K_i acting trivially and, hence, an action of $\overline{\mathfrak{D}_\Delta(n)_\mathbb{Q}}$ on $\Omega_\mathbb{Q}$. Thus, we obtain an action of $\overline{\mathfrak{D}_\Delta(n)_\mathbb{Q}}$ on $\Omega_\mathbb{Q}^{\otimes r}$ which gives an algebra homomorphism

$$\bar{\xi}_r : \overline{\mathfrak{D}_\Delta(n)_\mathbb{Q}} \longrightarrow \mathcal{S}_\Delta(n, r)_\mathbb{Q}.$$

This homomorphism can also be regarded as the reduction modulo $v = 1$ and $K_i = 1$ of the restriction $\xi_r|_{\mathfrak{D}_\Delta(n)} : \mathfrak{D}_\Delta(n) \to \mathcal{S}_\Delta(n, r)$ of the map in (3.5.5.4).

Though the map $\xi_r|_{\mathfrak{D}_\Delta(n)}$ is not surjective in general by Example 5.3.8, we will see that the map $\overline{\xi}_r$ is surjective. Assuming [**79**, Th. 6.6(i)], this can be seen from the following compatibility condition.

Proposition 6.1.2. *For any $r \geqslant 0$, we have $\eta_r \circ \phi = \overline{\xi}_r$.*

Proof. The action of u_i^+ (resp., u_i^-) on $\Omega_{\mathbb{Q}}^{\otimes r}$ is the same as the action of $E^{\Delta}_{i,i+1}$ (resp., $E^{\Delta}_{i+1,i}$) for all $i \in I$ and, by (3.5.4.1), the action of $\mathsf{z}_m^{\pm}$ is the same as the action of $\mathsf{z}_{\pm m}$ for all $m \geqslant 1$. This proves the compatibility condition. □

To establish the surjection directly, we want to understand the structure of $\overline{\mathfrak{D}_\Delta(n)_{\mathbb{Q}}}$. Recall that the idea used in the proof of Theorem 3.8.1 is to apply Proposition 3.7.4(1) to get the surjection on every triangular part of the affine Schur algebra. Thus, if we can prove that $\overline{\mathfrak{D}_\Delta(n)_{\mathbb{Q}}}$ contains Ringel–Hall algebras $\mathfrak{H}_\Delta(n)^{\pm}_{\mathbb{Q}} := \mathfrak{H}_\Delta(n)^{\pm} \otimes \mathbb{Q}$ (at $v = 1$), then the surjection follows. Note that this fact cannot be seen directly from the definition of $\overline{\mathfrak{D}_\Delta(n)_{\mathbb{Q}}}$.

Lemma 6.1.3. ([**59**, Cor. 4.1.1]) *For $1 \leqslant k \leqslant t$, let $A_k = (a^{(k)}_{i,j})$, $B = (b_{i,j}) \in \Theta^+_\Delta(n)$.*

(1) *If $\sigma(B) \geqslant \sum_{k=1}^t \sigma(A_k)$ and $B \neq \sum_{k=1}^t A_k$, then $(v^2 - 1) | \varphi^B_{A_1, A_2, \ldots, A_t}$.*
(2) *If $B = \sum_{k=1}^t A_k$, then*

$$(v^2 - 1) \left| \left(\varphi^B_{A_1, A_2, \ldots, A_t} - \prod_{\substack{1 \leqslant i \leqslant n,\, j \in \mathbb{Z} \\ i < j}} \frac{b_{i,j}!}{a^{(1)}_{i,j}! a^{(2)}_{i,j}! \cdots a^{(t)}_{i,j}!} \right) \right. .$$

Proposition 6.1.4. (1) *The Ringel–Hall algebra $\mathfrak{H}_\Delta(n)^{\pm}_{\mathbb{Q}}$ at $v = 1$ is generated by $u^{\pm}_{E^{\Delta}_{i,j}}$ for all $i < j$.*

(2) *If $\mathcal{U}(\widehat{\mathfrak{gl}}_n)^+$ (resp., $\mathcal{U}(\widehat{\mathfrak{gl}}_n)^-$) denotes the subalgebra of $\mathcal{U}(\widehat{\mathfrak{gl}}_n)$ generated by $E^{\Delta}_{i,j}$ (resp., $E^{\Delta}_{j,i}$) for all $i < j$, then there are algebra isomorphisms $f^{\pm}$: $\mathcal{U}(\widehat{\mathfrak{gl}}_n)^{\pm} \to \mathfrak{H}_\Delta(n)^{\pm}_{\mathbb{Q}}$ taking $E^{\Delta}_{i,j} \mapsto u^+_{E^{\Delta}_{i,j}}$ (resp., $E^{\Delta}_{j,i} \mapsto u^-_{E^{\Delta}_{i,j}}$), for all $i < j$.*

(3) *The Ringel–Hall algebras $\mathfrak{H}_\Delta(n)^{\pm}_{\mathbb{Q}}$ are subalgebras of $\overline{\mathfrak{D}_\Delta(n)_{\mathbb{Q}}}$ so that*

$$\overline{\mathfrak{D}_\Delta(n)_{\mathbb{Q}}} = \mathfrak{H}_\Delta(n)^+_{\mathbb{Q}} \otimes (0\text{-part}) \otimes \mathfrak{H}_\Delta(n)^-_{\mathbb{Q}}.$$

(4) *We have $f^{\pm} = \phi^{-1}|_{\mathcal{U}(\widehat{\mathfrak{gl}}_n)^{\pm}} : \mathcal{U}(\widehat{\mathfrak{gl}}_n)^{\pm} \to \mathfrak{H}_\Delta(n)^{\pm}_{\mathbb{Q}}$.*

Proof. It suffices to prove the + case.

(1) Let $\mathfrak{H}$ be the subalgebra generated by all $u_{E^{\Delta}_{i,j}}$. By induction on $\sigma(B)$ and applying Lemma 6.1.3, we can easily prove that every $u_B \in \mathfrak{H}$, for all $B \in \Theta^+_\Delta(n)$. Hence, $\mathfrak{H}_\Delta(n)_{\mathbb{Q}} \subseteq \mathfrak{H}$.

(2) Let $\widehat{\mathfrak{gl}}_n^+$ be the Lie subalgebra of $\widehat{\mathfrak{gl}}_n$ generated by $E^{\Delta}_{i,j}$ $(i < j)$. Then $\mathcal{U}(\widehat{\mathfrak{gl}}_n)^+$ is isomorphic to the enveloping algebra of $\widehat{\mathfrak{gl}}_n^+$ and in $\mathcal{U}(\widehat{\mathfrak{gl}}_n)^+$,

$$[E^{\Delta}_{i,j}, E^{\Delta}_{k,l}] = \delta_{\bar{j},\bar{k}} E^{\Delta}_{i,l+j-k} - \delta_{\bar{l},\bar{i}} E^{\Delta}_{k,j+l-i}, \quad \text{for } i, j, k, l \in \mathbb{Z}.$$

On the other hand, by Lemma 6.1.3, for $i < j$ and $k < l$, we have in $\mathfrak{H}_{\Delta}(n)^+_{\mathbb{Q}}$,

$$u^+_{E^{\Delta}_{i,j}} u^+_{E^{\Delta}_{k,l}} - u^+_{E^{\Delta}_{k,l}} u^+_{E^{\Delta}_{i,j}} = \delta_{\bar{j},\bar{k}} u^+_{E^{\Delta}_{i,l+j-k}} - \delta_{\bar{l},\bar{i}} u^+_{E^{\Delta}_{k,j+l-i}}. \tag{6.1.4.1}$$

Thus, there is an algebra homomorphism $f^+ : \mathcal{U}(\widehat{\mathfrak{gl}}_n)^+ \to \mathfrak{H}_{\Delta}(n)^+_{\mathbb{Q}}$ such that $f^+(E^{\Delta}_{i,j}) = u^+_{E^{\Delta}_{i,j}}$ for $i < j$.

Let $\mathcal{L}^+ = \{(i, j) \mid 1 \leqslant i \leqslant n,\ j \in \mathbb{Z},\ i < j\}$. By Lemma 6.1.3 again, we see, for any $A = \sum_{(i,j)\in\mathcal{L}^+} a_{i,j} E^{\Delta}_{i,j} \in \Theta^+_{\Delta}(n)$,

$$\prod_{(i,j)\in\mathcal{L}^+} (u^+_{E^{\Delta}_{i,j}})^{a_{i,j}} = \Big(\prod_{(i,j)\in\mathcal{L}^+} a_{i,j}!\Big) u^+_A + \sum_{B\in\Theta^+_{\Delta}(n),\, \sigma(B)<\sigma(A)} \varphi_B(1) u^+_B,$$

where $\varphi_B \in \mathcal{Z}$ and the products are taken with respect to a fixed total order on $\mathcal{L}^+$. Hence, the set

$$\Big\{ \prod_{(i,j)\in\mathcal{L}^+} (u^+_{E^{\Delta}_{i,j}})^{a_{i,j}} \,\Big|\, A = (a_{i,j}) \in \Theta^+_{\Delta}(n) \Big\}$$

is a $\mathbb{Q}$-basis of $\mathfrak{H}_{\Delta}(n)^+_{\mathbb{Q}}$. Thus, f sends a PBW-basis of $\mathcal{U}(\widehat{\mathfrak{gl}}_n)^+$ to a basis of $\mathfrak{H}_{\Delta}(n)^+_{\mathbb{Q}}$. Hence, f^+ is an isomorphism.

(3) By Remark 2.4.6 (see also Corollary 3.7.5), $\mathfrak{D}_{\Delta}(n)^+$ is a subalgebra of $\mathfrak{H}_{\Delta}(n)^+$. Thus, base change induces an algebra homomorphism $\iota_{\mathbb{Q}}$: $\mathfrak{D}_{\Delta}(n)^+_{\mathbb{Q}} \to \mathfrak{H}_{\Delta}(n)^+_{\mathbb{Q}}$, sending the basis $Y^+_{\mathbb{Q}}$ given in (6.1.1.1) to its image $Y^{+'}_{\mathbb{Q}}$. By (2.2.1.2), $\mathsf{z}^+_m = \sum_{i=1}^n u^+_{E^{\Delta}_{i,i+mn}}$ in $\mathfrak{H}_{\Delta}(n)^+_{\mathbb{Q}}$ equals $f^+(\mathsf{z}_m)$. Hence, $(f^+)^{-1}(Y^{+'}_{\mathbb{Q}})$ is a basis for $\mathcal{U}(\widehat{\mathfrak{gl}}_n)^+ = \mathcal{U}(\widehat{\mathfrak{sl}}_n)^+ \otimes \mathbb{Q}[\mathsf{z}_m]_{m\geqslant 1}$. This proves that $\iota_{\mathbb{Q}}$ is an isomorphism, and consequently, by Theorem 6.1.1, $\overline{\mathfrak{D}_{\Delta}(n)^+_{\mathbb{Q}}} \cong \mathfrak{H}_{\Delta}(n)^+_{\mathbb{Q}}$.

(4) Now, by (2.2.1.2) again, $\mathsf{z}^+_m = \sum_{l=1}^n u^+_{E^{\Delta}_{l,l+mn}}$ in $\overline{\mathfrak{D}_{\Delta}(n)_{\mathbb{Q}}}$. Hence, for $m \geqslant 1$,

$$(f^+)^{-1}(\mathsf{z}^+_m) = \mathsf{z}_m = \phi(\mathsf{z}^+_m).$$

Also, $(f^+)^{-1}(u^+_i) = \phi(u^+_i)$. Therefore, $(f^+)^{-1} = \phi|_{\mathfrak{H}_{\Delta}(n)^+_{\mathbb{Q}}}$. □

Let $\mathfrak{k}_i = 0(\boldsymbol{e}^{\Delta}_i, r) \in \mathcal{S}_{\Delta}(n, r)$, $1 \leqslant i \leqslant n$, be as considered in Proposition 3.7.4, and let $\mathcal{L} = \{(i, j) \mid 1 \leqslant i \leqslant n,\ j \in \mathbb{Z}\}$. Then the set $\{E^{\Delta}_{i,j}\}_{(i,j)\in\mathcal{L}}$

forms a basis for $\widehat{\mathfrak{gl}}_n(\mathbb{Q})$ and, hence, generates $\mathcal{U}(\widehat{\mathfrak{gl}}_n)$. We have almost proved the following result, which is independent of [**79**, Th. 6.6(i)]. The subscripts 1 indicate the elements in $\mathcal{S}_\Delta(n,r)_\mathbb{Q}$ obtained from elements in $\mathcal{S}_\Delta(n,r)$ by specializing v to 1.

Theorem 6.1.5. *The map η_r (or $\bar{\xi}_r$) is surjective. Moreover, we have*

$$\eta_r(E^\Delta_{i,i}) = \begin{bmatrix} \mathfrak{k}_i;0 \\ 1 \end{bmatrix}_1 \quad \textit{and} \quad \eta_r(E^\Delta_{j,k}) = E^\Delta_{j,k}(\mathbf{0},r)_1, \quad \textit{for all } i,\ j \neq k.$$

Proof. By Theorem 3.6.3, we have $\bar{\xi}_r(\left[\begin{smallmatrix} K_i;0 \\ t \end{smallmatrix}\right]) = \left[\begin{smallmatrix} \mathfrak{k}_i;0 \\ t \end{smallmatrix}\right]_1$, $\bar{\xi}_r(u^+_A) = A(\mathbf{0},r)_1$, and $\bar{\xi}_r(u^-_A) = {}^tA(\mathbf{0},r)_1$. Applying the isomorphism ϕ given in Theorem 6.1.1 yields the corresponding formulas for η_r. Now, by Proposition 6.1.2, the surjectivity of η_r follows from the surjectivity of $\bar{\xi}_r$ since $\bar{\xi}_r$ maps every triangular part of $\overline{\mathfrak{D}_\Delta(n)_\mathbb{Q}}$ onto the corresponding part of $\mathcal{S}_\Delta(n,r)_\mathbb{Q} = \mathcal{S}_\Delta(n,r)^+_\mathbb{Q}\mathcal{S}_\Delta(n,r)^0_\mathbb{Q}\mathcal{S}_\Delta(n,r)^-_\mathbb{Q}$ as given in Theorem 3.7.7. □

Remark 6.1.6. We now give a direct proof for the formula $\eta_r\left(\begin{smallmatrix} E^\Delta_{i,i} \\ t \end{smallmatrix}\right) = \left[\begin{smallmatrix} \mathfrak{k}_i;0 \\ t \end{smallmatrix}\right]_1$. For $\lambda \in \Lambda_\Delta(n,r)$, let $\mathbf{i}_\lambda$ be defined as in the proof of Proposition 3.3.1. Since, for all $w \in \mathfrak{S}_{\Delta,r}$, $E^\Delta_{i,i}(\omega_{\mathbf{i}_\lambda} w) = (E^\Delta_{i,i}\omega_{\mathbf{i}_\lambda})w = \lambda_i \omega_{\mathbf{i}_\lambda} w = (\left[\begin{smallmatrix} \mathfrak{k}_i;0 \\ 1 \end{smallmatrix}\right]_1 \omega_{\mathbf{i}_\lambda})w$ (cf. (6.1.0.1)), we obtain $\eta_r(E^\Delta_{i,i}) = \left[\begin{smallmatrix} \mathfrak{k}_i;0 \\ 1 \end{smallmatrix}\right]_1$. Hence,

$$\begin{aligned} \eta_r\begin{pmatrix} E^\Delta_{i,i} \\ t \end{pmatrix} &= \frac{1}{t!}\prod_{s=0}^{t-1}\Big(\sum_{\lambda\in\Lambda_\Delta(n,r)} (\lambda_i - s)[\mathrm{diag}(\lambda)]_1\Big) \\ &= \sum_{\lambda\in\Lambda_\Delta(n,r)} \binom{\lambda_i}{t}[\mathrm{diag}(\lambda)]_1 = \begin{bmatrix} \mathfrak{k}_i;0 \\ t \end{bmatrix}_1. \end{aligned}$$

The map $\bar{\xi}_r : \overline{\mathfrak{D}_\Delta(n)_\mathbb{Q}} \to \mathcal{S}_\Delta(n,r)_\mathbb{Q}$ is the classical ($v = 1$) version of the map $\xi_r : \mathfrak{D}_\Delta(n) \to \boldsymbol{\mathcal{S}}_\Delta(n,r)$. In the next two sections, we will describe explicitly the image of the map

$$\bar{\xi} = \prod_{r\geqslant 0} \bar{\xi}_r : \overline{\mathfrak{D}_\Delta(n)_\mathbb{Q}} \longrightarrow \widehat{\mathcal{K}}_\Delta(n)_\mathbb{Q},$$

or equivalently, the map

$$\eta = \prod_{r\geqslant 0} \eta_r : \mathcal{U}(\widehat{\mathfrak{gl}}_n) \longrightarrow \widehat{\mathcal{K}}_\Delta(n)_\mathbb{Q}. \tag{6.1.6.1}$$

Here we have identified the algebra $\widehat{\mathcal{K}}_\Delta(n)_\mathbb{Q}$ defined at the end of §5.4 with the direct product $\prod_{r\geqslant 0} \mathcal{S}_\Delta(n,r)_\mathbb{Q}$.

6.2. More multiplication formulas in affine Schur algebras

In order to prove Conjecture 5.4.2 in the classical case, we use the elements $A[\mathbf{j}, r]$ for $A \in \Theta_\Delta^\pm(n), \mathbf{j} \in \mathbb{N}_\Delta^n$ defined in (5.4.2.1):

$$A[\mathbf{j}, r] = \begin{cases} \sum_{\lambda \in \Lambda_\Delta(n, r-\sigma(A))} \lambda^{\mathbf{j}}[A + \mathrm{diag}(\lambda)]_1, & \text{if } \sigma(A) \leqslant r; \\ 0, & \text{otherwise.} \end{cases}$$

These elements cannot be obtained by specializing v to 1 from the elements $A(\mathbf{j}, r)$ defined in (3.4.0.2). However, we have $A[\mathbf{0}, r] = A(\mathbf{0}, r)_1$ for $A \in \Theta_\Delta^\pm(n)$ (assuming $0^0 = 1$). We also point out another difference when $\sigma(A) = r$. In this case, $A[\mathbf{j}, r] = \delta_{\mathbf{0},\mathbf{j}}[A]_1$ while $A(\mathbf{j}, r)_1 = [A]_1$ for all $\mathbf{j} \in \mathbb{N}_\Delta^n$.

We will first show that, for a given $r > 0$, the set $\{A[\mathbf{j}, r]\}_{A \in \Theta_\Delta^\pm(n), \mathbf{j} \in \mathbb{N}_\Delta^n}$ spans the affine Schur algebra $\mathcal{S}_\Delta(n, r)_\mathbb{Q}$. Then we derive some multiplication formulas between $A[\mathbf{j}, r]$ and certain generators corresponding to simple and homogeneous indecomposable representations of the cyclic quiver $\Delta(n)$. We will leave the proof of the conjecture to the next section.

The following result is the classical counterpart of [**24**, Prop. 4.1]. Its proof is similar to the proof there; cf. [**30**, 4.3,4.2].

Proposition 6.2.1. *For any fixed* $1 \leqslant i_0 \leqslant n$, *the set*

$$\{A[\mathbf{j}, r] \mid A \in \Theta_\Delta^\pm(n), \mathbf{j} \in \mathbb{N}_\Delta^n,\ j_{i_0} = 0,\ \sigma(A) + \sigma(\mathbf{j}) \leqslant r\}$$

is a basis for $\mathcal{S}_\Delta(n, r)_\mathbb{Q}$. *In particular, the set*

$$\{A[\mathbf{j}, r] \mid A \in \Theta_\Delta^\pm(n), \mathbf{j} \in \mathbb{N}_\Delta^n, \sigma(A) \leqslant r\}$$

forms a spanning set of $\mathcal{S}_\Delta(n, r)_\mathbb{Q}$.

The first two of the following multiplication formulas in affine Schur algebras are a natural generalization of the multiplication formulas for Schur algebras, given in [**30**, Prop. 3.1], which are the classical version of the quantum formulas in [**4**, Lem. 5.3]. The third formula is new and is the key to the proof of Conjecture 5.4.2 in the classical case. It would be interesting to find the corresponding formula for affine quantum Schur algebras. For the simplicity of the statement in the next result, we also set $A[\mathbf{j}, r] = 0$ if some off-diagonal entries of A are negative.

Theorem 6.2.2. *Assume* $1 \leqslant h, t \leqslant n$, $\mathbf{j} = (j_k) \in \mathbb{N}_\Delta^n$, *and* $A = (a_{i,j}) \in \Theta_\Delta^\pm(n)$. *The following multiplication formulas hold in* $\mathcal{S}_\Delta(n, r)_\mathbb{Q}$:

(1) $0[\boldsymbol{e}_t^\Delta, r]A[\mathbf{j}, r] = A[\mathbf{j} + \boldsymbol{e}_t^\Delta, r] + \big(\sum_{s \in \mathbb{Z}} a_{t,s}\big)A[\mathbf{j}, r]$;

(2) *for* $\varepsilon \in \{1, -1\}$,

$$E^{\triangle}_{h,h+\varepsilon}[\mathbf{0}, r]A[\mathbf{j}, r] = \sum_{\substack{a_{h+\varepsilon,i} \geqslant 1 \\ \forall i \neq h, h+\varepsilon}} (a_{h,i} + 1)(A + E^{\triangle}_{h,i} - E^{\triangle}_{h+\varepsilon,i})[\mathbf{j}, r]$$

$$+ \sum_{0 \leqslant i \leqslant j_h} (-1)^i \binom{j_h}{i} (A - E^{\triangle}_{h+\varepsilon,h})[\mathbf{j} + (1-i)\boldsymbol{e}^{\triangle}_h, r]$$

$$+ (a_{h,h+\varepsilon} + 1) \sum_{0 \leqslant i \leqslant j_{h+\varepsilon}} \binom{j_{h+\varepsilon}}{i} (A + E^{\triangle}_{h,h+\varepsilon})[\mathbf{j} - i\boldsymbol{e}^{\triangle}_{h+\varepsilon}, r];$$

(3) *for* $m \in \mathbb{Z}\backslash\{0\}$,

$$E^{\triangle}_{h,h+mn}[\mathbf{0}, r]A[\mathbf{j}, r] = \sum_{\substack{s \notin \{h, h-mn\} \\ a_{h,s} \geqslant 1}} (a_{h,s+mn} + 1)(A + E^{\triangle}_{h,s+mn} - E^{\triangle}_{h,s})[\mathbf{0}, r]$$

$$+ \sum_{0 \leqslant t \leqslant j_h} (a_{h,h+mn} + 1) \binom{j_h}{t} (A + E^{\triangle}_{h,h+mn})[\mathbf{j} - t\boldsymbol{e}^{\triangle}_h, r]$$

$$+ \sum_{0 \leqslant t \leqslant j_h} (-1)^t \binom{j_h}{t} (A - E^{\triangle}_{h,h-mn})[\mathbf{j} + (1-t)\boldsymbol{e}^{\triangle}_h, r].$$

It is natural to compare Theorem 6.2.2(1)–(2) with [**24**, (4.2.1-3)] or Theorem 3.4.2 which generalize the corresponding ones for quantum Schur algebras. They are *not* obtained from the quantum counterpart by specializing v to 1. For example, the second sum in the right-hand side of (2) above is slightly different from the quantum version. In the case where $\sigma(A) = r + 1$ and $a_{h+\varepsilon,h} \geqslant 1$, the left-hand side of Theorem 6.2.2(2) is zero. By the remark at the beginning of this section, the right-hand side is also zero since $\mathbf{j} + (1-i)\boldsymbol{e}^{\triangle}_h \neq \mathbf{0}$, for all $0 \leqslant i \leqslant j_h$.

The proof of the following result will be given at the end of the chapter as an appendix; see §6.4. It should be pointed out that the first formula can also be obtained from [**56**, 3.5] by specializing v to 1, while the second formula is the key to the proof of part (3) of the theorem above.

Proposition 6.2.3. *Let* $1 \leqslant h \leqslant n$, $B = (b_{i,j}) \in \Theta_{\triangle}(n, r)$, *and* $\lambda = \mathrm{ro}(B)$.

(1) *If* $\varepsilon \in \{1, -1\}$ *and* $\lambda \geqslant \boldsymbol{e}^{\triangle}_{h+\varepsilon}$, *then*

$$[E^{\triangle}_{h,h+\varepsilon} + \mathrm{diag}(\lambda - \boldsymbol{e}^{\triangle}_{h+\varepsilon})]_1 [B]_1 = \sum_{i \in \mathbb{Z}, b_{h+\varepsilon,i} \geqslant 1} (b_{h,i} + 1)[B + E^{\triangle}_{h,i} - E^{\triangle}_{h+\varepsilon,i}]_1.$$

(Here, by convention, $1 \leqslant h < n$ *for* $\varepsilon = 1$ *and* $1 < h \leqslant n$ *for* $\varepsilon = -1$.*)*

(2) *If* $m \in \mathbb{Z}\backslash\{0\}$ *and* $\lambda \geqslant e_h^\vartriangle$, *then*

$$[E^\vartriangle_{h,h+mn}+\mathrm{diag}(\lambda-e^\vartriangle_h)]_1[B]_1 = \sum_{\substack{s\in\mathbb{Z}\\ b_{h,s}\geqslant 1}} (b_{h,s+mn}+1)[B+E^\vartriangle_{h,s+mn}-E^\vartriangle_{h,s}]_1.$$

We now use these formulas to prove the theorem.

Proof. The proof of formula (1) is straightforward. Since, for any $A \in \Theta_\vartriangle(n,r)$ and $\lambda \in \Lambda_\vartriangle(n,r)$,

$$[\mathrm{diag}(\lambda)]_1[A]_1 = \begin{cases} [A]_1, & \text{if } \lambda = \mathrm{ro}(A);\\ 0, & \text{otherwise,}\end{cases}$$

it follows that

$$\begin{aligned}
0[e^\vartriangle_t, r]A[\mathbf{j}, r] &= \sum_{\lambda\in\Lambda(n,r-1)} \lambda_t[\mathrm{diag}(\lambda)]_1 \sum_{\mu\in\Lambda(n,r-\sigma(A))} \mu^{\mathbf{j}}[A+\mathrm{diag}(\mu)]_1\\
&= \sum_{\mu\in\Lambda(n,r-\sigma(A))} \lambda_t\mu^{\mathbf{j}}[A+\mathrm{diag}(\mu)]_1 \quad (\text{where } \lambda = \mathrm{ro}(A)+\mu)\\
&= \sum_{\mu\in\Lambda(n,r-\sigma(A))} (\sum_{j\in\mathbb{Z}} a_{t,j}+\mu_t)\mu^{\mathbf{j}}[A+\mathrm{diag}(\mu)]_1\\
&= \sum_{\mu\in\Lambda(n,r-\sigma(A))} \mu^{\mathbf{j}+e_t}[A+\mathrm{diag}(\mu)]_1 + \Big(\sum_{j\in\mathbb{Z}} a_{t,j}\Big)A[\mathbf{j}, r] = \mathrm{RHS}.
\end{aligned}$$

We now prove formula (2). For convenience, we set $[B]_1 = 0$ if one of the entries of B is negative. Since $[A]_1[B]_1 = 0$ whenever $\mathrm{co}(A) \neq \mathrm{ro}(B)$, we have

$$\begin{aligned}
&E^\vartriangle_{h,h+\varepsilon}[\mathbf{0}, r]A[\mathbf{j}, r]\\
&\quad= \sum_{\mu\in\Lambda(n,r-\sigma(A))} \mu^{\mathbf{j}}[E^\vartriangle_{h,h+\varepsilon}+\mathrm{diag}(\mu+\mathrm{ro}(A)-e^\vartriangle_{h+\varepsilon})]_1[A+\mathrm{diag}(\mu)]_1.
\end{aligned}$$

By Proposition 6.2.3(1),

$$\begin{aligned}
&[E^\vartriangle_{h,h+\varepsilon}+\mathrm{diag}(\mu+\mathrm{ro}(A)-e^\vartriangle_{h+\varepsilon})]_1[A+\mathrm{diag}(\mu)]_1\\
&\qquad= \sum_{\substack{i\neq h,h+\varepsilon\\ a_{h+\varepsilon,i}\geqslant 1}} (a_{h,i}+1)[A+E^\vartriangle_{h,i}-E^\vartriangle_{h+\varepsilon,i}+\mathrm{diag}(\mu)]_1\\
&\qquad\quad + (\mu_h+1)[A-E^\vartriangle_{h+\varepsilon,h}+\mathrm{diag}(\mu+e^\vartriangle_h)]_1\\
&\qquad\quad + (a_{h,h+\varepsilon}+1)[A+E_{h,h+\varepsilon}+\mathrm{diag}(\mu-e^\vartriangle_{h+\varepsilon})]_1.
\end{aligned}$$

Thus,

$$E^{\vartriangle}_{h,h+\varepsilon}[\mathbf{0}, r]A[\mathbf{j}, r] = \sum_{\substack{i\neq h,h+\varepsilon\\ a_{h+\varepsilon,i}\geqslant 1}} (a_{h,i}+1)(A+E^{\vartriangle}_{h,i}-E^{\vartriangle}_{h+\varepsilon,i})[\mathbf{j}, r]+\mathcal{Y}_h+\mathcal{Y}_{h+\varepsilon},$$

where

$$\begin{aligned}\mathcal{Y}_h &= \sum_{\mu\in\Lambda(n,r-\sigma(A))} \mu^{\mathbf{j}}(\mu_h+1)[A-E^{\vartriangle}_{h+\varepsilon,h}+\mathrm{diag}(\mu+\boldsymbol{e}^{\vartriangle}_h)]_1\\ &= \sum_{\mu\in\Lambda(n,r-\sigma(A))} \Big(\prod_{i\neq h}\mu_i^{j_i}\Big)(\mu_h+1-1)^{j_h}(\mu_h+1)\\ &\qquad\times[A-E^{\vartriangle}_{h+\varepsilon,h}+\mathrm{diag}(\mu+\boldsymbol{e}^{\vartriangle}_h)]_1\\ &= \sum_{0\leqslant i\leqslant j_h}(-1)^i\binom{j_h}{i}(A-E^{\vartriangle}_{h+\varepsilon,h})[\mathbf{j}+(1-i)\boldsymbol{e}^{\vartriangle}_h, r],\end{aligned}$$

and

$$\begin{aligned}\mathcal{Y}_{h+\varepsilon} &= \sum_{\mu\in\Lambda(n,r-\sigma(A))} \mu^{\mathbf{j}}(a_{h,h+\varepsilon}+1)[A+E_{h,h+\varepsilon}+\mathrm{diag}(\mu-\boldsymbol{e}^{\vartriangle}_{h+\varepsilon})]_1\\ &= (a_{h,h+\varepsilon}+1)\sum_{\mu\in\Lambda(n,r-\sigma(A))}\Big(\prod_{i\neq h+\varepsilon}\mu_i^{j_i}\Big)(\mu_{h+\varepsilon}-1+1)^{j_{h+\varepsilon}}\\ &\qquad\times[A+E_{h,h+\varepsilon}+\mathrm{diag}(\mu-\boldsymbol{e}^{\vartriangle}_{h+\varepsilon})]_1\\ &= (a_{h,h+\varepsilon}+1)\sum_{0\leqslant i\leqslant j_{h+\varepsilon}}\binom{j_{h+\varepsilon}}{i}\sum_{\mu\in\Lambda(n,r-\sigma(A)-1)}\mu^{\mathbf{j}-i\boldsymbol{e}^{\vartriangle}_{h+\varepsilon}}\\ &\qquad\times[A+E_{h,h+\varepsilon}+\mathrm{diag}(\mu)]_1\\ &= (a_{h,h+\varepsilon}+1)\sum_{0\leqslant i\leqslant j_{h+\varepsilon}}\binom{j_{h+\varepsilon}}{i}(A+E^{\vartriangle}_{h,h+\varepsilon})[\mathbf{j}-i\boldsymbol{e}^{\vartriangle}_{h+\varepsilon}, r].\end{aligned}$$

Substituting gives (2).

Finally, we prove formula (3). The proof is similar to that of (2). First, with the same reasoning,

$$\begin{aligned}&E^{\vartriangle}_{h,h+mn}[0, r]A[\mathbf{j}, r]\\ &\quad=\sum_{\mu\in\Lambda_{\vartriangle}(n,r-\sigma(A))}\mu^{\mathbf{j}}[E^{\vartriangle}_{h,h+mn}+\mathrm{diag}(\mu)+ro(A)-\boldsymbol{e}^{\vartriangle}_h]_1\cdot[A+\mathrm{diag}(\mu)]_1.\end{aligned}$$

Applying Proposition 6.2.3(2) yields

$$\begin{aligned}&[E^{\vartriangle}_{h,h+mn}+\mathrm{diag}(\mu)+ro(A)-\boldsymbol{e}^{\vartriangle}_h]_1\cdot[A+\mathrm{diag}(\mu)]_1\\ &\quad=\sum_{s\notin\{h,h-mn\}}(a_{h,s+mn}+1)[A+E^{\vartriangle}_{h,s+mn}-E^{\vartriangle}_{h,s}+\mathrm{diag}(\mu)]_1\end{aligned}$$

$$+(a_{h,h+mn}+1)[(A+E^{\triangle}_{h,h+mn})+\mathrm{diag}(\mu-e^{\triangle}_h)]_1$$
$$+(\mu_h+1)[(A-E_{h,h-mn})+\mathrm{diag}(\mu+e^{\triangle}_h)]_1.$$

Thus,

$$E^{\triangle}_{h,h+mn}[0,r]A[\mathbf{j},r]=$$
$$\sum_{s\notin\{h,h-mn\}}(a_{h,s+mn}+1)(A+E^{\triangle}_{h,s+mn}-E^{\triangle}_{h,s})[\mathbf{0},r]+\mathcal{X}_1+\mathcal{X}_2,$$

where

$$\begin{aligned}
\mathcal{X}_1&=(a_{h,h+mn}+1)\sum_{\mu\in\Lambda_\triangle(n,r-\sigma(A))}\mu^{\mathbf{j}}[(A+E^{\triangle}_{h,h+mn})+\mathrm{diag}(\mu-e^{\triangle}_h)]_1\\
&=(a_{h,h+mn}+1)\sum_{\mu\in\Lambda_\triangle(n,r-\sigma(A))}\prod_{\substack{s\neq h\\1\leqslant s\leqslant n}}\mu_s^{j_s}(\mu_h-1+1)^{j_h}\\
&\qquad\times[(A+E^{\triangle}_{h,h+mn})+\mathrm{diag}(\mu-e^{\triangle}_h)]_1\\
&=(a_{h,h+mn}+1)\sum_{\substack{\mu\in\Lambda_\triangle(n,r-\sigma(A))\\0\leqslant t\leqslant j_h}}\binom{j_h}{t}(\mu-e^{\triangle}_h)^{\mathbf{j}-te^{\triangle}_h}\\
&\qquad\times[A+E^{\triangle}_{h,h+mn}+\mathrm{diag}(\mu-e^{\triangle}_h)]_1\\
&=(a_{h,h+mn}+1)\sum_{0\leqslant t\leqslant j_h}\binom{j_h}{t}(A+E^{\triangle}_{h,h+mn})[\mathbf{j}-te^{\triangle}_h,r]
\end{aligned}$$

and

$$\begin{aligned}
\mathcal{X}_2&=\sum_{\mu\in\Lambda_\triangle(n,r-\sigma(A))}\mu^{\mathbf{j}}(\mu_h+1)[(A-E_{h,h-mn})+\mathrm{diag}(\mu+e^{\triangle}_h)]_1\\
&=\sum_{\mu\in\Lambda_\triangle(n,r-\sigma(A))}\prod_{s\neq h}\mu_s^{j_s}(\mu_h+1-1)^{j_h}(\mu_h+1)\\
&\qquad\times[(A-E_{h,h-mn})+\mathrm{diag}(\mu+e^{\triangle}_h)]_1\\
&=\sum_{\substack{\mu\in\Lambda_\triangle(n,r-\sigma(A))\\0\leqslant t\leqslant j_h}}(-1)^t\binom{j_h}{t}(\mu+e^{\triangle}_h)^{\mathbf{j}+(1-t)e^{\triangle}_h}\\
&\qquad\times[(A-E_{h,h-mn})+\mathrm{diag}(\mu+e^{\triangle}_h)]_1\\
&=\sum_{0\leqslant t\leqslant j_h}(-1)^t\binom{j_h}{t}(A-E_{h,h-mn})[\mathbf{j}+(1-t)e^{\triangle}_h,r],
\end{aligned}$$

proving (3). This completes the proof of the theorem. □

6.3. Proof of Conjecture 5.4.2 at $v = 1$

We now use Theorem 6.2.2 to prove Conjecture 5.4.2 in the classical case.

Recall from the proof of Proposition 6.1.4 that the specialized Ringel–Hall algebra $\mathfrak{H}_\Delta(n)_\mathbb{Q}$ is generated by $u_{i,j} := u_{E^\Delta_{i,j}}$ for all $i < j \in \mathbb{Z}$. As seen from Proposition 1.4.5, the Ringel–Hall algebra $\mathfrak{H}_\Delta(n)$ over $\mathbb{Q}(v)$ can be generated by the elements associated with simple and homogeneous indecomposable representations of $\Delta(n)$. We first prove that this is also true for $\mathfrak{H}_\Delta(n)_\mathbb{Q}$.

Lemma 6.3.1. *The $\mathfrak{H}_\Delta(n)_\mathbb{Q}$ is generated by the elements $u_i = u_{i,i+1}$ and $u_{i,i+mn}$, for $i \in \mathbb{Z}$ and $m \geqslant 1$. In particular, the subalgebra $\mathcal{S}_\Delta(n,r)^+_\mathbb{Q}$ (resp., $\mathcal{S}_\Delta(n,r)^-_\mathbb{Q}$) spanned by $A[\mathbf{0},r] = A(\mathbf{0},r)_1$, for all $A \in \Theta^+_\Delta(n)$ (resp., $A \in \Theta^-_\Delta(n)$), can be generated by $E^\Delta_{h,h'}[\mathbf{0},r]$ and $E^\Delta_{h,h+mn}[\mathbf{0},r]$, for all $1 \leqslant h \leqslant n$, $h - h' = -1$, and $m \geqslant 1$, (resp., $h - h' = 1$ and $m \leqslant -1$).*

Proof. Let $\mathfrak{K}$ be the subalgebra of $\mathfrak{H}_\Delta(n)_\mathbb{Q}$ generated by the elements $u_{i,i+1}$ and $u_{i,i+mn}$, for $i \in \mathbb{Z}$ and $m \geqslant 1$. It is enough to prove $u_{i,j} \in \mathfrak{K}$, for all $i < j$. Write $j - i = mn + k$, where $m \in \mathbb{Z}$ and $1 \leqslant k \leqslant n$. If $k = n$, then $u_{i,j} \in \mathfrak{K}$. Now assume $1 \leqslant k < n$. We apply induction on k. If $k = 1$, then by (6.1.4.1),

$$u_{i,mn+i+1} = u_{i,i+1}u_{i+1,mn+1+i} - u_{i+1,mn+i+1}u_{i,i+1} \in \mathfrak{K}.$$

Now suppose $k > 1$ and $u_{i,j} \in \mathfrak{K}$, for all $i < j$ with $j - i = mn + k - 1$. Then by (6.1.4.1) and the inductive hypothesis,

$$u_{i,mn+k+i} = u_{i,i+1}u_{i+1,mn+k+i} - u_{i+1,mn+k+i}u_{i,i+1} \in \mathfrak{K},$$

proving the first assertion.

The last assertion follows from Proposition 3.6.1. □

As in §5.4, let $\widehat{\mathcal{K}}_\Delta(n)_\mathbb{Q}$ be the vector space of all formal (possibly infinite) $\mathbb{Q}$-linear combinations $\sum_{A \in \Theta_\Delta(n)} \beta_A[A]_1$ satisfying (5.4.0.1). Recall from (5.4.2.2) that, for $A \in \Theta^\pm_\Delta(n)$ and $\mathbf{j} \in \mathbb{N}^n_\Delta$,

$$A[\mathbf{j}] = \sum_{r=0}^{\infty} A[\mathbf{j}, r] \in \widehat{\mathcal{K}}_\Delta(n)_\mathbb{Q}.$$

Furthermore, let $\leqslant$ and $\preccurlyeq$ be the orders on $\mathbb{Z}^n_\Delta$ and $\Theta_\Delta(n)$ defined in (1.1.0.3) and (3.7.1.1), respectively.

Proposition 6.3.2. *For $A \in \Theta^\pm_\Delta(n)$ and $\mathbf{j} \in \mathbb{N}^n_\Delta$, we have*

$$A^+[\mathbf{0}]0[\mathbf{j}]A^-[\mathbf{0}] = A[\mathbf{j}] + \sum_{\substack{\mathbf{j}' < \mathbf{j} \\ \mathbf{j}' \in \mathbb{N}^n_\Delta}} f^{\mathbf{j}'}_{A,\mathbf{j}} A[\mathbf{j}'] + \sum_{\substack{B \in \Theta^\pm_\Delta(n) \\ B \prec A,\, \mathbf{j}' \in \mathbb{N}^n_\Delta}} f^{B,\mathbf{j}'}_{A,\mathbf{j}} B[\mathbf{j}'],$$

where $f^{\mathbf{j}'}_{A,\mathbf{j}}, f^{B,\mathbf{j}'}_{A,\mathbf{j}} \in \mathbb{Q}$ and $A = A^+ + A^-$ with $A^+, {}^t(A^-) \in \Theta^+_\Delta(n)$.

Proof. For each $r \geqslant 0$, by Lemma 6.3.1, we may write $A^+[\mathbf{0}, r]$ as a linear combination of monomials in $E^{\Delta}_{h,h+1}[\mathbf{0}, r]$ and $E^{\Delta}_{h,h+mn}[\mathbf{0}, r]$. By Theorem 6.2.2, there exist $f^{B,\mathbf{j}'}_{A,\mathbf{j}} \in \mathbb{Q}$ (independent of r) such that

$$A^+[\mathbf{0}, r]0[\mathbf{j}, r]A^-[\mathbf{0}, r] = \sum_{\substack{B \in \Theta^{\pm}_{\Delta}(n) \\ \mathbf{j}' \in \mathbb{N}^n_{\Delta}}} f^{B,\mathbf{j}'}_{A,\mathbf{j}} B[\mathbf{j}', r], \qquad (6.3.2.1)$$

for all $r \geqslant 0$. On the other hand, using an argument similar to the second display for the computation of $\mathfrak{p}_A$ in the proof of Theorem 3.7.7 yields

$$\begin{aligned} A^+[\mathbf{0}, r]0[\mathbf{j}, r]A^-[\mathbf{0}, r] &= \sum_{\lambda \in \Lambda_{\Delta}(n,r)} \lambda^{\mathbf{j}} A^+[\mathbf{0}, r][\mathrm{diag}(\lambda)]_1 A^-[\mathbf{0}, r] \\ &= \sum_{\substack{\lambda \in \Lambda_{\Delta}(n,r) \\ \lambda \geqslant \sigma(A)}} \lambda^{\mathbf{j}} [A + \mathrm{diag}(\lambda - \sigma(A))]_1 + g, \end{aligned}$$

where $\sigma(A) = \mathrm{co}(A^+) + \mathrm{ro}(A^-)$ and g is a $\mathbb{Q}$-linear combination of $[B]_1$ with $B \in \Theta_{\Delta}(n, r)$ and $B \prec A$. Since

$$\lambda^{\mathbf{j}} = (\lambda - \sigma(A) + \sigma(A))^{\mathbf{j}} = (\lambda - \sigma(A))^{\mathbf{j}} + \sum_{\substack{\mathbf{j}' < \mathbf{j} \\ \mathbf{j}' \in \mathbb{N}^n_{\Delta}}} f^{\mathbf{j}'}_{A,\mathbf{j}} (\lambda - \sigma(A))^{\mathbf{j}'},$$

where $f^{\mathbf{j}'}_{A,\mathbf{j}} \in \mathbb{Z}$ are independent of λ and r, it follows that

$$A^+[\mathbf{0}, r]0[\mathbf{j}, r]A^-[\mathbf{0}, r] = A[\mathbf{j}, r] + \sum_{\substack{\mathbf{j}' < \mathbf{j} \\ \mathbf{j}' \in \mathbb{N}^n_{\Delta}}} f^{\mathbf{j}'}_{A,\mathbf{j}} A[\mathbf{j}', r] + g.$$

Combining this with (6.3.2.1) proves the assertion. □

For integers $a \geqslant -1$ and $l \geqslant 1$, let $\mathbb{Z}_{a,l} = \{a + i \mid i = 1, 2, \ldots, l\}$ and

$$(\mathbb{Z}_{a,l})^n = \{\lambda \in \mathbb{N}^n \mid \lambda_i \in \mathbb{Z}_{a,l}, \forall i\}.$$

Lemma 6.3.3. *For fixed integers $a \geqslant -1$ and $n, l \geqslant 1$, if we order $(\mathbb{Z}_{a,l})^n$ lexicographically and form an $l^n \times l^n$ matrix $B_n = (\lambda^{\mu})_{\lambda,\mu \in (\mathbb{Z}_{a,l})^n}$, where $\lambda^{\mu} = \lambda_1^{\mu_1} \lambda_2^{\mu_2} \cdots \lambda_n^{\mu_n}$, then $\det(B_n) \neq 0$.*

Proof. Write $(\mathbb{Z}_{a,l})^n = \{\boldsymbol{a}_1, \boldsymbol{a}_2, \ldots, \boldsymbol{a}_{l^n}\}$ with $\boldsymbol{a}_i <_{\mathrm{lx}} \boldsymbol{a}_{i+1}$ for all i under the lexicographical order $<_{\mathrm{lx}}$. If the (i, j) entry of B_n is $\boldsymbol{a}_i^{\boldsymbol{a}_j}$, then B_n has the form

$$B_n = \begin{pmatrix} (a+1)^{(a+1)}B_{n-1} & (a+1)^{a+2}B_{n-1} & \cdots & (a+1)^{a+l}B_{n-1} \\ (a+2)^{(a+1)}B_{n-1} & (a+2)^{(a+2)}B_{n-1} & \cdots & (a+2)^{(a+l)}B_{n-1} \\ \vdots & \vdots & \ddots & \vdots \\ (a+l)^{(a+1)}B_{n-1} & (a+l)^{(a+2)}B_{n-1} & \cdots & (a+l)^{(a+l)}B_{n-1} \end{pmatrix}.$$

Thus,

$$\begin{aligned} &\det(B_n) \\ &= \prod_{i=1}^{l}(a+i)^{(a+1)l^{n-1}} \det \begin{pmatrix} B_{n-1} & (a+1)B_{n-1} & \cdots & (a+1)^{l-1}B_{n-1} \\ B_{n-1} & (a+2)B_{n-1} & \cdots & (a+2)^{l-1}B_{n-1} \\ \vdots & \vdots & \ddots & \vdots \\ B_{n-1} & (a+l)B_{n-1} & \cdots & (a+l)^{l-1}B_{n-1} \end{pmatrix} \\ &= \prod_{i=1}^{l}(a+i)^{(a+1)l^{n-1}} \\ &\quad \times \det \begin{pmatrix} B_{n-1} & 0 & \cdots & 0 \\ B_{n-1} & (2-1)B_{n-1} & \cdots & ((a+2)^{l-1}-(a+2)^{l-2}(a+1))B_{n-1} \\ \vdots & \vdots & \ddots & \vdots \\ B_{n-1} & (l-1)B_{n-1} & \cdots & ((a+l)^{l-1}-(a+l)^{l-2}(a+1))B_{n-1} \end{pmatrix} \\ &= \prod_{i=1}^{l}(a+i)^{(a+1)l^{n-1}} \det(B_{n-1}) \prod_{j=2}^{l}(j-1)^{l^{n-1}} \\ &\quad \times \det \begin{pmatrix} B_{n-1} & \cdots & (a+2)^{l-2}B_{n-1} \\ \vdots & \ddots & \vdots \\ B_{n-1} & \cdots & (a+l)^{l-2}B_{n-1} \end{pmatrix}. \end{aligned}$$

Hence,

$$\det(B_n) = \det(B_{n-1})^l \prod_{i=1}^{l}(a+i)^{(a+1)l^{n-1}} \prod_{1\leqslant j<i\leqslant l}(i-j)^{l^{n-1}} \neq 0,$$

by induction. □

Note that, in the proof above, if $a = -1$, then

$$\mathbb{Z}_l := \mathbb{Z}_{-1,l} = \{0, 1, \ldots, l-1\}$$

and so the product $\prod_{i=1}^{l}(a+i)^{(a+1)l^{n-1}} = 1$.

As introduced at the end of §5.4, let $\mathfrak{A}_{\triangle,\mathbb{Q}}[n]$ be the subspace of $\widehat{\mathcal{K}}_\triangle(n)_\mathbb{Q}$ spanned by the elements $A[\mathbf{j}]$ for all $A \in \Theta^{\pm}_\triangle(n)$ and $\mathbf{j} \in \mathbb{N}^n_\triangle$. The following theorem gives a realization of the universal enveloping algebra $\mathcal{U}(\widehat{\mathfrak{gl}}_n)$. Let $(\mathbb{Z}_l)^n_\triangle = \flat_2^{-1}((\mathbb{Z}_l)^n)$, where $\flat_2$ is defined in (1.1.0.2).

Theorem 6.3.4. *The $\mathbb{Q}$-space $\mathfrak{A}_{\triangle,\mathbb{Q}}[n]$ is a subalgebra of $\widehat{\mathcal{K}}_{\triangle}(n)_{\mathbb{Q}}$ with $\mathbb{Q}$-basis*

$$\mathfrak{B} = \{A[\mathbf{j}] \mid A \in \Theta_{\triangle}^{\pm}(n),\ \mathbf{j} \in \mathbb{N}_{\triangle}^{n}\}.$$

Moreover, the map $\eta := \prod_{r\geqslant 0} \eta_r$ defined in (6.1.6.1) *is injective and induces a $\mathbb{Q}$-algebra isomorphism $\mathcal{U}(\widehat{\mathfrak{gl}}_n) \overset{\eta}{\cong} \mathfrak{A}_{\triangle,\mathbb{Q}}[n]$.*

Proof. We first prove the linear independence of $\mathfrak{B}$. Suppose

$$\sum_{A\in\Theta_{\triangle}^{\pm}(n),\,\mathbf{j}\in\mathbb{N}_{\triangle}^{n}} f_{A,\mathbf{j}} A[\mathbf{j}] = 0$$

for some $f_{A,\mathbf{j}} \in \mathbb{Q}$. Then

$$0 = \sum_{\substack{A\in\Theta_{\triangle}^{\pm}(n)\\ \mathbf{j}\in\mathbb{N}_{\triangle}^{n}}} f_{A,\mathbf{j}} A[\mathbf{j}] = \sum_{r\geqslant 0} \sum_{\substack{A\in\Theta_{\triangle}^{\pm}(n)\\ \lambda\in\Lambda_{\triangle}(n,r-\sigma(A))}} \Big(\sum_{\mathbf{j}\in\mathbb{N}_{\triangle}^{n}} \lambda^{\mathbf{j}} f_{A,\mathbf{j}}\Big)[A + \mathrm{diag}(\lambda)].$$

Thus, $\sum_{\mathbf{j}\in\mathbb{N}_{\triangle}^{n}} \lambda^{\mathbf{j}} f_{A,\mathbf{j}} = 0$ for all $A \in \Theta_{\triangle}^{\pm}(n)$, $\lambda \in \Lambda_{\triangle}(n, r - \sigma(A))$, and $r \geqslant \sigma(A)$. So, when A is arbitrarily fixed, there is a finite subset J of $\mathbb{N}_{\triangle}^{n}$ satisfying $f_{A,\mathbf{j}} \neq 0$ for all $\mathbf{j} \in J$. Choose $\ell \geqslant 1$ such that J is a subset of $(\mathbb{Z}_l)_{\triangle}^{n}$ and set $f_{A,\mathbf{j}} = 0$ if $\mathbf{j} \in (\mathbb{Z}_l)_{\triangle}^{n}\backslash J$. Since $\bigcup_{r\geqslant\sigma(A)}\Lambda_{\triangle}(n, r - \sigma(A)) = \mathbb{N}_{\triangle}^{n}$ contains $(\mathbb{Z}_l)_{\triangle}^{n}$, it follows that $\sum_{\mathbf{j}\in(\mathbb{Z}_l)_{\triangle}^{n}} \lambda^{\mathbf{j}} f_{A,\mathbf{j}} = 0$ for all $\lambda \in (\mathbb{Z}_l)_{\triangle}^{n}$. Applying Lemma 6.3.3 gives $f_{A,\mathbf{j}} = 0$ for all $\mathbf{j}$. Hence, $\mathfrak{B}$ forms a basis for $\mathfrak{A}_{\triangle,\mathbb{Q}}[n]$.

Since $\mathcal{U}(\widehat{\mathfrak{gl}}_n) = \mathcal{U}(\widehat{\mathfrak{gl}}_n)^{+}\mathcal{U}(\widehat{\mathfrak{gl}}_n)^{0}\mathcal{U}(\widehat{\mathfrak{gl}}_n)^{-}$, by identifying $\mathcal{U}(\widehat{\mathfrak{gl}}_n)^{\pm}$ with $\mathfrak{H}_{\triangle}(n)_{\mathbb{Q}}^{\pm}$, the set

$$\{u_A^{+}(E_{1,1}^{\triangle})^{j_1}\cdots(E_{n,n}^{\triangle})^{j_n}u_B^{-} \mid A, B \in \Theta_{\triangle}^{+}(n), \mathbf{j} \in \mathbb{N}_{\triangle}^{n}\}$$

forms a basis of $\mathcal{U}(\widehat{\mathfrak{gl}}_n)$. Now Theorem 6.1.5 implies

$$\eta(u_{A^+}^{+}(E_{1,1}^{\triangle})^{j_1}\cdots(E_{n,n}^{\triangle})^{j_n}u_{{}^tA^-}^{-}) = A^{+}[\mathbf{0}]0[\mathbf{j}]A^{-}[\mathbf{0}].$$

By Proposition 6.3.2, the set

$$\{A^{+}[\mathbf{0}]0[\mathbf{j}]A^{-}[\mathbf{0}] \mid A \in \Theta_{\triangle}^{\pm}(n),\ \mathbf{j} \in \mathbb{N}_{\triangle}^{n}\}$$

forms another basis for $\mathfrak{A}_{\triangle,\mathbb{Q}}[n]$. Hence, η is injective and $\mathfrak{A}_{\triangle,\mathbb{Q}}[n]$ is exactly the image of η. This completes the proof of the theorem. □

This theorem together with Theorem 6.2.2 implies immediately the following multiplication formulas in $\mathcal{U}(\widehat{\mathfrak{gl}}_n)$.

Corollary 6.3.5. *The universal enveloping algebra $\mathcal{U}(\widehat{\mathfrak{gl}}_n)$ of the loop algebra $\widehat{\mathfrak{gl}}_n(\mathbb{Q})$ has a basis $\{A[\mathbf{j}] \mid A \in \Theta_{\triangle}^{\pm}(n), \mathbf{j} \in \mathbb{N}_{\triangle}^{n}\}$ which satisfies the following multiplication formulas: for $1 \leqslant h, t \leqslant n$, $\mathbf{j} = (j_k) \in \mathbb{N}_{\triangle}^{n}$, $A = (a_{i,j}) \in \Theta_{\triangle}^{\pm}(n)$, $\varepsilon \in \{1, -1\}$, and $m \in \mathbb{Z}\backslash\{0\}$,*

$$
\begin{aligned}
0[e_t^{\triangle}]A[\mathbf{j}] =& A[\mathbf{j}+e_t^{\triangle}] + \Big(\sum_{s\in\mathbb{Z}} a_{t,s}\Big)A[\mathbf{j}],\\
E^{\triangle}_{h,h+\varepsilon}[\mathbf{0}]A[\mathbf{j}] =& \sum_{\substack{a_{h+\varepsilon,i}\geqslant 1\\ \forall i\neq h,h+\varepsilon}} (a_{h,i}+1)(A+E^{\triangle}_{h,i}-E^{\triangle}_{h+\varepsilon,i})[\mathbf{j}]\\
&+\sum_{0\leqslant i\leqslant j_h}(-1)^i\binom{j_h}{i}(A-E^{\triangle}_{h+\varepsilon,h})[\mathbf{j}+(1-i)e_h^{\triangle}]\\
&+(a_{h,h+\varepsilon}+1)\sum_{0\leqslant i\leqslant j_{h+\varepsilon}}\binom{j_{h+\varepsilon}}{i}(A+E^{\triangle}_{h,h+\varepsilon})[\mathbf{j}-ie^{\triangle}_{h+\varepsilon}],
\end{aligned}
$$

and

$$
\begin{aligned}
E^{\triangle}_{h,h+mn}[\mathbf{0}]A[\mathbf{j}] =& \sum_{\substack{s\notin\{h,h-mn\}\\ a_{h,s}\geqslant 1}} (a_{h,s+mn}+1)(A+E^{\triangle}_{h,s+mn}-E^{\triangle}_{h,s})[\mathbf{0}]\\
&+\sum_{0\leqslant t\leqslant j_h}(a_{h,h+mn}+1)\binom{j_h}{t}(A+E^{\triangle}_{h,h+mn})[\mathbf{j}-te_h^{\triangle}]\\
&+\sum_{0\leqslant t\leqslant j_h}(-1)^t\binom{j_h}{t}(A-E^{\triangle}_{h,h-mn})[\mathbf{j}+(1-t)e_h^{\triangle}].
\end{aligned}
$$

Remark 6.3.6. There should be applications of these multiplication formulas. For example, one may define the $\mathbb{Z}$-subalgebra $\mathcal{U}(\widehat{\mathfrak{gl}}_n)_{\mathbb{Z}}$ generated by the divided powers of $E^{\triangle}_{h,h+\varepsilon}[\mathbf{0}]$ together with $E^{\triangle}_{h,h+mn}[\mathbf{0}]$ for all $1\leqslant h\leqslant n$, $\varepsilon\in\{1,-1\}$ and $m\in\mathbb{Z}\backslash\{0\}$. This should serve as the Kostant $\mathbb{Z}$-form of $\mathcal{U}(\widehat{\mathfrak{gl}}_n)$.

6.4. Appendix: Proof of Proposition 6.2.3

For a finite subset $X\subseteq\mathfrak{S}_{\triangle,r}$, define $\underline{X}=\sum_{x\in X}x\in\mathbb{Q}\mathfrak{S}_{\triangle,r}$. If $A=\jmath_\triangle(\lambda,d,\mu)$ is the matrix corresponding to the double coset $\mathfrak{S}_\lambda d\mathfrak{S}_\mu$ for $\lambda,\mu\in\Lambda_\triangle(n,r)$, $d\in\mathscr{D}^{\triangle}_{\lambda,\mu}$, then the element $[A]_1\in\mathcal{S}_\triangle(n,r)_{\mathbb{Q}}$ is the map $\phi^d_{\lambda,\mu}$ (at $v=1$) as defined in (3.2.6.1). Note also that, if $v=1$, then $x_\lambda=\underline{\mathfrak{S}_\lambda}$. Thus, $[A]_1(\underline{\mathfrak{S}_\nu})=0$ for $\nu\neq\mu$ and $[A]_1(\underline{\mathfrak{S}_\mu})=\underline{\mathfrak{S}_\lambda d\mathfrak{S}_\mu}=\underline{\mathfrak{S}_\lambda d\mathscr{D}^{\triangle}_\nu\cap\mathfrak{S}_\mu}$ by Lemma 3.2.5, where ν is the composition defined by $\mathfrak{S}_\nu=d^{-1}\mathfrak{S}_\lambda d\cap\mathfrak{S}_\mu$ (see Corollary 3.2.3 for a precise description of ν). Lemma 3.2.5 implies also that for $\lambda,\mu\in\Lambda_\triangle(n,r)$ and $w\in\mathfrak{S}_{\triangle,r}$,

$$
\underline{\mathfrak{S}_\lambda}w\underline{\mathfrak{S}_\mu}=|w^{-1}\mathfrak{S}_\lambda w\cap\mathfrak{S}_\mu|\underline{\mathfrak{S}_\lambda w\mathfrak{S}_\mu}. \tag{6.4.0.1}
$$

This fact will be frequently used in the proofs below.

Recall also from (3.2.1.4) the sets

$$R^{\lambda}_{i+kn} = \{\lambda_{k,i-1}+1, \lambda_{k,i-1}+2, \ldots, \lambda_{k,i-1}+\lambda_i\}$$

associated with $\lambda \in \Lambda_\triangle(n,r)$, where $\lambda_{k,i-1} := kr + \sum_{j=1}^{i-1}\lambda_j$.

PROPOSITION 6.2.3(1). *Let* $A = (a_{i,j}) = \jmath_\triangle(\lambda, d, \mu) \in \Theta_\triangle(n,r)$ *with* $\mathrm{ro}(A) = \lambda$. *If* $\varepsilon \in \{1,-1\}$ *and* $\lambda \geqslant \boldsymbol{e}^\triangle_{h+\varepsilon}$, *then*

$$[E^\triangle_{h,h+\varepsilon} + \mathrm{diag}(\lambda - \boldsymbol{e}^\triangle_{h+\varepsilon})]_1[A]_1 = \sum_{i\in\mathbb{Z},\ a_{h+\varepsilon,i}\geqslant 1}(a_{h,i}+1)[A+E^\triangle_{h,i}-E^\triangle_{h+\varepsilon,i}]_1,$$

where $1 \leqslant h, h+\varepsilon \leqslant n$.

Proof. Observe that $\mathrm{ro}(E^\triangle_{h,h+\varepsilon}) = \boldsymbol{e}^\triangle_h$ and $\mathrm{co}(E^\triangle_{h,h+\varepsilon}) = \boldsymbol{e}^\triangle_{h+\varepsilon}$. Thus, applying (3.2.2.2) yields

$$\jmath_\triangle(\lambda + \boldsymbol{\alpha}^\triangle_{h,\varepsilon}, 1, \lambda) = E^\triangle_{h,h+\varepsilon} + \mathrm{diag}(\lambda - \boldsymbol{e}^\triangle_{h+\varepsilon}),$$

where $\boldsymbol{\alpha}^\triangle_{h,\varepsilon} = \boldsymbol{e}^\triangle_h - \boldsymbol{e}^\triangle_{h+\varepsilon}$. In other words, the matrix $E^\triangle_{h,h+\varepsilon} + \mathrm{diag}(\lambda - \boldsymbol{e}^\triangle_{h+\varepsilon})$ is defined by the double coset $\mathfrak{S}_{\lambda-\boldsymbol{\alpha}^\triangle_{h,\varepsilon}}\mathfrak{S}_\lambda$. Hence, putting $\mathcal{L} = [1,n]\times\mathbb{Z}$, we have by Corollary 3.2.3 that

$$\begin{aligned}
&[E^\triangle_{h,h+\varepsilon} + \mathrm{diag}(\lambda - \boldsymbol{e}^\triangle_{h+\varepsilon})]_1[A]_1(\underline{\mathfrak{S}_\mu})\\
&= [E^\triangle_{h,h+\varepsilon} + \mathrm{diag}(\lambda - \boldsymbol{e}^\triangle_{h+\varepsilon})]_1(\underline{\mathfrak{S}_\lambda d\mathfrak{S}_\mu})\\
&= \prod_{s,t\in\mathcal{L}}\frac{1}{a_{s,t}!}\underline{\mathfrak{S}_{\lambda+\boldsymbol{\alpha}^\triangle_{h,\varepsilon}}\mathfrak{S}_\lambda d\mathfrak{S}_\mu}\\
&= \prod_{s,t\in\mathcal{L}}\frac{1}{a_{s,t}!}\underline{\mathfrak{S}_{\lambda+\boldsymbol{\alpha}^\triangle_{h,\varepsilon}}\mathscr{D}^\triangle_\gamma\cap\mathfrak{S}_\lambda d}\underline{\mathfrak{S}_\mu},
\end{aligned}$$

where

$$\gamma = \gamma(\varepsilon) = \begin{cases}(\lambda_1,\ldots,\lambda_h,1,\lambda_{h+1}-1,\lambda_{h+2},\ldots,\lambda_n), & \text{if } \varepsilon = 1;\\ (\lambda_1,\ldots,\lambda_{h-1}-1,1,\lambda_h,\lambda_{h+1},\ldots,\lambda_n), & \text{if } \varepsilon = -1.\end{cases}$$

For $i \in R^\lambda_{h+\varepsilon} = \{\lambda_{0,h+\varepsilon}+1, \lambda_{0,h+\varepsilon}+2, \ldots, \lambda_{0,h+\varepsilon}+\lambda_{0,h+\varepsilon+1}\}$, define $y_{\varepsilon,i} \in \mathfrak{S}_\lambda$ by setting

$$y_{1,i} = \begin{pmatrix} 1\cdots\lambda_{0,h} & \lambda_{0,h}+1 & \cdots & i-1 & i & i+1 & \cdots\lambda_{0,h+1}\cdots r\\ 1\cdots\lambda_{0,h} & \lambda_{0,h}+2 & \cdots & i & \lambda_{0,h}+1 & i+1 & \cdots\lambda_{0,h+1}\cdots r\end{pmatrix}$$

and

$$y_{-1,i} = \begin{pmatrix} 1\cdots\lambda_{0,h-2}\cdots i-1 & i & i+1 & \cdots & \lambda_{0,h-1} & \lambda_{0,h-1}+1\cdots r\\ 1\cdots\lambda_{0,h-2}\cdots i-1 & \lambda_{0,h-1} & i & \cdots & \lambda_{0,h-1}-1 & \lambda_{0,h-1}+1\cdots r\end{pmatrix}.$$

Then $\mathscr{D}^{\vartriangle}_{\gamma} \cap \mathfrak{S}_\lambda = \{y_{\varepsilon,i} \mid i \in R^\lambda_{h+\varepsilon}\}$. Hence,

$$[E^{\vartriangle}_{h,h+\varepsilon} + \mathrm{diag}(\lambda - e^{\vartriangle}_{h+\varepsilon})]_1[A]_1(\underline{\mathfrak{S}_\mu}) = \prod_{s,t\in\mathcal{L}} \frac{1}{a_{s,t}!} \sum_{i\in R^\lambda_{h+\varepsilon}} \underline{\mathfrak{S}_{\lambda+\alpha^{\vartriangle}_{h,\varepsilon}} y_{\varepsilon,i} d \mathfrak{S}_\mu}.$$

Let $B^{(\varepsilon,i)} = (b^{(\varepsilon,i)}_{s,t}) \in \Theta_\vartriangle(n,r)$ be the matrix associated with $\lambda + \boldsymbol{\alpha}^{\vartriangle}_{h,\varepsilon}$, μ and the double coset $\mathfrak{S}_{\lambda+\alpha^{\vartriangle}_{h,\varepsilon}} y_{\varepsilon,i} d\mathfrak{S}_\mu$. Since

$$y^{-1}_{\varepsilon,i}(R^{\lambda+\alpha^{\vartriangle}_{h,\varepsilon}}_s) = \begin{cases} R^\lambda_s, & \text{if } 1 \leqslant s \leqslant n \text{ but } s \neq h, h+\varepsilon; \\ R^\lambda_h \cup \{i\}, & \text{if } s = h; \\ R^\lambda_{h+\varepsilon}\backslash\{i\}, & \text{if } s = h+\varepsilon, \end{cases}$$

it follows that, for $i \in R^\lambda_{h+\varepsilon}$,

$$\begin{aligned} b^{(\varepsilon,i)}_{s,t} &= |d^{-1} y^{-1}_{\varepsilon,i} R^{\lambda+\alpha^{\vartriangle}_{h,\varepsilon}}_s \cap R^\mu_t| \\ &= \begin{cases} a_{s,t}, & \text{if } 1 \leqslant s \leqslant n \text{ but } s \neq h, h+\varepsilon; \\ a_{h,t} + |\{d^{-1}(i) \cap R^\mu_t\}|, & \text{if } s = h; \\ a_{h+\varepsilon,t} - |\{d^{-1}(i) \cap R^\mu_t\}|, & \text{if } s = h+\varepsilon. \end{cases} \end{aligned}$$

If $t_i \in \mathbb{Z}$ is the unique integer such that $d^{-1}(i) \in R^\mu_{t_i}$ (and $a_{h+\varepsilon,t_i} \geqslant 1$), then

$$b^{(\varepsilon,i)}_{s,t} = \begin{cases} a_{s,t}, & \text{if } 1 \leqslant s \leqslant n, s \neq h, h+\varepsilon \text{ or } t \neq t_i; \\ a_{h,t} + 1, & \text{if } s = h, t = t_i; \\ a_{h+\varepsilon,t} - 1, & \text{if } s = h+\varepsilon, t = t_i. \end{cases}$$

This implies that $B^{(\varepsilon,i)} = A + E^{\vartriangle}_{h,t_i} - E^{\vartriangle}_{h+\varepsilon,t_i}$, for all $i \in R^\lambda_{h+\varepsilon}$. By Corollary 3.2.3 again,

$$\begin{aligned} &[E^{\vartriangle}_{h,h+\varepsilon} + \mathrm{diag}(\lambda - e^{\vartriangle}_{h+\varepsilon})]_1[A]_1(\underline{\mathfrak{S}_\mu}) \\ &= \prod_{s,t\in\mathcal{L}} \frac{1}{a_{s,t}!} \sum_{i\in R^\lambda_{h+\varepsilon}} \prod_{s,t\in\mathcal{L}} b^{(\varepsilon,i)}_{s,t}!\underline{\mathfrak{S}_{\lambda+\alpha^{\vartriangle}_{h,\varepsilon}} y_{\varepsilon,i} d\mathfrak{S}_\mu} \\ &= \sum_{i\in R^\lambda_{h+\varepsilon}} \prod_{s,t\in\mathcal{L}} \frac{b^{(\varepsilon,i)}_{s,t}!}{a_{s,t}} [B^{(\varepsilon,i)}]_1(\underline{\mathfrak{S}_\mu}) \\ &= \sum_{i\in R^\lambda_{h+\varepsilon}} \frac{a_{h,t_i}+1}{a_{h+\varepsilon,t_i}} [B^{(\varepsilon,i)}]_1(\underline{\mathfrak{S}_\mu}). \end{aligned}$$

Finally,

$$[E^{\Delta}_{h,h+\varepsilon} + \mathrm{diag}(\lambda - e^{\Delta}_{h+\varepsilon})]_1[A]_1$$
$$= \sum_{t\in\mathbb{Z},a_{h+\varepsilon,t}\geqslant 1} |\{i \in \mathbb{Z} \mid i \in R^{\lambda}_{h+\varepsilon}, t = t_i\}| \frac{a_{h,t}+1}{a_{h+\varepsilon,t}}[A + E^{\Delta}_{h,t} - E^{\Delta}_{h+\varepsilon,t}]_1$$
$$= \sum_{t\in\mathbb{Z},a_{h+\varepsilon,t}\geqslant 1} (a_{h,t}+1)[A + E^{\Delta}_{h,t} - E^{\Delta}_{h+\varepsilon,t}]_1,$$

as $|\{i \in \mathbb{Z} \mid i \in R^{\lambda}_{h+\varepsilon}, t = t_i\}| = |d^{-1}R^{\lambda}_{h+\varepsilon} \cap R^{\mu}_t| = a_{h+\varepsilon,t}$. □

We need some preparation before proving Proposition 6.2.3(2). We follow the notation used in §3.2. Thus, for $1 \leqslant i \leqslant r$, $\boldsymbol{e}_i = (0, \ldots, 0, \underset{(i)}{1}, 0, \ldots, 0) \in \mathfrak{S}_{\Delta,r}$ is the permutation sending i to $i + r$ and j to j, for all $1 \leqslant j \leqslant r$ with $j \neq i$, and $\boldsymbol{e}_i = \rho s_{r+i-2} \cdots s_{r+1}s_r \cdots s_{i+1}s_i$ as seen in the proof of Proposition 3.2.1. Note that $s_{j+1}\rho = \rho s_j$ for all $j \in \mathbb{Z}$.

To $\lambda \in \Lambda_\Delta(n, r)$, $h \in [1, n]$ with $\lambda_h \neq 0$, and $m > 0$, we associate the element

$$u^{\lambda}_{m,h} =$$
$$\begin{pmatrix} 1\cdots\lambda_{0,h-1} & \lambda_{0,h-1}+1 & \lambda_{0,h-1}+2 & \cdots & \lambda_{0,h}-1 & \lambda_{0,h} & \lambda_{0,h}+1\cdots r \\ 1\cdots\lambda_{0,h-1} & \lambda_{0,h-1}+2 & \lambda_{0,h-1}+3 & \cdots & \lambda_{0,h} & mr+\lambda_{0,h-1}+1 & \lambda_{0,h}+1\cdots r \end{pmatrix}$$

in $\mathfrak{S}_{\Delta,r}$ and let $u^{\lambda}_{-m,h} = (u^{\lambda}_{m,h})^{-1}$. Then

$$u^{\lambda}_{-m,h} = \begin{pmatrix} 1\cdots\lambda_{0,h-1} & \lambda_{0,h-1}+1 & \lambda_{0,h-1}+2 & \cdots & \lambda_{0,h} & \lambda_{0,h}+1\cdots r \\ 1\cdots\lambda_{0,h-1} & -mr+\lambda_{0,h} & \lambda_{0,h-1}+1 & \cdots & \lambda_{0,h}-1 & \lambda_{0,h}+1\cdots r \end{pmatrix}.$$

For any $i \in R^{\lambda}_h$, removing those simple reflections s_j from $\boldsymbol{e}_i$ indexed by the numbers $i, i+1, \ldots, \lambda_{0,h}-1$ and $r+\lambda_{0,h-1}, r+\lambda_{0,h-1}+1, \ldots, r+i-2$ yields the shortest representative $\rho s_{r+\lambda_{0,h-1}-1} \cdots s_{\lambda_{0,h}+1}s_{\lambda_{0,h}}$ of the double coset $\mathfrak{S}_\lambda \boldsymbol{e}_i \mathfrak{S}_\lambda$. Clearly, $u^{\lambda}_{1,h} = \rho s_{r+\lambda_{0,h-1}-1} \cdots s_{\lambda_{0,h}+1}s_{\lambda_{0,h}}$. This observation has the following generalization.

Lemma 6.4.1. *Maintain the notation introduced above. Suppose* $\lambda \in \Lambda_\Delta(n, r)$, $m \in \mathbb{Z}\backslash\{0\}$, *and* $1 \leqslant h \leqslant n$.

(1) $u^{\lambda}_{m,h}$ *is the shortest element in* $\mathfrak{S}_\lambda \boldsymbol{e}^m_i \mathfrak{S}_\lambda$ *for all* $i \in R^{\lambda}_h$;
(2) $\jmath_\Delta(\lambda, u^{\lambda}_{m,h}, \lambda) = E^{\Delta}_{h,h-mn} + \mathrm{diag}(\lambda - \boldsymbol{e}^{\Delta}_h)$;
(3) $(u^{\lambda}_{m,h})^{-1}\mathfrak{S}_\lambda u^{\lambda}_{m,h} \cap \mathfrak{S}_\lambda = \mathfrak{S}_\nu$, *where*

$$\nu = \begin{cases} (\lambda_1, \ldots, \lambda_{h-1}, \lambda_h - 1, 1, \lambda_{h+1}, \ldots, \lambda_n), & \text{if } m > 0; \\ (\lambda_1, \ldots, \lambda_{h-1}, 1, \lambda_h - 1, \lambda_{h+1}, \ldots, \lambda_n), & \text{if } m < 0. \end{cases}$$

Proof. For $m > 0$, it is straightforward to check that

$$e_i^m = s_{i-1} \cdots s_{\lambda_{0,h-1}+2} s_{\lambda_{0,h-1}+1} \cdot u^\lambda_{m,h} \cdot s_{\lambda_{0,h}-1} \cdots s_{i+1} s_i.$$

Thus, $\mathfrak{S}_\lambda e_i^m \mathfrak{S}_\lambda = \mathfrak{S}_\lambda u^\lambda_{m,h} \mathfrak{S}_\lambda$. Now, by definition, $u^\lambda_{m,h}(j) < u^\lambda_{m,h}(j+1)$, for any j with $s_j \in \mathfrak{S}_\lambda$ and $m \in \mathbb{Z}\backslash\{0\}$. Hence, by (3.2.1.5), $u^\lambda_{m,h} \in \mathscr{D}^\vartriangle_{\lambda,\lambda}$, proving (1). The assertion (2) follows from Lemma 3.2.2, noting that $e_h^\vartriangle = e^\vartriangle_{h-mn}$ and $|R^\lambda_s \cap u^\lambda_{m,h} R^\lambda_t| = \big(E^\vartriangle_{h,h-mn} + \mathrm{diag}(\lambda - e_h^\vartriangle)\big)_{s,t}$. The assertion (3) is a consequence of (2) and Corollary 3.2.3. □

PROPOSITION 6.2.3(2). *Let $h \in [1,n]$ and $A = (a_{i,j}) \in \Theta_\vartriangle(n,r)$ with $\mathrm{ro}(A) = \lambda$. If $m \in \mathbb{Z}\backslash\{0\}$ and $\lambda \geqslant e_h^\vartriangle$, then*

$$[E^\vartriangle_{h,h+mn} + \mathrm{diag}(\lambda - e_h^\vartriangle)]_1 [A]_1 = \sum_{s\in\mathbb{Z},\ a_{h,s}\geqslant 1} (a_{h,s+mn}+1)[A + E^\vartriangle_{h,s+mn} - E^\vartriangle_{h,s}]_1.$$

Proof. As above, let

$$\lambda_{0,h} = \lambda_1 + \cdots + \lambda_h \quad (\text{and } \lambda_{0,0} = 0).$$

Assume $\mu = \mathrm{co}(A)$ and $d \in \mathscr{D}^\vartriangle_{\lambda,\mu}$ such that $\jmath_\vartriangle(\lambda, d, \mu) = A$. By Lemma 6.4.1 and Corollary 3.2.3 (and (6.4.0.1)),

$$\begin{aligned}
&[E^\vartriangle_{h,h+mn} + \mathrm{diag}(\lambda - e_h^\vartriangle)]_1 [A]_1 (\underline{\mathfrak{S}_\mu}) \\
&= \underline{\mathfrak{S}_\lambda (u^\lambda_{m,h})^{-1} \mathfrak{S}_\lambda} \cdot d \cdot \underline{\mathscr{D}^\vartriangle_\alpha \cap \mathfrak{S}_\mu} \quad (\text{noting } \lambda_h > 0) \\
&= \frac{1}{|\mathfrak{S}_\lambda|} \underline{\mathfrak{S}_\lambda (u^\lambda_{m,h})^{-1} \mathfrak{S}_\lambda} \cdot \underline{\mathfrak{S}_\lambda d \mathfrak{S}_\mu} \\
&= \frac{1}{|\mathfrak{S}_\lambda|} \prod_{\substack{1\leqslant s\leqslant n\\ t\in\mathbb{Z}}} \frac{1}{a_{s,t}!} \underline{\mathfrak{S}_\lambda (u^\lambda_{m,h})^{-1} \mathfrak{S}_\lambda} \cdot \underline{\mathfrak{S}_\lambda} \cdot d \cdot \underline{\mathfrak{S}_\mu} \\
&= \prod_{\substack{1\leqslant s\leqslant n\\ t\in\mathbb{Z}}} \frac{1}{a_{s,t}!} \underline{\mathfrak{S}_\lambda (u^\lambda_{m,h})^{-1} \mathfrak{S}_\lambda} \cdot d \cdot \underline{\mathfrak{S}_\mu} \\
&= \prod_{\substack{1\leqslant s\leqslant n\\ t\in\mathbb{Z}}} \frac{1}{a_{s,t}!} \underline{\mathfrak{S}_\lambda} \cdot (u^\lambda_{m,h})^{-1} \cdot \underline{\mathscr{D}^\vartriangle_\beta \cap \mathfrak{S}_\lambda} \cdot d \cdot \underline{\mathfrak{S}_\mu},
\end{aligned}$$

where $\mathfrak{S}_\alpha = d^{-1}\mathfrak{S}_\lambda d \cap \mathfrak{S}_\mu$ and $\mathfrak{S}_\beta = u^\lambda_{m,h} \mathfrak{S}_\lambda (u^\lambda_{m,h})^{-1} \cap \mathfrak{S}_\lambda$ with

$$\beta = \begin{cases} (\lambda_1, \ldots, \lambda_{h-1}, 1, \lambda_h - 1, \lambda_{h+1}, \ldots, \lambda_n), & \text{if } m > 0; \\ (\lambda_1, \ldots, \lambda_{h-1}, \lambda_h - 1, 1, \lambda_{h+1}, \ldots, \lambda_n), & \text{if } m < 0. \end{cases}$$

We now compute $\mathscr{D}_\beta^\Delta \cap \mathfrak{S}_\lambda$. For $m \neq 0$ and $\lambda_{0,h-1}+1 \leqslant i \leqslant \lambda_{0,h}$ (i.e., $i \in R_h^\lambda$), define $w_{m,i} \in \mathfrak{S}_\lambda$ as follows. If $m > 0$, then $w_{m,\lambda_{0,h-1}+1} = 1$ and, for $\lambda_{0,h-1}+1 < i$,

$$w_{m,i} := \begin{pmatrix} 1\cdots\lambda_{0,h-1} & \lambda_{0,h-1}+1 & \cdots & i-1 & i & i+1 & \cdots\lambda_{0,h}\cdots r \\ 1\cdots\lambda_{0,h-1} & \lambda_{0,h-1}+2 & \cdots & i & \lambda_{0,h-1}+1 & i+1 & \cdots\lambda_{0,h}\cdots r \end{pmatrix}.$$

If $m < 0$, then $w_{m,\lambda_{0,h}} = 1$ and, for $i < \lambda_{0,h}$,

$$w_{m,i} := \begin{pmatrix} 1\cdots\lambda_{0,h-1}\cdots i-1 & i & i+1 & \cdots & \lambda_{0,h} & \lambda_{0,h}+1\cdots r \\ 1\cdots\lambda_{0,h-1}\cdots i-1 & \lambda_{0,h} & i & \cdots & \lambda_{0,h}-1 & \lambda_{0,h}+1\cdots r \end{pmatrix}.$$

It is also clear that

$$\mathscr{D}_\beta^\Delta \cap \mathfrak{S}_\lambda = \{w_{m,i} \mid \lambda_{0,h-1}+1 \leqslant i \leqslant \lambda_{0,h}\} \text{ for all } m \in \mathbb{Z}\backslash\{0\}.$$

Hence,

$$[E_{h,h+mn}^\Delta + \mathrm{diag}(\lambda - e_h^\Delta)]_1[A]_1(\underline{\mathfrak{S}_\mu}) = \prod_{\substack{1\leqslant s\leqslant n\\ t\in\mathbb{Z}}} \frac{1}{a_{s,t}!} \sum_{i\in R_h^\lambda} \underline{\mathfrak{S}_\lambda}\cdot(u_{m,h}^\lambda)^{-1} w_{m,i} d\cdot\underline{\mathfrak{S}_\mu}.$$

Let $B^{(m,i)} = (b_{s,t}^{(m,i)}) \in \Theta_\Delta(n,r)$ be the matrix associated with λ, μ and the double coset $\mathfrak{S}_\lambda(u_{m,h}^\lambda)^{-1}w_{m,i}d\mathfrak{S}_\mu$, where $m \in \mathbb{Z}\backslash\{0\}$ and $i \in R_h^\lambda$. Since

$$w_{m,i}^{-1}u_{m,h}^\lambda(R_s^\lambda) = \begin{cases} R_s^\lambda, & \text{if } 1 \leqslant s \leqslant n \text{ but } s \neq h; \\ (R_h^\lambda\backslash\{i\}) \cup \{mr+i\}, & \text{if } s = h, \end{cases}$$

it follows that, for $i \in R_h^\lambda$,

$$\begin{aligned} b_{s,t}^{(m,i)} &= |d^{-1}w_{m,i}^{-1}u_{m,h}^\lambda R_s^\lambda \cap R_t^\mu| \\ &= \begin{cases} a_{s,t}, & \text{if } 1 \leqslant s \leqslant n \text{ but } s \neq h; \\ a_{s,t} - |\{\{d^{-1}(i)\} \cap R_t^\mu\}| + |\{d^{-1}(mr+i)\} \cap R_t^\mu|, & \text{if } s = h. \end{cases} \end{aligned}$$

If $t_i \in \mathbb{Z}$ is the unique integer such that $d^{-1}(i) \in R_{t_i}^\mu$ (and thus, $a_{h,t_i} \geqslant 1$), then $d^{-1}(mr+i) \in R_{mn+t_i}^\mu$ and

$$b_{h,t}^{(m,i)} = \begin{cases} a_{h,t}, & \text{if } t \notin \{t_i, mn+t_i\}; \\ a_{h,t}-1, & \text{if } t = t_i; \\ a_{h,t}+1, & \text{if } t = mn+t_i. \end{cases}$$

This implies that $B^{(m,i)} = A + E^{\Delta}_{h,mn+t_i} - E^{\Delta}_{h,t_i}$ for all $i \in R^{\lambda}_h$. Thus, applying Corollary 3.2.3 again yields

$$\begin{aligned}
&[E^{\Delta}_{h,h+mn} + \mathrm{diag}(\lambda - e^{\Delta}_h)]_1[A]_1(\underline{\mathfrak{S}_\mu})\\
&\quad= \prod_{\substack{1\leqslant s\leqslant n\\ t\in\mathbb{Z}}} \frac{1}{a_{s,t}!} \sum_{i\in R^{\lambda}_h} \prod_{\substack{1\leqslant s\leqslant n\\ t\in\mathbb{Z}}} b^{(m,i)}_{s,t}!\,\underline{\mathfrak{S}_\lambda (u^{\lambda}_{m,h})^{-1} w_{m,i} d\mathfrak{S}_\mu}\\
&\quad= \sum_{i\in R^{\lambda}_h} \prod_{\substack{1\leqslant s\leqslant n\\ t\in\mathbb{Z}}} \frac{b^{(m,i)}_{s,t}!}{a_{s,t}!}[B^{(m,i)}]_1(\underline{\mathfrak{S}_\mu})\\
&\quad= \sum_{i\in R^{\lambda}_h, a_{h,t_i}\geqslant 1} \frac{a_{h,mn+t_i}+1}{a_{h,t_i}}[B^{(m,i)}]_1(\underline{\mathfrak{S}_\mu}).
\end{aligned}$$

Therefore,

$$\begin{aligned}
&[E^{\Delta}_{h,h+mn} + \mathrm{diag}(\lambda - e^{\Delta}_h)]_1[A]_1\\
&= \sum_{\substack{t\in\mathbb{Z}\\ a_{h,t}\geqslant 1}} |\{i\in\mathbb{Z} \mid i\in R^{\lambda}_h,\ t = t_i\}| \frac{a_{h,mn+t}+1}{a_{h,t}}[A + E^{\Delta}_{h,mn+t} - E^{\Delta}_{h,t}]_1\\
&= \sum_{\substack{t\in\mathbb{Z}\\ a_{h,t}\geqslant 1}} (a_{h,t+mn}+1)[A + E^{\Delta}_{h,t+mn} - E^{\Delta}_{h,t}]_1,
\end{aligned}$$

since $a_{h,t} = |d^{-1}R^{\lambda}_h \cap R^{\mu}_t| = |\{i\in\mathbb{Z} \mid i\in R^{\lambda}_h,\ t = t_i\}|$ by Lemma 3.2.2. □

Proposition 6.2.3(2) is the key to the establishment of the multiplication formulas in Theorem 6.2.2, which in turn play a decisive role in the proof of the conjecture in the classical case (see Proposition 6.3.2). It would be natural to raise the following question as a finer version of Problem 3.4.3.

Problem 6.4.2. Find the quantum version of the multiplication formulas given in Proposition 6.2.3(2) for affine quantum Schur algebras.

Bibliography

[1] M. Auslander, I. Reiten and S. O. Smalø, *Representation Theory of Artin Algebras*, Cambridge Studies in Advanced Mathematics, No. 36, Cambridge University Press, Cambridge, 1995.

[2] J. Beck, *Braid group action and quantum affine algebras*, Commun. Math. Phys. **165** (1994), 555–568.

[3] J. Beck, V. Chari and A. Pressley, *An algebraic characterization of the affine canonical basis*, Duke Math. J. **99** (1999), 455–487.

[4] A. A. Beilinson, G. Lusztig and R. MacPherson, *A geometric setting for the quantum deformation of GL_n*, Duke Math. J. **61** (1990), 655–677.

[5] R. Borcherds, *Generalized Kac–Moody algebras*, J. Algebra **115** (1988), 501–512.

[6] V. Chari and A. Pressley, *Quantum affine algebras*, Commun. Math. Phys. **142** (1991), 261–283.

[7] V. Chari and A. Pressley, *A Guide to Quantum Groups*, Cambridge University Press, Cambridge, 1994.

[8] V. Chari and A. Pressley, *Quantum affine algebras and their representations*, Representations of Groups (Banff, AB, 1994), CMS Conf. Proc., **16**, American Mathematical Society, Providence, RI, 1995, 59–78.

[9] V. Chari and A. Pressley, *Quantum affine algebras and affine Hecke algebras*, Pacific J. Math. **174** (1996), 295–326.

[10] V. Chari and A. Pressley, *Quantum affine algebras at roots of unity*, Represen. Theory **1** (1997), 280–328.

[11] B. Deng and J. Du *Monomial bases for quantum affine $\mathfrak{sl}_n$*, Adv. Math. **191** (2005), 276–304.

[12] B. Deng, J. Du, B. Parshall and J. Wang, *Finite Dimensional Algebras and Quantum Groups*, Mathematical Surveys and Monographs Volume 150, American Mathematical Society, Providence, RI, 2008.

[13] B. Deng, J. Du and J. Xiao, *Generic extensions and canonical bases for cyclic quivers*, Can. J. Math. **59** (2007), 1260–1283.

[14] B. Deng and J. Xiao, *On double Ringel–Hall algebras*, J. Algebra **251** (2002), 110–149.

[15] R. Dipper and G. James, *The q-Schur algebra*, Proc. London Math. Soc. **59** (1989), 23–50.

[16] R. Dipper and G. James, *q-Tensor spaces and q-Weyl modules*, Trans. Amer. Math. Soc. **327** (1991), 251–282.

[17] S. Donkin, *The q-Schur algebra*, LMS Lecture Note Series **253**, Cambridge University Press, Cambridge, 1998.

[18] S. Doty and A. Giaquinto, *Presenting Schur algebras,* Internat. Math. Res. Not. **36** (2002), 1907–1944.

[19] S. Doty and R. M. Green, *Presenting affine q-Schur algebras*, Math. Z. **256** (2007), 311–345.

[20] V. G. Drinfeld, *A new realization of Yangians and quantized affine algebras*, Soviet Math. Dokl. **32** (1988), 212–216.

[21] J. Du, *Kahzdan–Lusztig bases and isomorphism theorems for q-Schur algebras*, Contemp. Math. **139** (1992), 121–140.

[22] J. Du, *A note on the quantized Weyl reciprocity at roots of unity*, Alg. Colloq. **2** (1995), 363–372.

[23] J. Du and Q. Fu, *Quantum* $\mathfrak{gl}_\infty$*, infinite q-Schur algebras and their representations*, J. Algebra **322** (2009), 1516–1547.

[24] J. Du and Q. Fu, *A modified BLM approach to quantum affine* $\mathfrak{gl}_n$, Math. Z. **266** (2010), 747–781.

[25] J. Du, Q. Fu and J. Wang, *Infinitesimal quantum* $\mathfrak{gl}_n$ *and little q-Schur algebras,* J. Algebra **287** (2005), 199–233.

[26] J. Du and B. Parshall, *Monomial bases for q-Schur algebras,* Trans. Amer. Math. Soc. **355** (2003), 1593–1620.

[27] J. Du, B. Parshall and L. Scott, *Quantum Weyl reciprocity and tilting modules*, Commun. Math. Phys. **195** (1998), 321–352.

[28] E. Frenkel and E. Mukhin, *The Hopf algebra* Rep $U_q(\widehat{\mathfrak{gl}}_\infty)$, Sel. Math., New Ser. **8** (2002), 537–635.

[29] E. Frenkel and N. Reshetikhin, *The q-characters of representations of quantum affine algebras and deformations of W-algebras*, Contemp. Math., AMS **248** (1998), 163–205.

[30] Q. Fu, *On Schur algebras and little Schur algebras*, J. Algebra **322** (2009), 1637–1652.

[31] V. Ginzburg, N. Reshetikhin and E. Vasserot, *Quantum groups and flag varieties*, Contemp. Math. **175** (1994), 101–130.

[32] V. Ginzburg and E. Vasserot, *Langlands reciprocity for affine quantum groups of type* A_n, Internat. Math. Res. Not. **3** (1993), 67–85.

[33] J. A. Green, *Polynomial Representations of* GL_n, 2nd edn, with an appendix on Schensted correspondence and Littelmann paths by K. Erdmann, J. A. Green and M. Schocker, Lecture Notes in Mathematics, No. 830, Springer-Verlag, Berlin, 2007.

[34] J. A. Green, *Hall algebras, hereditary algebras and quantum groups*, Invent. Math. **120** (1995), 361–377.

[35] R. M. Green, *The affine q-Schur algebra*, J. Algebra **215** (1999), 379–411.

[36] J. Y. Guo, *The Hall polynomials of a cyclic serial algebra*, Commun. Algebra **23** (1995) 743–751.

[37] J. Y. Guo and L. Peng, *Universal PBW-basis of Hall–Ringel algebras and Hall polynomials*, J. Algebra **198** (1997) 339–351.

[38] J. Hua and J. Xiao, *On Ringel–Hall algebras of tame hereditary algebras*, Algebra Represent. Theory **5** (2002), 527–550.

[39] A. Hubery, *Symmetric functions and the center of the Ringel–Hall algebra of a cyclic quiver*, Math. Z. **251** (2005), 705–719.

[40] A. Hubery, *Three presentations of the Hopf algebra $\mathcal{U}_v(\widehat{\mathfrak{gl}}_n)$*, preprint, 2009 (available at http://www1.maths.leeds.ac.uk/~ahubery).

[41] N. Iwahori and H. Matsumoto, *On some Bruhat decomposition and the structure of the Hecke rings of $\mathfrak{p}$-adic Chevalley groups*, Inst. Hautes Études Sci. Publ. Math. **25** (1965), 5–48.

[42] K. Jeong, S.-J. Kang and M. Kashiwara, *Crystal bases for quantum generalized Kac–Moody algebras*, Proc. London Math. Soc. **90** (2005), 395–438.

[43] M. Jimbo, *A q-analogue of $U(\mathfrak{gl}(N+1))$, Hecke algebra, and the Yang–Baxter equation*, Lett. Math. Physics **11** (1986), 247–252.

[44] N. Jing, *On Drinfeld realization of quantum affine algebras*, in: The Monster and Lie algebras (Columbus, OH, 1996), Ohio State University Mathematical Research Institute Publications, **7**, de Gruyter, Berlin, 1998, 195–206.

[45] V. F. R. Jones, *A quotient of the affine Hecke algebra in the Brauer algebra*, Enseign. Math. **40** (1994), 313–344.

[46] A. Joseph, *Quantum Groups and Their Primitive Ideals*, Ergebnisse der Mathematik und ihrer Grenzgebiete, Vol. **29**, Springer-Verlag, Berlin, 1995.

[47] V. Kac, *Infinite Dimensional Lie Algebras*, 3rd edn, Cambridge University Press, Cambridge, 1990.

[48] S.-J. Kang, *Quantum deformations of generalized Kac–Moody algebras and their modules*, J. Algebra **175** (1995), 1041–1066.

[49] S.-J. Kang and O. Schiffmann, *Canonical bases for quantum generalized Kac–Moody algebras*, Adv. Math. **200** (2006), 445–478.

[50] D. Kazhdan and G. Lusztig, *Representations of Coxeter groups and Hecke algebras*, Invent. Math. **53** (1979), 165–184.

[51] G. Lusztig, *Some examples of square integrable representations of semisimple p-adic groups*, Trans. Amer. Math. Soc. **277** (1983), 623–653.

[52] G. Lusztig, *Finite dimensional Hopf algebras arising from quantized universal enveloping algebras*, J. Amer. Math. Soc. **3** (1990), 257–296.

[53] G. Lusztig, *Quivers, perverse sheaves, and quantized enveloping algebras*, J. Amer. Math. Soc. **4** (1991), 365-421.

[54] G. Lusztig, *Introduction to Quantum Groups*, Progress in Math. **110**, Birkhäuser, Boston, 1993.

[55] G. Lusztig, *Canonical bases and Hall algebras*, in *Representation Theories and Algebraic Geometry*, A. Braer and A. Daigneault (eds.), Kluwer, Dordrecht, 1998, 365–399.

[56] G. Lusztig, *Aperiodicity in quantum affine $\mathfrak{gl}_n$*, Asian J. Math. **3** (1999), 147–177.

[57] G. Lusztig, *Transfer maps for quantum affine $\mathfrak{sl}_n$*, in: Representations and quantizations (Shanghai, 1998), China Higher Education Press, Beijing, 2000, 341–356.

[58] K. McGerty, *Generalized q-Schur algebras and quantum Frobenius*, Adv. Math. **214** (2007), 116–131.

[59] L. Peng, *Some Hall polynomials for representation-finite trivial extension algebras*, J. Algebra **197** (1997), 1–13.

[60] L. Peng and J. Xiao, *Triangulated categories and Kac–Moody algebras*, Invent. Math. **140** (2000), 563–603.

[61] G. Pouchin, *A geometric Schur–Weyl duality for quotients of affine Hecke algebras*, J. Algebra **321** (2009), 230–247.

[62] M. Reineke, *Generic extensions and multiplicative bases of quantum groups at* $q = 0$, Represent. Theory **5** (2001), 147–163.

[63] C. M. Ringel, *Hall algebras and quantum groups*, Invent. Math. **101** (1990), 583–592.

[64] C. M. Ringel, *Hall algebras revisited*, Israel Math. Conf. Proc. **7** (1993), 171–176.

[65] C. M. Ringel, *The composition algebra of a cyclic quiver*, Proc. London Math. Soc. **66** (1993), 507–537.

[66] J. D. Rogawski, *On modules over the Hecke algebra of a p-adic group*, Invent. Math. **79** (1985), 443–465.

[67] O. Schiffmann, *The Hall algebra of a cyclic quiver and canonical bases of Fock spaces*, Internat. Math. Res. Not. **8** (2000), 413–440.

[68] O. Schiffmann, *Noncommutative projective curves and quantum loop algebras*, Duke Math. J. **121** (2004) 113–168.

[69] I. Schur, *Über eine Klasse von Matrizen, die sich einer gegebenen Matrix zuordnen lassen*, Dissertation, Berlin 1902.

[70] I. Schur, *Über die rationalen Darstellungen der allgemeinen linearen Gruppe*, Sitzber Königl. Preuß. Ak. Wiss., Physikal.-Math. Klasse, pages 58–75, 1927.

[71] B. Sevenhant and M. Van den Bergh, *On the double of the Hall algebra of a quiver*, J. Algebra **221** (1999), 135–160

[72] M. E. Sweedler, *Hopf Algebras*, Benjamin, New York, 1969.

[73] M. Varagnolo and E. Vasserot, *On the decomposition matrices of the quantized Schur algebra*, Duke Math. J. **100** (1999), 267–297.

[74] M. Varagnolo and E. Vasserot, *From double affine Hecke algebras to quantized affine Schur algebras*, Internat. Math. Res. Not. **26** (2004), 1299–1333.

[75] E. Vasserot, *Affine quantum groups and equivariant K-theory*, Transf. Groups **3** (1998), 269–299.

[76] M.-F. Vignéras, *Schur algebras of reductive p-adic groups. I*, Duke Math. J. **116** (2003), 35–75.

[77] H. Weyl, *The Classical Groups*, Princeton University Press, Princeton, NJ, 1946.

[78] J. Xiao, *Drinfeld double and Ringel–Green theory of Hall algebras*, J. Algebra **190** (1997), 100–144.

[79] D. Yang, *On the affine Schur algebra of type A*, Commun. Algebra **37** (2009), 1389–1419.

[80] D. Yang, *On the affine Schur algebra of type A. II*, Algebr. Represent. Theory **12** (2009), 63–75.

[81] A. V. Zelevinsky, *Induced representations of reductive p-adic groups II. On irreducible representations of* GL_n, Ann. Sci. Ec. Norm. Sup. 4e Sér. 13 (1980), 165–210.

[82] G. Zwara, *Degenerations for modules over representation-finite biserial algebras*, J. Algebra **198** (1997), 563–581.

Index

Printed in the United States
By Bookmasters